Molecular and Cell Biology of Muscular Dystrophy

Molecular and Cell Biology of Human Diseases Series

Series Editors

D.J.M. WRIGHT MD FRCPath
Reader in Medical Microbiology,
Charing Cross and Westminster Medical School, London, UK

L.C. ARCHARD PhD
Senior Lecturer in Biochemistry,
Charing Cross and Westminster Medical School, London, UK

The continuing developments in molecular biology have made possible a new approach to a whole range of different diseases. The books in this series each concentrate on a disease or group of diseases where real progress is being made in understanding the pathogenesis, diagnosis and management. Experts on aspects of each area provide a text accessible to scientists and clinicians in a form which records advances and points to the potential application of these advances in a clinical setting.

Other titles in this series

1 *Molecular and Cell Biology of Sexually Transmitted Diseases*
Edited by David Wright and Leonard Archard

2 *Molecular and Cell Biology of Opportunistic Infections in AIDS*
Edited by Steven Myint and Alan Cann

4 *Molecular and Cell Biology of Neuropsychiatric Diseases*
Edited by Frank Owen and Ruth Itzhaki

Molecular and Cell Biology of Muscular Dystrophy

EDITED BY

Terence Partridge
Charing Cross & Westminster Medical School,
University of London, UK.

SPRINGER-SCIENCE+BUSINESS MEDIA, B.V.

A catalogue record for this book is available from the British Library

Library of Congress Cataloging-in-Publication data

Molecular and cell biology of muscular dystrophy / edited by Terence Partridge. -- 1st ed.

p. cm.

Includes bibliographical references and index.

ISBN 978-94-010-4667-1 ISBN 978-94-011-1528-5 (eBook)
DOI 10.1007/978-94-011-1528-5

1. Becker muscular dystrophy--Molecular aspects. 2. Becker muscular dystrophy--Genetic aspects. 3. Duchenne muscular dystrophy--Molecular aspects. 4. Duchenne muscular dystrophy--Genetic aspects. I. Partridge, Terrence.

[DNLM: 1. Genetics, Biochemical. 2. Muscular Dystrophy--physiopathology. WE 559 M718 1993]
RC935.M7M65 1993
616.7'48--dc20
DNLM/DLC
for Library of Congress 92–48393
CIP

Contents

Contributors

Kevin P. Campbell
Howard Hughes Medical Institute
University of Iowa College of Medicine
400 EMRB
Iowa City
Iowa 52242
USA

Jeffrey S. Chamberlain
The Department of Human Genetics
University of Michigan Medical School
Ann Arbor
Michigan 48109–0618
USA

Gregory A. Cox
The Department of Human Genetics
University of Michigan Medical School
Ann Arbor
Michigan 48109–0618
USA

George Dickson
Department of Experimental Pathology
UMDS Guy's Hospital Medical School
London Bridge
London SE1 9RT
UK

Matthew Dunckley
Department of Experimental Pathology
UMDS Guy's Hospital Medical School
London Bridge
London SE1 9RT
UK

James M. Ervasti
Department of Physiology
University of Wisconsin Medical School
Madison
Wisconsin 53706
USA

Rune R. Frants
MGC-Department of Human Genetics
Leiden University
The Netherlands

Alex D. Greenwood
The Department of Human Genetics
University of Michigan Medical School
Ann Arbor
Michigan 48109–0618
USA

Miranda D. Grounds
Department of Pathology
University of Western Australia
Queen Elizabeth II Medical Centre
Perth
Western Australia 6009

Eric P. Hoffman
Department of Molecular Genetics and Biochemistry
Department of Human Genetics
and
Department of Pediatrics
University of Pittsburgh School of Medicine
Pittsburgh
Pennsylvania 15261
USA

Marten H. Hofker
MCG-Department of Human Genetics
Leiden University
The Netherlands

Malcom J. Jackson
Muscle Research Centre
Department of Medicine
University of Liverpool
P.O. Box 147
Liverpool L69 3BX
UK

Keith Johnson
Department of Anatomy
Charing Cross and Westminster Medical School
St Dunstan's Road
London W6 8RP
UK

Nigel G. Laing
Chief of Research and Diagnostic Molecular Neurogenetics
Australian Neuromuscular Research Institute
QEII Medical Centre
Nedlands
Western Australia 6009

Andrea J. Maichele
The Department of Human Genetics
University of Michigan Medical School
Ann Arbor
Michigan 48109–0618
USA

Kathleen M. McCormick
Department of Anatomy
University of Wisconsin
Madison
Wisconsin 53706
USA

Anthony P. Monaco
Human Genetics Laboratory
Imperial Cancer Research Fund
Institute of Molecular Medicine
John Radcliffe Hospital
Headington
Oxford OX3 9DU
UK

Jennifer E. Morgan
Department of Histopathology
Charing Cross and Westminster Medical School
St Dunstan's Road
London W6 8RF
UK

George W. Padberg
Department of Neurology
Leiden University
The Netherlands

Stephanie F. Phelps
The Department of Human Genetics
University of Michigan Medical School
Ann Arbor
Michigan 48109–0618
USA

Edward Schultz
Department of Anatomy
University of Wisconsin
Madison
Wisconsin 53706
USA

Diana J. Watt
Department of Anatomy
Charing Cross and Westminster Medical School
St Dunstan's Road
London W6 8RF
UK

Cisca Wijmenga
MGC-Department of Human Genetics
Sylvius Laboratory
Wassenaarseweg 72
2333 AL Leiden
The Netherlands

Zipora Yablonka-Reuveni
Department of Biological Structure
School of Medicine
University of Washington
Seattle
Washington 98195
USA

Preface

From the time of the original descriptions by Meryon (1852) and Duchenne (1868), the severe X-linked childhood muscular dystrophy, now termed Duchenne or Becker muscular dystrophy depending on its severity, has been particularly problematical. For more than a century it has proven refractory to all attempts at therapy and, until recently, to investigation of its cause. However, it had become clear by the early 1980s, that this would be the first genetic disease to succumb to the strategy of 'reverse genetics' or 'positional cloning' as it is more properly known. Isolation of the full length cDNA from the disease locus followed rapidly by preliminary characterization of dystrophin, the protein product of this gene, set off an exponential expansion in research activity across a surprisingly broad front. Experience of this prototype is now at a stage where some general lessons may be drawn as to both the power and the limitations of the molecular genetic approach to inherited diseases. This volume is intended to catch some of the zest and enthusiasm engendered by these discoveries and to anticipate the prospects for other muscular dystrophies which are on the verge of elucidation at the molecular genetic level.

Identification of the Duchenne gene had the immediate effect of concentrating research efforts on a number of obviously relevant topics; for example, the relationships between genetic defect, gene product, and clinical disease were early beneficiaries of this breakthrough. The resultant increase in the scope and accuracy of genetic counselling is the major practical gain to date. Rather unexpected perhaps, was the revelation that we were already in possession of a genuine animal model of Duchenne dystrophy, in the form of the *mdx* mouse; an animal which had previously achieved some disrepute as a model of any sort of dystrophy because of its tendency to become sturdy and strong rather than wasted and weak. This mouse, although not a faithful model of all aspects of Duchenne muscular dystrophy, has been the mainstay of research into a number of features which it does share with the human disease; namely the perturbing effects of lack of dystrophin on the molecular biology, macromolecular permeability, calcium ion homeostasis and mechanical stability of skeletal muscle fibres. Identification of an X-linked dystrophic dog has provided us with what is widely regarded as a more appropriate clinical model of DMD, although the relatively high cost and slow breeding rate of dogs leave the *mdx* mouse as the principal animal model, at least for preliminary investigations of mechanisms of pathogenesis and therapeutic strategies.

In this latter respect, the *mdx* mouse has already proved its worth, for it has provided the test system on which two ideas for therapy, namely myoblast transplantation and gene therapy, have been validated. In each case, the degree of success has been tantalizing; confirming the potentials of both approaches but, at the same time, showing them to be in need of technical improvement if either is to be converted to practical use in man. Current reports of experiments on myoblast transplantation in DMD patients have tended to confirm this view; having produced little to encourage the idea that this will be a simple approach to therapy. Both the mouse and the dog should prove important test systems in which to examine ideas for achieving the required improvements in efficiency. The essential role of these animals at the early stages of investigation of potential therapies serves to illustrate the need for animal experiments in a field where both technical and ethical barriers impose severe limitations on the conduct of effective investigations in human volunteers.

Notwithstanding the advent of experimental tools in the form of nucleotide probes, antibodies and pertinent animal models, there has, as yet, been no satisfactory disclosure of the precise functional defect underlying Duchenne dystrophy. This is not too surprising, for dystrophin appears to belong to the 'cytoskeletal' category of proteins, the exact functions of most of which are still largely a matter of speculation, despite the fact that many of them have been known and characterized for many years. Moreover, the immediate functional consequences of the absence of this protein, when considered at the single cell level, appear to be quite subtle and probably sporadic and fleeting; a combination of attributes as defiant of resolution in muscle fibres as it is in television receivers. All the same, the availability of animal models of Duchenne dystrophy has permitted valid inter-species comparison as a firm basis for the gathering of data which addresses the pathogenetic mechanisms and has allowed a more critical evaluation of these data than was previously possible. Here again the *mdx* mouse has proved of value, both as a definitive system in which to seek accurate detailed data on abnormalities of muscle fibre function associated with the absence of dystrophin and as the most accessible experimental system in which to test theories linking these abnormalities to myonecrosis. By imaginative use of these resources, considerable inroads have already been made into the biochemical consequences of the absence of dystrophin and a convincing picture has emerged of a complex of proteins and glycoproteins which mediates the association between intra-cellular dystrophin and the extra-cellular matrix of the muscle fibre. At present, this biochemical approach appears to be the firmest foundation for constructing testable models of the pathogenetic mechanism of muscle fibre necrosis in this disease.

These direct consequences of the identification of the dystrophin gene were manifested as a distinct awakening of a largely dormant topic, for there were ideas and techniques awaiting the precise questions which could be posed once the nature of the Duchenne gene was known. Much as the whole castle was re-vivified by the Prince's kiss in the story of the *Sleeping Beauty*, this arousal set off a bustle of activity in distant provinces of muscle research as it became obvious that any radical approach to therapy would entail some strengthening of our understanding of the cell biology of this tissue, especially of the mechanisms underlying muscle degeneration and regeneration.

Over the five years or so in which we have been able to observe the repercussions of the discovery of the Duchenne gene, other inherited myopathies classified as dystrophies have been following parallel paths towards elucidation. The two most advanced of these investigations, which have recently revealed the chromosomal location of Facio-scapulo-humeral dystrophy and the nature of the genetic mutation in myotonic dystrophy, are described. Quite apart from their intrinsic interest as tales of exploration, they impart some savour of the vitality of a field in which the technological innovations are often as spectacular as the knowledge they generate.

A general aim in assembling the articles in this volume was to catch some of the mood of elation at having reached a historic turning-point, tapping first-hand knowledge of active scientists of world-standing, themselves intimately involved in the work they describe. The topics were chosen to interlock or overlap with one another and it has been gratifying to note that even where a given subject, such as the satellite cell, has been addressed by a number of authors, the different perspectives have tended to complement rather than to reiterate one another.

So entrancing were the immediate and practical consequences of the molecular genetic revolution on research into the muscular dystrophies that there was, initially, a widespread feeling that the dystrophin gene itself had yielded all its important secrets. This is now, clearly, a mistaken view; a number of facets of this remarkable gene; its conservation of non-coding sequence, its enormous introns and its rapidly increasing catalogue of splicing isoforms in different tissues, are difficult to reconcile with the rather restricted range of clinical phenotype in Duchenne dystrophy arising from a huge variety of individual mutations. Such observations also raise questions as to the nature of selective pressures which permit (or cause) the dystrophin gene to conserve so much genetic sequence which, as far as is known, does not emerge as phenotype.

Investigations of these aspects of the gene, together with the discovery of a closely related autosomal gene of similar size and complexity, continue to develop very rapidly and are only touched upon in this volume but promise to give us insights into areas of basic molecular biology which were not at all envisaged by the teams who originally sought the genetic cause of DMD.

REFERENCES

Duchenne, G.B. (1868) Recherches sur la paralysie musculaire pseudo-hypertrophique ou paralysie myosclerotique. *Archives génèrales de médicine* **11**, 5, 178, 305, 421, 552.

Meryon, E. (1852). On granular and fatty degeneration of the voluntary muscles. *Medico-Chirurgical Transactions* (London) **35**, 73.

1

Molecular human genetics and the Duchenne/Becker muscular dystrophy gene

ANTHONY P. MONACO

1.1 INTRODUCTION

To isolate the gene involved in a human inherited disorder without prior knowledge of the defective protein product is a direct application of molecular human genetics and has been termed 'reverse genetics' (Orkin, 1986) or 'positional cloning'. The phenotype associated with the disease, its segregation in families and linkage to genetic markers, and any associated cytogenetic abnormalities provide the information necessary to localize the disease gene to a precise chromosome map position. This map position then focuses experimental approaches to isolate the gene by collecting genomic DNA from the region and analysing it for candidate genes for the disease. Once the correct gene has been identified and the nucleotide sequence obtained, information can be gained on the structure and possible function of its protein product. The gene can be used to construct fusion proteins for antibody production thus allowing the cell biology of the protein to be studied and related to the pathophysiology of the disease. In this chapter the various aspects of molecular human genetics will be discussed in relation to the cloning and analysis of the Duchenne/Becker muscular dystrophy gene and study of its protein product dystrophin. The relationship between phenotype and genotype will be discussed in the light of the unusual organization and size of the dystrophin gene.

1.2 DMD AND BMD PHENOTYPE

The phenotype of X-linked muscular dystrophy varies between severe and mild forms (Moser, 1984). Duchenne muscular dystrophy (DMD) is the

Molecular and Cell Biology of Muscular Dystrophy
Edited by Terence Partridge
Published in 1993 by Chapman & Hall, London. ISBN 0 412 43440 7

most severe form with progressive muscle wasting and weakness requiring a wheelchair for ambulation by the age of twelve. Death usually occurs from respiratory or cardiac failure by the late second or early third decade. DMD affects 1/3500 live male births worldwide and one-third of cases are sporadic without any prior family history. Becker muscular dystrophy (BMD) is a rarer and milder form of muscle weakness. BMD patients show a much wider spectrum of disease phenotype than DMD, usually requiring a wheelchair sometime after 15 years of age, yet some never lose the ability to walk. Approximately one-third of both DMD and BMD patients exhibit a non-progressive mental retardation.

1.3 LOCALIZATION TO Xp21

DMD and BMD phentoypes segregating in families were used to pinpoint the gene's location in the cytogenetic band Xp21 by genetic linkage analysis (for review see Goodfellow *et al.*, 1985). DNA probes from the human X chromosome that detected restriction fragment length polymorphisms (RFLPs, Botstein *et al.*, 1980) were used to follow the X chromosome carrying the disease gene in each family. RFLPs are usually recognized as different length alleles on separate chromosomes due to variation in nucleotide sequences at sites for endonuclease restriction enzymes. By analysis of meiotic recombinations between DNA markers recognizing RFLPs and the DMD or BMD phenotype, the most likely position of the gene for both disorders was reported to be in Xp21. This confirmed that DMD and BMD were probably different alleles of the same gene or two genes very close on the X chromosome.

Rare cases of females exhibiting DMD or BMD also provided information for gene location in the band Xp21. Cytogenetic analysis of their chromosomes revealed balanced translocations, in each case between an autosome and the X chromosome (Boyd *et al.*, 1986; 1987). The translocation breakpoints on the X chromosome were always located in the band Xp21, although there was some degree of heterogeneity of the exact location within Xp21. The mechanism giving rise to DMD or BMD in these females was thought to involve the translocated X chromosome that disrupted the gene escaping X-inactivation when adjacent to an autosomal sequence. This gives rise to non-random inactivation of the normal X chromosome and therefore lack of expression of the only normal copy of the DMD/BMD gene.

The location of DMD in Xp21 was confirmed with the cytogenetic and molecular analysis of a male patient (B.B.) with an Xp21 interstitial deletion (Francke *et al.*, 1985). B.B. exhibited four X-linked disorders, DMD, chronic granulomatous disease (CGD), retinitis pigmentosa and McLeod syndrome, the red blood cell Kell antigen defect. This was a contiguous gene syndrome associated with an Xp21 deletion detected both cytogenetically and by the absence of the DNA probe 754 (*DXS84*). This Xp21 deletion was to prove useful in the isolation of the genes for DMD and CGD (Royer-Pokora *et al.*, 1986) and hopefully the future isolation of the genes for retinitis pigmentosa and McLeod syndrome.

1.4 ISOLATION OF Xp21 SEQUENCES

As mentioned above, the DNA probe 754 (*DXS84*) was shown to be deleted from the patient B.B., confirming the cytogenetic deletion in this patient and localizing all four disorders to the band Xp21. However, 754 did not detect deletions in unrelated males with only the DMD phenotype, and exhibited approximately a 5–10% recombination rate with DMD and BMD mutations segregating in families. There was a definite need to obtain more sequences from the Xp21 region, hopefully within the DMD or BMD gene. Two experimental approaches were taken by several different laboratories that provided DNA probes within the DMD gene in Xp21.

One approach was to use differential hybridization of genomic DNA between normal X chromosomes and the deleted X chromosome of the patient B.B. to enrich for DNA sequences missing from the Xp21 deletion. Kunkel *et al.* (1985) used a phenol emulsion reassociation technique (PERT) and a plasmid cloning procedure to generate a library with a 20-fold enrichment of sequences missing from the Xp21 deletion of B.B. Nine DNA probes missing in B.B. were generated from this library of which one, pERT87 (*DXS164*), were found to be deleted in approximately 9% of unrelated DMD males (Monaco *et al.*, 1985). This probe was potentially within the DMD gene itself since deletions had been found with cloned genes in other X-linked disorders.

The second approach was to utilize the rare translocations giving rise to DMD or BMD in females to isolate sequences from within Xp21. One translocation was found to break within Xp21 and the ribosomal cistrons on chromosome 21. A genomic bacteriophage library was constructed from the DNA of this translocated chromosome and screened with a DNA probe from chromosome 21 which was very close to the translocation breakpoint. This led to the isolation of a genomic clone containing the translocation breakpoint with DNA from both chromosome 21 and Xp21 (Ray *et al.*, 1985). The DNA sequences from Xp21 (pXJ, *DXS206*) were also found to be deleted in some unrelated DMD patients, thus showing that this sequence may be within the DMD gene.

1.5 CHROMOSOME WALKING AND JUMPING IN Xp21

Both pERT87 (*DXS164*) and pXJ (*DXS206*) were found to be deleted in approximately 5–10% of DMD males and some BMD males. The next step was to isolate the surrounding genomic DNA to search for polymorphic markers and possible expressed sequences, a process termed 'chromosome walking' (Bender *et al.*, 1983). Both pERT87 (*DXS164*) and pXJ (*DXS206*) were expanded in bacteriophage and cosmid genomic libraries to over 200 kilobasepairs (kb) of genomic DNA each. The surrounding genomic DNA was screened to identify single copy DNA probes to test the extent of deletions identified in DMD and BMD males (Kunkel *et al.*, 1986). The deletion breakpoints were spread heterogeneously throughout the *DXS164* locus with large deletions extending towards both the centromere and telomere. Single copy DNA probes from the *DXS164* locus were also used to identify RFLPs for genetic linkage analysis in families segregating DMD or BMD mutations.

Although the *DXS164* locus was deleted in 6.5% of DMD and BMD males it recombined with mutations at a rate of 4–6% (Kunkel *et al.*, 1986). Thus from the heterogeneity of deletions and the high rate of recombination, the DMD/BMD gene seemed to be outside the cloned region or possibly was extremely large and was spread throughout the Xp21 region. To investigate these possibilities, four patients with one breakpoint of their deletion in the *DXS164* locus were used to jump to the other side of their deletions. This involved constructing a genomic bacteriophage library from the DNA of each patient and finding DNA probes within the *DXS164* locus that were quite close to the deletion breakpoints. From the deletion junction cloning four new loci were isolated, two centromeric and two telomeric to the *DXS164* locus (Monaco *et al.*, 1987; Van Ommen *et al.*, 1987). These new deletion jump loci were mapped relative to X;autosome translocation breakpoints in DMD and BMD females and used to study known recombinants in DMD and BMD families. This analysis showed that mutations and translocations giving rise to DMD and BMD were still mapping outside the cloned region of approximately 500 kb, thus giving indirect evidence that the genomic locus containing the DMD/BMD gene was extremely large.

1.6 IDENTIFICATION OF EXPRESSED SEQUENCES

To find potential expressed sequences of the DMD/BMD gene in this large stretch of genomic DNA cloned from Xp21, individual DNA fragments were analysed for conservation on 'zoo' blots of genomic DNA isolated from a variety of mammalian species and birds (Monaco *et al.*, 1986). The premise for this experiment was that exons encoding amino acids should be conserved through evolution whereas intron sequences would have diverged. For practical reasons, it was much easier to test multiple small DNA fragments for conservation on zoo blots than to test each fragment for expression by hybridization to RNA or cDNA libraries. Two DNA fragments from over 50 tested from the *DXS164* locus cross-hybridized to other mammalian DNA samples. The mouse equivalents were isolated from a mouse genomic DNA library and their nucleotide sequence determined and compared between mouse and human. Both conserved DNA fragments were shown to contain short open reading frames (ORF) of triplet codons for amino acids surrounded by appropriate sequences for splicing of introns to produce mature transcripts. When tested in RNA from fetal skeletal muscle, one of the conserved fragments hybridized to a large (14 kb) transcript and was used to isolate short cDNA clones. Similar approaches led to the identification of a similar large transcript and cDNA clones from the *DXS206* locus (Burghes *et al.*, 1987).

1.7 GENE SEQUENCE AND ORGANIZATION

The complete 14 kb representation of the DMD gene was isolated and found to be partially deleted in 50–65% of DMD patients (Koenig *et al.*, 1987; Forrest *et al.*, 1987; Den Dunnen *et al.*, 1987; Darras *et al.*, 1988) and partially duplicated in about 5% of cases (Hu *et al.*, 1990). The gene was organized into a minimum of 65 small exons spread over a large region with very large introns (Koenig *et al.*, 1987). The size of the genomic locus of the DMD/BMD gene was estimated to be 2300 kb using pulsed field gel electrophoresis (PFGE; Van Ommen *et al.*, 1986, 1987; Burmeister and Lehrach, 1986; Kenwrick *et al.*, 1987; Burmeister *et al.*, 1988). The large size of the DMD/BMD gene was consistent with the high mutation frequency, the heterogeneity of deletion and translocation breakpoints and the high recombination rate of intragenic probes with mutations segregating in families.

The relation of exon-containing *Hind*III fragments relative to the long-range *Sfi*I map has been completed by Den Dunnen *et al.*, (1989). The genomic locus for the DMD/BMD gene has recently been extended to 2400 kb with the analysis of a human brain specific promoter located 100 kb proximal to the muscle specific promoter (Boyce *et al.*, 1991). Using the new cloning system of yeast artificial chromosomes (YACs, Burke *et al.*, 1987) the complete 2400 kb genomic locus of the DMD/BMD gene has been isolated in overlapping YAC clones by two groups (Monaco *et al.*, 1992; Coffey *et al.*, 1992). The DMD/BMD gene has been reconstructed into a single large YAC using individual YAC clones and homologous recombination in yeast (Den Dunnen *et al.*, 1992). These YAC clones should be useful for studies of exon/intron organization of the gene, isolation of regulatory sequences, alternative promoters and exons, and for possible gene therapy approaches.

The complete nucleotide sequence of the DMD/BMD gene was determined and predicted a protein product of 3685 amino acids organized into four domains (Koenig *et al.*, 1988). The N-terminal domain (240 amino acids) is similar to the actin binding portion of α-actinin. The large second domain consists of 26 triple helical segments in succession that are similar to the rod-shaped domains of spectrin and α-actinin. The third domain is cysteine rich with similarity to the C-terminal domain of α-actinin. The last domain of dystrophin (420 amino acids) has no similarity to known protein sequences but has the potential to form multiple isoforms due to alternate splicing of exons at the 3′ end of the gene (Feener *et al.*, 1989). The DMD/BMD gene product, termed 'dystrophin', is therefore most likely a member of a family of flexible rod-shaped proteins that includes spectrin and α-actinin.

1.8 GENOTYPE AND PHENOTYPE

From the analysis of partial deletions in the DMD/BMD gene a paradox emerged that deletion size had no direct correlation with the severity of the resulting phenotype (DMD versus BMD). To explain this paradox at the molecular level, a model was proposed which suggested that the severity of the disease would be directly correlated with the effect of partial gene deletions on the open reading frame (ORF) of triplet codons in the mRNA transcript (Monaco *et al.*, 1988). Deletions predicted to shift the ORF of triplet codons for amino acids would result in the more severe DMD phenotype. Due to an eventual stop codon, the frameshifted mRNA would prematurely halt protein synthesis during translation. The resulting truncated dystrophin molecules would most likely be non-functional or degraded. Deletions that maintain the ORF for amino acids despite the deletion event would be predicted to result in the less severe BMD phenotype. Translation of the mRNA transcripts with internal non-frameshifting deletions would produce shorter, lower molecular weight dystrophin molecules presumed to be semifunctional.

After genomic sequence analysis of exon:intron borders in the *DXS164* locus relative to deletions in three DMD and three BMD patients, the ORF model was found to be consistent (Monaco *et al.*, 1988). However, studies limited to the 5′-end of the gene suggested there were exceptions to this model especially for deletions of exons 3 to 7 giving rise to BMD, intermediate and DMD phenotypes (Malhotra *et al.*, 1988). There were also several exceptions found with deletions around the middle portion of the gene near the hotspot for deletions (Baumbach *et al.*, 1989; Blonden *et al.*, 1989). When a large study of 256 deletions was analysed along the entire length of the gene, the ORF model was found to be consistent with phenotype in 92% of cases (Koenig *et al.*, 1989). Exceptions to the ORF model to explain genotype and phenotype of DMD and BMD patients may be due to alternate splicing of exons in some patients to overcome the effect on the ORF of the original deletion mutation (Chelly *et al.*, 1991).

1.9 DYSTROPHIN

Polyclonal antibodies were raised against two different N-terminal domains of mouse cardiac dystrophin. These antibodies recognized a 400 kd protein in skeletal, smooth and cardiac muscle and brain (Hoffman *et al.*, 1987a). Dystrophin was localized to the muscle plasma membrane and t-tubules by subcellular fractionation, immunofluorescence and immunogold electron microscopy (Hoffman *et al.*, 1987b; Zubrzycka-Gaarn *et al.*, 1988; Bonilla *et al.*, 1988; Arahata *et al.*, 1988; Watkins *et al.*, 1988). In brain, dystrophin has been localized in neurons of the cerebral and cerebellar cortices and specifically at postsynaptic membrane specializations (Lidov *et al.*, 1990). In the *mdx* mouse, dystrophin has been found to be missing in both muscle and brain (Hoffman *et al.*, 1987b; Watkins *et al.*, 1988; Lidov *et al.*, 1990) thus substantiating this mouse model of DMD.

In skeletal muscle samples from most DMD patients dystrophin has been found to be absent yet BMD patients exhibit altered size dystrophin in about 80% of cases (Hoffman *et al.*, 1988). These results are consistent with the ORF model for the clinical difference between DMD and BMD patients bearing partial gene deletions.

Dystrophin has been shown in biochemical studies to be tightly bound to a complex of glycoproteins isolated from skeletal muscle membranes (Campbell and Kahl, 1989). The large oligomeric complex containing dystrophin was purified and four glycoproteins (156K, 50K, 43K, and 35K) were identified (Ervasti *et al.*, 1990). In more detailed studies, the dystrophin–glycoprotein complex has been shown to consist of cytoskeletal (dystrophin and 59K), transmembrane (50K, 43K, 35K, and 25K) and extracellular (156K) components (Ervasti and Campbell, 1991). In DMD and *mdx* skeletal muscle there is complete absence of dystrophin and almost all of the glycoprotein complex. More recently, the primary structure has been identified for two components of the dystrophin–glycoprotein complex (43K and 156K) and they have been shown to be encoded by a single gene which after post-translational modification of its predicted 97K precursor product results in two mature proteins (Ibraghimov-Beskrovnaya *et al.*, 1992). Furthermore, the 156K dystrophin-associated glycoprotein binds laminin, thus indicating that the dystrophin–glycoprotein complex may link the subsarcolemmal cytoskeleton to the extracellular matrix.

1.10 CONCLUSIONS

The process of molecular human genetics, going from phenotype to gene and then to protein, has been outlined for the DMD/BMD gene. Isolation and characterization of the large DMD/BMD gene has allowed accurate prenatal diagnosis and carrier detection of DMD and BMD mutations segregating in families. Recent progress using multiplex polymerase chain reaction amplification of exon sequences has identified almost all deletion cases, thus making diagnosis more efficient (Chamberlain *et al.*, 1988; Beggs *et al.*, 1990). Identification and subcellular localization of the protein product dystrophin and recent identification of glycoproteins which interact as a complex with dystrophin may direct therapeutic interventions. The study of the DMD/BMD gene locus and its protein product dystrophin by molecular human genetics should be a good model for the molecular understanding of other inherited human disorders where there is no prior knowledge of the defective protein involved.

REFERENCES

Arahata, K., Ishiura, S., Ishiguro, T. *et al.* (1988) Immunostaining of skeletal and cardiac muscle surface membrane with antibody against Duchenne muscular dystrophy peptide. *Nature*, **333**, 861–3.

Baumbach, L.L., Chamberlain, J.S., Ward, P.A. *et al.* (1989) Molecular and clinical correlations of deletions leading to Duchenne and Becker muscular dystrophies. *Neurology*, **39**, 465–74.

Beggs, A.H., Koenig, M., Boyce, F.M. and Kunkel, L.M. (1990) Detection of 98% of DMD/BMD deletions by PCR. *Hum. Genet.*, **86**, 45–8.

Bender, W., Arkam, M., Karch, F. *et al.* (1983) Molecular genetics of the bithorax complex in *Drosophila melanogaster*. *Science*, **221**, 23–9.

Blonden, L.A.J., Den Dunnen, J.T., Van Paassen, H.M.B. *et al.* (1989) High resolution deletion breakpoint mapping in the DMD-gene by whole cosmid hybridization. *Nucl. Acids Res.*, **17**, 5611–21.

Bonilla, E., Samitt, C.E., Miranda, A.F. *et al.* (1988) Duchenne muscular dystrophy: Deficiency of dystrophin at the muscle cell surface. *Cell*, **54**, 447–52.

Botstein, D., White, R.L., Skolnick, M. and Davis, R.W. (1980) Construction of a genetic linkage map in man using restriction fragment length polymorphisms. *Am. J. Hum. Genet.*, **32**, 314–31.

Boyce, F.M., Beggs, A.H., Feener, C. and Kunkel, L.M. (1991) Dystrophin is transcribed in brain from a distant upstream promoter. *Proc. Natl. Acad. Sci. USA*, **88**, 1276–80.

Boyd, Y., Buckle, V., Holt, S. *et al.* (1986) Muscular dystrophy in girls with X;autosome translocations. *J. Med. Genet.*, **23**, 484–90.

Boyd, Y., Munro, E., Ray, P. *et al.* (1987) Molecular heterogeneity of translocations associated with muscular dystrophy. *Clin. Genet.*, **31**, 265–72.

Burghes, A.H.M., Logan, C., Hu, X. *et al.* (1987) Isolation of a cDNA clone from the region of an X;21 translocation that breaks within the Duchenne/Becker muscular dystrophy gene. *Nature*, **328**, 434–6.

Burke, D.T., Carle, G.F. and Olson, M.V. (1987) Cloning of large segments of exogenous DNA into yeast by means of artificial chromosome vectors. *Science*, **236**, 806–12.

Burmeister, M. and Lehrach, H. (1986) Long-range restriction map around the Duchenne muscular dystrophy gene. *Nature*, **324**, 582–4.

Burmeister, M., Monaco, A.P., Gillard, E.F. *et al.* (1988) A 10 megabase map of human Xp21 including the Duchenne muscular dystrophy gene. *Genomics*, **2**, 189–202.

Campbell, K.P. and Kahl, S.D. (1989) Association of dystrophin and an integral membrane glycoprotein. *Nature*, **338**, 259–62.

Chamberlain, J.S., Gibbs, R.A., Ranier, J.E. *et al.* (1988) Deletion screening of the Duchenne muscular dystrophy locus via multiplex DNA amplification. *Nucl. Acids Res.*, **16**, 11141–56.

Chelly, J., Gilgenkrantz, H., Lambert, M. *et al.* (1991) Effect of dystrophin gene deletions on mRNA levels and processing in Duchenne and Becker muscular dystrophies. *Cell*, **63**, 1239–48.

Coffey, A.J., Roberts, R.G., Green, E.D. *et al.* (1992) Construction of a 2.6-Mb contig in yeast artificial chromosomes spanning the human dystrophin gene using an STS-based approach. *Genomics*, **12**, 474–84.

Darras, B.T., Blattner, P., Harper, J.F. *et al.* (1988) Intrageneic deletions in 21 Duchenne muscular dystrophy (DMD)/Becker muscular dystrophy (BMD) families studied with the dystrophin cDNA: location of breakpoints on *Hind*III and *Bgl*II exon-containing fragment maps, meiotic and mitotic origin of the mutations. *Am. J. Hum. Genet.*, **43**, 620–9.

Den Dunnen, J.T., Bakker, E., Klein-Breteler, E.G. *et al.* (1987) Direct detection of more than 50% Duchenne muscular dystrophy mutations by field inversion gel electrophoresis. *Nature*, **329**, 640–2.

Den Dunnen, J.T., Grootscholten, P.M., Bakker, E. *et al.* (1989) Topography of the Duchenne muscular dystrophy gene: FIGE and cDNA analysis of 194 cases reveals 115 deletions and 13 duplications. *Am. J. Hum. Genet.*, **45**, 835–47.

Den Dunnen, J.T., Grootscholten, P.M., Dauwerse, J.G. *et al.* (1992) Reconstruction of the 2.4 Mb human DMD-gene by homologous YAC recombination. *Hum. Molec. Genet.*, **1**, 19–28.

Ervasti, J.M. and Campbell, K.P. (1991) Membrane organization of the dystrophin-glycoprotein complex. *Cell*, **66**, 1121–31.

Ervasti, J.M., Ohlendieck, K., Kahl, S.D. *et al.* (1990) Deficiency of a glycoprotein component of the dystrophin complex in dystrophic muscle. *Nature*, **345**, 315–19.

Feener, C.A., Koenig, M. and Kunkel, L.M. (1989) Alternative splicing of human dystrophin mRNA generates isoforms at the carboxy terminus. *Nature*, **338**, 509–11.

Forrest, S.M., Cross, G.S., Speer, A. *et al.* (1987) Preferential deletion of exons in Duchenne and Becker muscular dystrophies. *Nature*, **329**, 638–40.

Francke U., Ochs H.D., de Martinville B. *et al.* (1985) Minor Xp21 chromosome deletion in a male associated with expression of Duchenne muscular dystrophy, chronic granulomatous disease, retinitis pigmentosa and McLeod syndrome. *Am. J. Hum. Genet.*, **37**, 250–67.

Goodfellow, P.N., Davies K.E. and Ropers, H.H. (1985) Report of the committee on the genetic constitution of the X and Y chromosomes. *Cytogenet. Cell Genet.*, **40**, 296–352.

Hoffman, E.P., Brown, R.H., Jr. and Kunkel, L.M. (1987a) Dystrophin: The protein product of the Duchenne muscular dystrophy locus. *Cell*, **51**, 919–28.

Hoffman, E.P., Knudson, C.M., Campbell, K.P. and Kunkel, L.M. (1987b) Subcellular fractionation of dystrophin to the triads of skeletal muscle. *Nature*, **330**, 754–8.

Hoffman, E.P., Fischbeck, K., Brown, R.H. *et al.* (1988) Dystrophin quality and quantity determines the clinical severity of Duchenne/Becker muscular dystrophies. *N. Engl. J. Med.*, **318**, 1363–8.

Hu, X., Ray, P.N., Murphy, E.G. *et al.* (1990) Duplicational mutation at the Duchenne muscular dystrophy locus: Its frequency, distribution, origin, and phenotype genotype correlation. *Am. J. Hum. Genet.*, **46**, 682–95.

Kenwrick, S., Patterson, M., Speer, A. *et al.* (1987) Molecular analysis of the Duchenne muscular dystrophy region using pulsed-field gel electrophoresis. *Cell*, **48**, 351–7.

Koenig, M., Hoffman, E.P., Bertelson, C.J. *et al.* (1987) Complete cloning of the Duchenne muscular dystrophy (DMD) cDNA and preliminary genomic organization of the DMD gene in normal and affected individuals. *Cell*, **50**, 509–17.

Koenig, M., Monaco, A.P. and Kunkel, L.M. (1988) The complete sequence of dystrophin predicts a rod-shaped cytoskeletal protein. *Cell*, **53**, 219–28.

Koenig, M., Beggs, A.H., Moyer, M. *et al.* (1989) The molecular basis for Duchenne versus Becker muscular dystrophy: correlation of severity with type of deletion. *Am. J. Hum. Genet.*, **45**, 498–506.

Kunkel, L.M., Monaco, A.P., Middlesworth, W. *et al.* (1985) Specific cloning of DNA fragments absent from the DNA of a male patient with an X chromosome deletion. *Proc. Natl. Acad. Sci. USA*, **82**, 4778–82.

Kunkel, L.M. *et al.* (1986) Analysis of deletions in the DNA of patients with Becker and Duchenne muscular dystrophy. *Nature*, **322**, 73–7.

Ibraghimov-Beskrovnaya, O., Ervasti, J.M., Leveille, C.J. *et al.* (1992) Primary structure of dystrophin-associated glycoproteins linking dystrophin to the extracellular matrix. *Nature*, **355**, 696–702.

Lidov, H.G.W., Byers, T.J., Watkins, S.C. and Kunkel, L.M. (1990) Localization of dystrophin to postsynaptic regions of central nervous system cortical neurons. *Nature*, **348**, 725–8.

Malhotra, S.B., Hart, K.A., Klamut, H.J. *et al.* (1988) Frame-shift deletions in patients with Duchenne and Becker muscular dystrophy. *Science*, **242**, 756–9.

Monaco, A.P., Bertelson, C.J., Middlesworth, W. *et al.* (1985) Detection of deletions spanning the Duchenne muscular dystrophy locus using a tightly linked DNA segment. *Nature*, **316**, 842–5.

Monaco, A.P., Neve, R.L., Colletti-Feener, C. *et al.* (1986) Isolation of candidate cDNAs for portions of the Duchenne muscular dystrophy gene. *Nature*, **323**, 646–50.

Monaco, A.P., Bertelson, C.J., Colletti-Feener, C. and Kunkel, L.M. (1987) Localization and cloning of Xp21 deletion breakpoints involved in muscular dystrophy. *Hum. Genet.*, **75**, 221–7.

Monaco, A.P., Bertelson, C.J., Liechti-Gallati, S. *et al.* (1988) An explanation for the phenotypic difference between patients bearing partial deletions of the DMD locus. *Genomics*, **2**, 90–95.

Monaco, A.P., Walker, A.P., Millwood, I. *et al.* (1992) A yeast artificial chromosome contig containing the complete Duchenne muscular dystrophy gene. *Genomics*, **12**, 465–73.

Moser, H. (1984) Duchenne muscular dystrophy: Pathogenetic aspects and genetic prevention. *Hum. Genet.*, **66**, 17–40.

Orkin, S.A. (1986) Reverse genetics and human disease. *Cell*, **47**, 845.

Ray, P.N., Belfall, B., Duff, C. *et al.* (1985) Cloning of the breakpoint of an X;21 translocation associated with Duchenne muscular dystrophy. *Nature*, **318**, 672–5.

Royer-Pokora, B., Kunkel, L.M., Monaco, A.P. *et al.* (1986) Cloning the gene for an inherited human disorder – chronic granulomatous disease – on the basis of its chromosomal location. *Nature*, **322**, 32–8.

Van Ommen, G.J.B., Verkerk, J.M.H., Hofker, M.H. *et al.* (1986) A physical map of 4 million base pairs around the Duchenne muscular dystrophy gene on the human X-chromosome. *Cell*, **47**, 499–504.

Van Ommen, G.J.B., Bertelson, C.E., Ginjaar, H.B. *et al.* (1987) Long-range genomic map of the Duchenne muscular dystrophy (DMD) gene: isolation and use of J66 (DXS268), a distal intragenic marker. *Genomics*, **1**, 329–36.

Watkins, S.C., Hoffman, E.P., Slayter, H.S. and Kunkel, L.M. (1988) Immunoelectron microscopic localization of dystrophin in myofibres. *Nature*, **333**, 863–6.

Zubrzycka-Gaarn, E.E., Bulman, D.E., Karpati, G. *et al.* (1988) The Duchenne muscular dystrophy gene product is localized in sarcolemma of human skeletal muscle. *Nature*, **333**, 466–9.

2

Genotype/phenotype correlations in Duchenne/Becker dystrophy

ERIC P. HOFFMAN

2.1 INTRODUCTION

Genotype/phenotype correlations in Duchenne/Becker muscular dystrophy can be attempted at many levels: molecular genetics (dystrophin gene mutations), biochemistry (dystrophin protein abnormalities), cell biology (muscle fibre dysfunction), tissue physiology (muscle dysfunction and pathology), and clinical phenotype (disease presentation and progression). In many ways, Duchenne/Becker dystrophy represents an ideal genetic disorder in which to make such correlations (Table 2.1). The plethora of mutations of the massive and complex dystrophin gene are known to cause a dramatic range of clinical disorders, and the X-linked recessive expression enables study of solitary loss-of-function or change-of-function mutations in hemizygous males, and in the mosaic situation caused by X-inactivation in heterozygous females.

The engaging complexity of dystrophin gene structure and expression also makes Duchenne/Becker one of the more difficult disorders in which to study genotype/phenotype correlations (Table 2.1). The dystrophin gene was one of the first disease genes identified by 'positional cloning'. Thus, there was no antecedent biochemical data on dystrophin protein structure or function. It has been only four years since the identification of the gene and protein, and questions still remain. Does dystrophin simply stabilize muscle plasma membranes, or is the pathophysiology of dystrophin-deficiency more complicated? Dystrophin-deficiency in vascular smooth muscle and some neurones begets complicating variables: are there vasculogenic or neurogenic components involved in the progressive skeletal muscle pathology? Finally, amino acid substitutions caused by point mutations may yield the most important data.

Molecular and Cell Biology of Muscular Dystrophy
Edited by Terence Partridge
Published in 1993 by Chapman & Hall, London. ISBN 0 412 43440 7

Table 2.1 Advantages and disadvantages of dystrophinopathy as a paradigm for gene/protein/clinical correlations

Advantages
Complex at the molecular genetic, biochemical, physiological, and clinical levels
Highest germ-line mutation rate known, most cases represent independent mutation events
Gene cloned, protein identified, molecular tools available
Both loss-of-function and change-of-function mutations in single copy hemizygous background
The effect of mosaicism can be studied: X-inactivation in female heterozygotes
Extensive biomedical network documenting patient symptoms, including subjective symptoms such as fatigue, muscle pain, stiffness
Disadvantages
Complex at the molecular genetic, biochemical, physiological, and clinical levels
40% of mutations remain unidentified (point mutations in very large gene)
Low quantities of dystrophin make biochemical studies of abnormal dystrophin proteins impossible
The highly differentiated and specialized state of mature muscle *in vivo* is impossible to duplicate experimentally
The connections between muscle membrane instability and muscle pathology (progressive muscle loss) are not understood

However, screening the 2.5 million base pair gene for point mutations is, at best, problematic.

There is undoubtedly a wealth of information entombed in mutant dystrophin genes and proteins that seem in unending supply in both humans and animals. Efforts to unearth these data have been successful to a limited extent: deletion mutations have been defined for over 1000 Duchenne and Becker patients, but in nearly 700 other patients studied no mutation has been found. Even when the molecular genetic basis for the disease is understood in a cohort of patients, attempts at genotype/phenotype correlations reveal our relatively shallow understanding of variables contributing to the pathophysiology of dystrophinopathies. To quote a recent study by Beggs *et al.*, (1991), 'Unfortunately, the intra-group variation among patients with similar deletions is so great that it is difficult to draw firm conclusions . . .'.

Reviews and hypotheses on the pathophysiology of Duchenne/Becker dystrophy can be found elsewhere in this volume, or in other recent publications (Hoffman and Gorospe, 1991; Rojas and Hoffman, 1991). This chapter is intended to summarize the studies on gene/protein/phenotype correlations, and emphasize the documented generalizations that can be made, as well as what remains to be learned.

2.2 DUCHENNE MUSCULAR DYSTROPHY

Duchenne muscular dystrophy is, in general, a clinically homogeneous disorder. Boys with Duchenne dystrophy typically present at age 4 to 5 years, with difficulty keeping up with peers. The disease is progressive, with proximal muscle weakness leading to loss of ambulation around 11 years and death due to respiratory failure usually by 20 years. If ventilated, some men with Duchenne dystrophy live into their 40s.

Boys with Duchenne dystrophy can be ascertained as infants due to delayed motor milestones, although there is overlap with normal children. Dramatic elevations of serum creatine kinase (CK) is an invariant laboratory finding in boys with Duchenne dystrophy at any age, even in the absence of obvious clinical symptoms. Assay for CK in blood spots has proved an effective screen for Duchenne in neonates (Greenberg *et al.*, 1991; Naylor *et al.*, 1992). Muscle histopathology is evident from birth onwards, though it is progressive and generally parallels the clinical disease expression (Table 2.2).

There is strong correlation between biochemical and clinical findings in Duchenne dystrophy: > 99% of boys show marked dystrophin deficiency at all ages (Tables 2.2, 2.3). The biochemical deficiency is neither age-dependent nor progressive: dystrophin appears in normal skeletal muscle during early differentiation (~12 weeks gestational age), and is deficient in fetuses with Duchenne dystrophy (Figure 2.1) (Bieber *et al.*, 1989; Clerk *et al.*, 1992b). Fetal muscle biopsies are diagnostic of Duchenne dystrophy *in utero* (Evans *et al.*, 1991; Kuller *et al.*, 1992).

All reports agree that dystrophin-deficiency is an invariant feature of Duchenne dystrophy (Table 2.3). Early reports described a few boys carrying a Duchenne dystrophy diagnosis with dystrophin of normal size and amount. These patients are now thought to have unrelated disorders: of the 1363 patients studied at the biochemical level, no case has yet been described which shows X-linked inheritance of the disease, a clinical picture consistent with Duchenne dystrophy, and normal dystrophin. Most deletion mutations are expected to produce truncated dystrophin molecules, but these are infrequently seen and when present show considerable reductions in quantity (Bulman *et al.*, 1991b; Hoffman *et al.*, 1991; Lupski *et al.*, 1991). The correlation of dystrophin-deficiency with Duchenne dystrophy is extraordinary: many other inherited metabolic diseases show high levels of the culpable protein, however it lacks functional activity. The findings in Duchenne dystrophy suggest that there is tight control of dystrophin protein or mRNA stability, or both: dystrophin mRNA which is not efficiently or completely translated, or dystrophin proteins which do not correctly integrate into the membrane cytoskeleton, are catabolized (Hoffman and Kunkel, 1989; Chelly *et al.*, 1990). Point mutations resulting in amino-acid

Table 2.2 Gene, protein, histopathology, and clinical correlations

Disease Type: Gene Mutation Type	*Gene-correlated Dystrophin Protein Abnormality*	*Histopathology. Generally correlated with age, not gene or protein abnormality*	*Clinical Presentation/ Progression. Generally correlated with histopathology*
Duchenne: 55% gene deletion	> 99% Dystrophin deficiency (< 5% protein)	Fetal – 1 yr: hypercontracted fibres, loose fibrosis	Extremely high serum CK
Duchenne: 5% gene duplication	> 99%Dystrophin deficiency (< 5%)	1–5 yr: fibre size variation, progressive fibrosis	Extremely high CK, muscle hypertrophy, delayed motor milestones
Duchenne: 40% presumed point mutations	> 99% Dystrophin deficiency (< 5%)	> 5 yr: failed regeneration, fibro/fatty replacement	Extremely high CK, progressive muscle wasting, weakness
Becker: 68% gene deletion	95% dystrophin of smaller molecular weight. 5% normal size, low quantity	Variable. Generally less dramatic than Duchenne	Very high CK
Becker: 5% gene duplication	100% dystrophin of larger molecular weight	At all ages, fibre hypertrophy, mild to severe endomysial fibrosis, degen./regen.	Variable muscle hypertrophy Variable muscle wasting
Becker: 27% presumed point mutation	80% dystrophin of smaller molecular weight. 20% normal size, reduced quantity	Progressive, often with failed regeneration and focal fibro/fatty replacement	Variable progression, often with proximal muscle weakness
Dystrophinopathy in females (manifesting carriers)	Mosaic immunostaining pattern in biopsy	Variable, though similar to either Duchenne or Becker. Often focal	Variable, though similar to either Duchenne or Becker
Unrelated disorders: 100% no detectable mutation	100% dystrophin of normal size Rare patients with 2° partial deficiency	Duchenne-like: rare Becker-like: frequent Carrier-like: very frequent	Duchenne-like: rare Becker-like: frequent Carrier-like: very frequent

Table 2.3 Dystrophin protein findings. Summary of dystrophin protein findings from different laboratories. Patients included in more than one publication are listed only once

Part A Muscle biopsies from postnatal males and females

Citation	*No. patients studied*	*Antibody*	*Technique used*[1]	*Postnatal males: DMD*		*Postnatal males: Becker*		*Postnatal: unrelated disorders*		*Females*		
				No.	No. dystr. def.	No.	No. dystr. abnorm.	No.	No. dystr. normal	No. familial	No. familial mosaic	No. isolated mosaic
Hoffman *et al.*, 1988	104	30 kDa	B	38	35	23	18	40	38			2
Arahata *et al.*, 1988	85	DMDP-I	IF	27	27	9	9	49	49			1
Arahata *et al.*, 1989b	68	DMDP-I,II; 30 kDa, 60 kDa	IF, B	18	18	14	14	32	29	4	4	
Hoffman *et al.*, 1989b	97	30 kDa	B			54	54	29	29[2]			
Patel *et al.*, 1988[3]	26	60 kDa	B	6	6	4	2	16	16			
Bonilla *et al.*, 1988	8	60 kDa	IF							8	7	
Burrow *et al.*, 1991	33	Prit	IF, B, DNA	30	30	3	3					
Arahata *et al.*, 1989a	3	DMDP-II	IF							3	3	

Nicholson *et al.*, 1990[4]	226	dys-I, (Dy4/6D3)	IF, B	85	85	55	55	83	83	3	3	
Voit *et al.*, 1991b[5]	86	60 kDa; d10; dysII		62	62	26	26			2	2	
Uchino *et al.*, 1989	50	30 kDa; 60 kDa	If, B	6	6	5	5	38	38	1	1	
Bulman *et al.*, 1991b[6]	29	9219; 1461	B	25	25	4	2					
Byers *et al.*, 1992[7]	25	6/1 & 2/6	ELISA, DNA	5	5	9	9	11	11			
Arahata *et al.*, 1991	75	4C5; DMDP I, II	IF, B, DNA	21	21	30	30	24	24			
Clerk *et al.*, 1991	16	60 kDa; 2814	IF, B							16	11	
Hoffman *et al.*, 1992[8]	505	DMDP-II; 60 kDa; Torp	IF					459	459	21	21	25
Totals	1436			323	320	236	227	781	776	58	52	28
Percentage					99%		96%		99%		90%	

Part B Muscle from fetuses and neonatal screening patients

Citation	*Antibody used*	*Technique used*	*Fetal muscle*				*Neonatal screening: Duchenne*
			Normal necropsy	No. known or probable Duchenne necropsy	No. Duchenne dystrophin deficient necropsy	No. fetal biopsy	No. dystrophin deficient from CK test
Evans *et al.*, 1991	30 kDa	B				1 normal	
Clerk *et al.*, 1992b	60 kDa; H1; 2P 6; dys I, II	IF, B, DNA	4	12	12		
Kuller *et al.*, 1992	60 kDa; d10; 30 kDa	IF, B, DNA	8	4	4	3 2 normal, 1 DMD	
Greenberg *et al.*, 1991	60 kDa	B, DNA					8
Naylor *et al.*, 1992	60 kDa, d10, 30 kDa	IF, B, DNA					12

[1]Immunoblot, B; immunofluorescence, IF; DNA, dystrophin gene analysis also done.
[2]There was no evidence of an X-linked family history in patients with normal dystrophin, however this does not exclude the possibility that a few of these patients have Becker dystrophy.
[3]Ten fetal samples were also studied. These are included in Part B, Clerk *et al.*, 1992b.
[4]This report did not give specific numbers for the number of 'dystrophin-deficient' Duchenne or 'abnormal dystrophin' Becker patients,however it states that the patients generally fell into these categories with overlap and some exceptions.
[5]Also studied, but not shown here,were five cases of intermediate severity (severe-Becker), and four cases of female carriers of Becker dystrophy.
[6]Two of the four Becker patients were wheelchair bound at age 14 yrs, and both showed dystrophin deficiency.
[7]This included three severe-Becker (intermediate) patients who were found to have levels of dystrophin intermediate between Duchenne and Becker. These patients were previously described in Beggs *et al.* (1991).
[8]There were 25 cases of limb-girdle/hyper-CKaemia which were re-classified as isolated female dystrophinopathy patients.

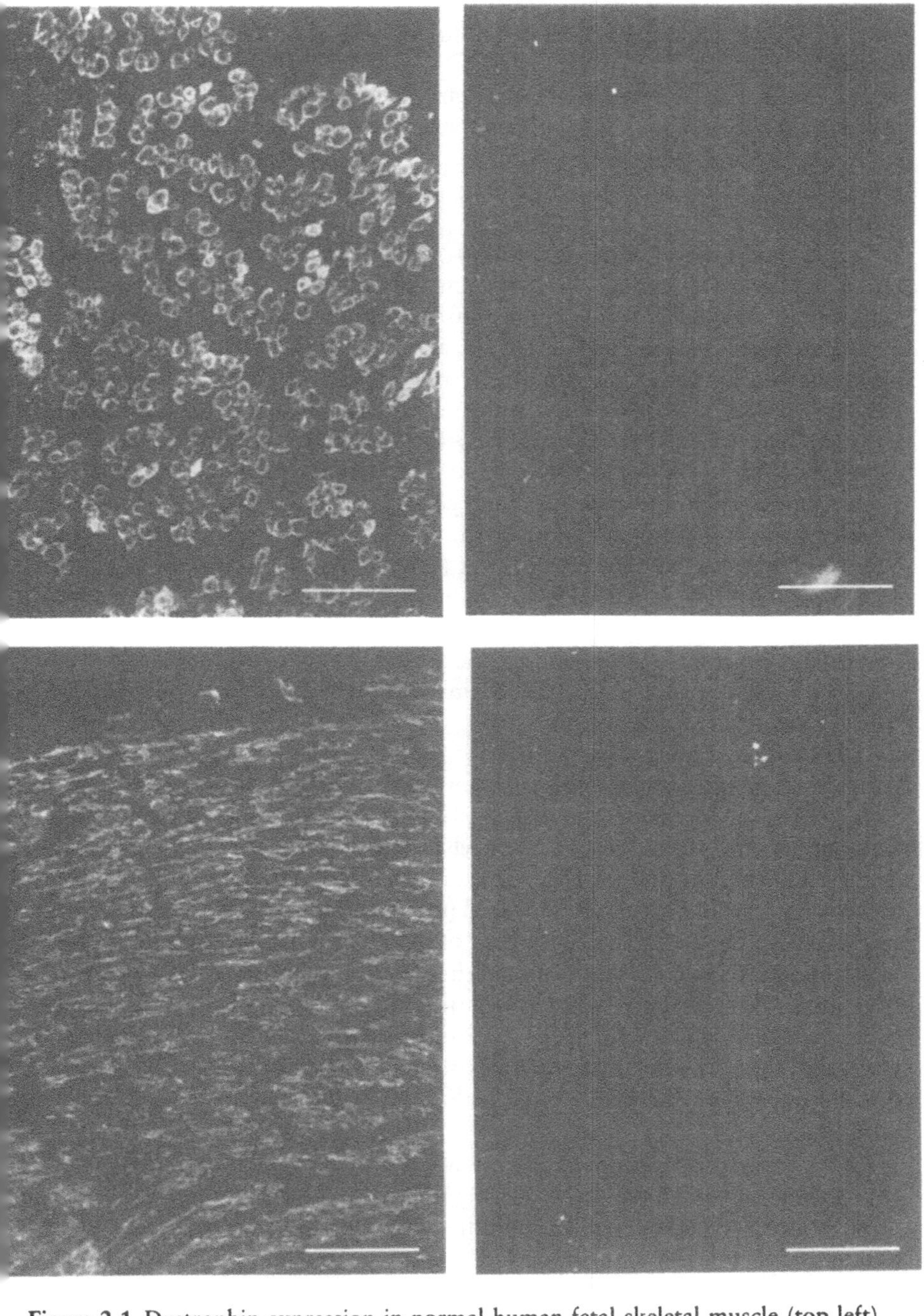

Figure 2.1 Dystrophin expression in normal human fetal skeletal muscle (top left) and heart (bottom left), and dystrophin deficiency in Duchenne fetal skeletal muscle (top right) and heart (bottom right). Shown are cryosections from 18 week fetopsy specimens. The antibody used was 60 kDa, and all experiments were done in parallel, with the photographic exposure times identical. Dystrophin deficiency is clearly manifested biochemically from the point at which muscle fibres first form in fetal life, however the clinical phenotype does not become apparent until much later in life. Bar = 100 μm.

substitutions which perturb interactions between dystrophin and other proteins (Ervasti *et al.*, 1990) probably exist, and these would also be expected to cause dystrophin catabolism. The future identification of these point mutations is critical for the understanding of dystrophin structure and function.

Precise quantitation of the extent of dystrophin deficiency is difficult, and reports vary concerning the amount of dystrophin which defines 'dystrophin-deficiency' in Duchenne muscle. All quantitations of dystrophin are done using immunological methods, and antibodies and techniques vary between laboratories. Quantitation is made more difficult because many anti-dystrophin antibodies show some cross-reaction with chromosome-6-encoded 'dystrophin-related protein' (DRP) (Hoffman *et al.* 1989a; Love *et al.*, 1989; Khurana *et al.*, 1990; Voit *et al.*, 1991a). DRP is indistinguishable from true dystrophin in immunoblots, yet is present at normal amounts in Duchenne muscle. DRP is expressed in most tissues, but is localized at the neuromuscular junction in muscle tissue (Love *et al.*, 1991; Takemitsu *et al.*, 1991; Khurana *et al.*, 1992). Also, somatic reversion of dystrophin gene mutations occurs in many Duchenne patients, and this causes a low level of dystrophin expression in patient muscle (Shimizu *et al.*, 1988; Nicholson *et al.*, 1989; Hoffman *et al.*, 1990; Klein *et al.*, 1992; Fanin *et al.*, 1992). Thirty percent of Duchenne boys show some reversion in their muscle, but the prevalence of revertant myofibres rarely exceeds 1% of fibres in young patients. Patients with somatic reversion or truncated abnormal dystrophin proteins are still 'dystrophin-deficient', yet they are not 'dystrophin-absent'. The ability to detect very low levels of either normal or abnormal dystrophin depends on the specificity and sensitivity of the antibodies used, and the immunological assay employed (immunofluorescence, immunoblotting, ELISA). The most accurate report of dystrophin quantitation in muscle biopsies has recently been accomplished by Byers *et al.* (1992) using a two antibody sandwich ELISA technique. The findings of this study agreed with previous subjective determinations of dystrophin quantity, with non-DMD/BMD (unrelated disorders) showing 100 ± 14.1% of normal dystrophin ($n = 11$), and Duchenne muscle showing 0.3 ± 0.1% ($n = 5$).

Molecular genetic studies of the 2.5 million base pair dystrophin gene in Duchenne patients have been largely limited to finding deletions and duplications of the 70 exons. While it is often stated that 65% of DMD patients have a deletion of one or more exons, compilation of studies using the entire cDNA shows the actual frequency to be lower (Table 2.4). Some reports do not discriminate between the results for Duchenne versus Becker dystrophy, while others do. In the 687 DMD patients for whom DNA data has been detailed, 375 (54.6%) have a deletion mutation (Table 2.4). Duplication mutations are more difficult to detect, and many molecular

Table 2.4 Genotype/phenotype correlations in 1689 cases of Duchenne/Becker dystrophy

Citation	*No. DMD +BMD*	*No. deletions*	*No. duplication*	*No. nothing*	*No. DMD*	*No. DMD deletion*	*No. DMD duplication*	*No. DMD nothing*	*No. BMD*	*No. BMD deletion*	*No. BMD duplication*	*No. BMD nothing*
Asano *et al.*, 1991, Japan					28	9	0	19	22	11	0	11
Hu *et al.*, 1990, Canada	181	109	10	62								
Koenig *et al.*, 1987, USA					104	53	0	51				
Den Dunnen *et al.*, 1989, Dutch					160	98	12	50	34	17	1	16
Upadhaya *et al.*, 1990, England					164	82	0	82				
Medori *et al.*, 1989, USA					32	14	0	18	6	5	0	1
Hodgson *et al.*, 1989, England	287	163		124								
Darras *et al.*, 1988, USA					29	18	0	11	3	3	0	0
Baumbach *et al.*, 1989, USA	160	90	8	62								
Beggs *et al.*, 1991, USA/Japan/Italy									68	53	5	10

Table 2.4 *(con'd)*

Citation	*No. DMD +BMD*	*No. deletions*	*No. duplication*	*No. nothing*	*No. DMD*	*No. DMD deletion*	*No. DMD duplication*	*No. DMD nothing*	*No. BMD*	*No. BMD deletion*	*No. BMD duplication*	*No. BMD nothing*
Ballo *et al.*, 1991, South Africa					60	30		30				
Laing *et al.*, 1991, Australia	60	39	2	19								
Cooke *et al.*, 1990, Scotland					110	71	1	38	25	18	1	6
Covone *et al.*, 1991, Italy	127	73		54								
Soong *et al.*, 1991, China	29	13		16								
No. Patients	844	487	20	337	687	375	13	299	158	107	7	44
% Mutation		57.7	2.4	40.0		54.6	1.9	43.5		67.7	4.4	27.8

genetic studies have not looked for this type of genetic alteration. The 2% prevalence of duplications shown in the compiled data (Table 2.4) is an underestimate of the true frequency: careful studies have shown a duplication frequency of approximately 5% (Hu *et al.*, 1990).

Most intragenic deletion and duplication mutations probably do not preclude expression of mRNA from the dystrophin gene. In the few cases which have been studied at the mRNA level, the remaining exons are transcribed and spliced, though the abnormal mRNA seems to be unstable or inefficiently expressed (Chelly *et al.*, 1990). The majority (90%) of deletions and duplications in Duchenne dystrophy cause a frame-shift in the translational reading frame of the transcribed mRNA (Chelly *et al.*, 1990; Monaco *et al.*, 1988; Koenig *et al.*, 1989): when the remaining exons are spliced together, translation cannot continue over the new exon boundary and dystrophin protein translation is prematurely terminated. The abnormal truncated dystrophin proteins also seem to be unstable. The truncated protein has been visualized in some Duchenne muscle, although the quantity of the abnormal dystrophin is always quite low (Bulamn *et al.*, 1990b; Hoffman *et al.*, 1991; Lupski *et al.*, 1991). Approximately 10% of Duchenne patients have in-frame deletions or duplications, and are exceptions to the reading frame rule (Byers *et al.*, 1992; Malhotra *et al.*, 1988). Both exon-skipping (Matsuo *et al.*, 1991) and protein instability (Chelly *et al.*, 1990) have been hypothesized to be mechanisms for dystrophin-deficiency seen in Duchenne boys having in-frame deletions/duplications of their dystrophin gene.

About 40% of patients with Duchenne muscular dystrophy have no detectable deletion or duplication (Table 2.4). It is presumed that these patients have point mutations of the dystrophin gene affecting transcription, mRNA processing, translation, or protein stability. To date, only a few point mutations have been identified (Bulman *et al.*, 1991a; Roberts *et al.*, 1992). From the compiled published data, about 600 Duchenne boys have point mutations which remain to be identified.

What has been learned from genotype/phenotype correlations in Duchenne dystrophy, and what remains to be learned? We have learned that the genotype/phenotype correlations are quite straightforward in Duchenne dystrophy: dystrophin-deficiency causes a remarkably uniform disease in humans. However, we are only partially aware of the possible genotypes. While a few non-deletion mutations which inactivate the dystrophin gene have recently been found, those point mutations which render the dystrophin protein non-functional and unstable remain to be found. These will tell us a considerable amount about dystrophin protein structure and function.

We know that there must be additional aspects of the human pathophysiology which may not be a direct and immediate consequence of dystrophin deficiency: affected fetuses and neonates do not express the profound weakness that a 10-year-old patient does, yet the extent of dystrophin deficiency is the same. Does the chronic leakage of muscle fibres cause progressive fibrosis which inhibits the ability of the muscle to regenerate? This question must be answered if we hope to develop rational therapies for Duchenne dystrophy. We have learned that somatic reversion takes place, though this seems to have little consequence on the clinical expression of the disease.

2.3 DYSTROPHINOPATHY IN GIRLS AND WOMEN

Female cells inactivate one of their two X chromosomes early in embryonic development. In a girl or woman who is a carrier of Duchenne dystrophy, approximately half the nuclei have the abnormal, non-functional dystrophin gene active, while the other half have the normal, functional gene active. Since muscle fibres are syncytial, both dystrophin-positive and dystrophin-negative myonuclei exist in the same fibre. In dogs and mice, evidence has accumulated which suggests that the positive nuceli seem to compensate for the negative ones within the fibre causing biochemical normalization of dystrophin content over time (Watkins *et al.*, 1989; Cooper *et al.*, 1990; Tanaka *et al.*, 1990; Weller *et al.*, 1991): this may explain the decrease of serum creatine kinase levels in human female carriers as they grow older.

If X-inactivation in a carrier is skewed, then the majority of cells may become unable to produce dystrophin, and dystrophin-deficiency in females becomes possible. In the late 1970s, isolated cases of Duchenne dystrophy in females with X-autosome translocations were found, and all cases showed the X chromosome translocation breakpoint to occur in the Xp21 region (see Boyd *et al.*, 1986). Indeed, this observation was the first data pointing to Xp21 as the site of the Duchenne gene. In these cases, the translocated X chromosome, with its defective dystrophin gene, was kept active, thereby forcing the inactivation of the normal X chromosome. The translocation females are always isolated cases with no previous history of Duchenne dystrophy in males. 'Manifesting carriers' also occasionally occur in families with boys with typical Duchenne dystrophy, and these cases show skewed inactivation of the normal X chromosome. Since X-inactivation is thought to be a random process, one expects statistical outliers which inactivate a greater proportion of the normal X chromosome. In keeping with this logic, the clinical expression of familial cases of manifesting carriers is quite broad, with many showing only minor symptoms (Moser and Emery, 1974).

Bonilla *et al.* (1988) and Arahata *et al.* (1989a) showed that manifesting carriers could be biochemically detected by the mosaic pattern of dystrophin protein expression in muscle: populations of dystrophin-positive and dystrophin-negative fibres could be seen using dystrophin immunofluorescence of muscle cryosections. A single fibre from a manifesting carrier can show normal dystrophin over patches of membrane and dystrophin-deficiency over neighbouring regions: these regions correspond to nuclear domains of the dystrophin-positive and dystrophin-negative nuclei. This finding led to the identification of another developmental genetic mechanism responsible for skewing of X-inactivation, monozygotic twinning (Richards *et al.*, 1990; Bonilla *et al.*, 1990; Lupski *et al.*, 1991). The twinning process in early embryonic development leads to a 'sampling error' of the inner cell mass, such that one of the twins inherits cells where the majority have the normal X chromosome inactivated, thereby manifesting dystrophin-deficiency (Nance, 1990). Additional studies proved valuable at the clinical diagnostic level: isolated cases of dystrophinopathy in girls and women were found carrying different diagnoses and showing different presentations and progressions (Minetti *et al.*, 1991; Arikawa *et al.*, 1991). Finally, the ability to examine both dystrophin-positive and dystrophin-negative muscle fibres in the same muscle biopsy has enabled elegant comparative studies of muscle fibre membrane pathophysiology in Duchenne dystrophy (Morandi *et al.*, 1990). Indeed, the observed influx of large molecules (albumin) into non-necrotic dystrophin-deficient fibres and not into adjacent dystrophin-positive fibres has provided some of the most supportive evidence for a generalized 'leaky' membrane as the primary cellular consequence of dystrophin-deficiency.

The study of manifesting carrier girls and women could potentially answer some important questions. What is the degree of correlation between dystrophin quantities and clinical phenotype? Is there a minimum amount of dystrophin needed to prevent overt clinical symptoms? Does dystrophin-deficiency correlate with histopathology in muscle? What is the range of clinical phenotypes caused by quantitative abnormalities of dystrophin?

A recent large collaborative study strove to answer these questions (Hoffman *et al.*, 1992). Muscle biopsies from 505 girls and women were studied for dystrophin content. Forty-six patients were found to have mosaic immunostaining denoting a carrier status. Twenty-one had a family history for Duchenne dystrophy in males. Twenty-five were isolated cases. The isolated cases showed a correlation between the biochemical and clinical findings; the patients with the lowest percentage of dystrophin generally had the highest CK levels and the most severe clinical phenotype. In addition, the findings suggest that dystrophin levels must be relatively high before precluding clinical expression of disease. A patient with 70% dystrophin by immunoblot (52% negative fibres by immunofluorescence) presented with myalgia at 8 years and had striking elevations of serum CK

(8060–12660 IU/l; normal < 200). Another patient who had only 6.2% negative fibres by immunofluorescence and 64% dystrophin by immunoblot presented at 20 years with easy fatigue, and had CK levels of 1834. Thus, muscle needs at least half of its normal complement of dystrophin to preclude overt clinical symptoms. The findings of elevated CKs in 70% of female carriers (without clinical symptoms) suggests that the muscle requires considerably more than 50% dystrophin to preclude muscle fibre abnormalities (leakage). This is in marked contrast to enzymatic deficiencies where as little as 10% of enzyme activity can preclude clinical and biochemical symptoms.

At the histopathological level, the pathology seen in the muscle biopsies from isolated cases of manifesting carriers often correlated with dystrophin expression: dystrophin-negative regions of the muscle tended to show more pronounced histopathology (Morandi *et al.*, 1990). This finding bolsters the expected cell-autonomy of dystrophin-deficiency, and also underscores the localized nature of the hypothesized fibrotic reaction of muscle tissue to fibre leakage in dystrophin-deficient regions (Hoffman and Gorospe, 1991). However, none of the correlations are complete, and the complicating variables of clonality and sampling error (small muscle biopsy) in human manifesting carriers make controlled and conclusive studies very difficult. Unfortunately, the occurrence of manifesting carriers is rare enough in humans so that the study of manifesting carriers in the dystrophin-deficient dog, cat, or mouse is probably impossible.

While the biochemical/clinical correlations in the isolated cases were satisfying, similar correlations were not found in the familial cases (positive family history of Duchenne in males). The authors conclude that there was substantial ascertainment bias in the familial cases which was not encountered in the isolated cases. This latter finding has an important bearing on all studies of genotype/phenotype correlations in dystrophinopathy: if ascertainment bias on the part of the patient, her family and doctors has such an influence on clinical expression, objective studies become very difficult.

What remains to be learned from the study of manifesting carriers? Aside from girls with translocations and monozygotic twins, the genetic or developmental mechanisms leading to skewed lyonization in manifesting carriers is unknown. Is this simply expected statistical deviation of X-inactivation, or could other developmental or genetic triggers cause this? The biochemical normalization taking place in carriers suggests that dystrophin-positive nuclei are somehow compensating for the dystrophin-negative nuclei. Is this due to a positive up-regulation of dystrophin gene expression or simply due to a lack of down-regulation because of insufficient saturation of the dystrophin protein in the membrane cytoskeleton?

2.4 BECKER MUSCULAR DYSTROPHY

Phenotypes that seemed similar to, yet milder than Duchenne dystrophy were recognized as early as 1879 (Gowers, 1879). The finding of families with clear X-linked inheritance established 'Becker muscular dystrophy' as a clinically distinct entity (Becker and Liener, 1955). The lower incidence of Becker muscular dystrophy (approximately 10-fold less frequent than Duchenne), and the clinical overlap with other dystrophies limited previous studies of Becker dystrophy to only about 20 families showing a clear X-linked family history (Emery and Skinner, 1976; Bradley *et al.*, 1978).

At least 96% of Becker dystrophy patients show dystrophin in their muscle which is abnormal in quantity, quality (molecular weight), or both (Hoffman *et al.* 1988; Hoffman *et al.*, 1989b) (Tables 2.2, 2.3). Early reports mentioned a few Becker patients with no detectable dystrophin in their muscle, however most if not all such cases can be explained by technical problems of the dystrophin assay (poor condition of biopsy, or 'false-negative' due to deletion of antigenic regions of the dystrophin protein). A few additional patients carrying a Becker dystrophy diagnosis were found to have normal dystrophin, however no such patient has been found to have an X-linked family history: this suggests that the clinical diagnosis of Becker may be erroneous in most or all of these cases (Hoffman *et al.*, 1989b). An additional category of patients, 'severe-Becker dystrophy' (wheelchair-bound 14–20 years) has been proposed which includes cases with a clinical phenotype intermediate between that of typical Duchenne dystrophy (wheelchair bound < 14 years) and Becker dystrophy (wheelchair-bound > 20 years). These patients have low amounts (~10%) of dystrophin in their muscle (Hoffman *et al.*, 1989b; Byers *et al.*, 1992), however the number of patients studied is small, and the clinical and biochemical overlap with both Duchenne and Becker dystrophies may be substantial. Dystrophin quantitation is thus able to offer a prognosis to young male patients by assigning them into the biochemical groups of Duchenne (< 5% dystrophin), severe-Becker (tentatively 5%–12%), and Becker (> 12%).

Genetic analyses of Becker patients showed a high proportion of deletion mutations of the dystrophin gene (Tables 2.2, 2.4). Compilation of the data for 158 patients reported to date shows 68% of Becker patients to have a gene deletion mutation of one or more exons, and approximately 5% to have duplication mutations. The remaining 28% of patients are presumed to have point mutations. The frequency of deletion mutations in Becker dystrophy (68%) is substantially higher than in Duchenne dystrophy (55%).

The clinical range of disorders caused by abnormal dystrophin is quite dramatic, and continues to expand. In the first extensive biochemical study of Becker patients, 30% of the 54 patients with abnormal dystrophin presented with clinical manifestations not previously considered diagnostic of Becker dystrophy (Hoffman *et al.*, 1988). Traditionally, the diagnostic criteria for Becker dystrophy were a presentation much like Duchenne dystrophy, but at a later age. In this study, patient presentations included muscle pain (myalgia), incidental findings of elevated serum CK, myoglobinuria, malignant hyperthermia, and ankle tendon contractures. Most of these atypical patients did not have detectable proximal limb-weakness. The clinical criteria for consideration of a Becker dystrophy diagnosis were changed to male sex and CK levels usually in excess of 1000 U/l, with primarily myopathic changes of both muscle histology and electromyography. Becker patients were also found to have a high incidence of cardiac abnormalities detected by ECG (50%) compared to patients with normal dystrophin but similar clinical phenotypes (7%).

Subsequent studies continued to expand the clinical phenotypes resulting from abnormal dystrophin. Weakness isolated to the quadriceps muscles (Sunohara *et al.*, 1990), patients with myalgias and cramps (Gospe *et al.*, 1989), patients with clinical phenotypes more consistent with limb-girdle muscular dystrophy (Arikawa *et al.*, 1991), and patients presenting with cardiomyopathy (Palmucci *et al.*, 1992) all were shown to have a dystrophinopathy at both the gene and protein levels.

The large range of clinical presentations and progressions in Becker dystrophy provides a unique opportunity to study genotype/phenotype correlations. All patients produce some dystrophin, however the dystrophin is abnormal, and most (68%) have missing blocks of amino acids due to a deletion mutation. Thus, a series of nested deletions exists *in vivo* in Becker dystrophy: the structure and function of the dystrophin protein can be studied by correlating specific gene mutations with the resulting phenotype at the biochemical, cellular (histopathological) and clinical levels.

To make such correlations, a Becker patient's gene must be fully analysed and the deletion defined. If a deletion is not present, then the patient must be excluded from the study because the genotype is not identifiable. Of those Becker patients with a defined deletion, a muscle biopsy must be obtained for biochemical and histopathological studies. In many cases the diagnosis of Becker dystrophy can be established by clinical and gene deletion data, precluding the requirement for a 'diagnostic' muscle biopsy: this functionally reduces the number of patients available for study. Finally, definitive clinical characterization of Becker patients can be difficult: the disease is often mild, and fulminant symptoms may not be evident until old age. Thus, genotype/phenotype studies must be limited to patients with a previous X-linked family history so that the disease progression of the proband can be predicted based upon that of an older relative (an assumption which is often invalid because of intrafamilial variability in Becker dystrophy), or be limited to the study of older patients whose clinical disease is sufficiently characterized. Given these difficulties, together with the lower incidence of Becker dystrophy, it is easy to understand why only a single large genotype/phenotype study in Becker dystrophy has been published (Beggs *et al.*, 1991) (Figure 2.2). In addition to the single large study many case reports have appeared which have correlated the gene and protein defect with the clinical phenotypes of one or a few patients. Taken together these reports have suggested that deletion of regions of the central rod domain (Figure 2.2) often lead to mild phenotypes, while deletion of either the amino-terminal or carboxyl-terminal domains leads to more severe phenotypes. There is substantial clinical variability, and patients with similar deletions in the 'hot spot' near the centre of the rod domain can show a dramatic range of clinical phenotypes. For example, patients with deletions in the exon 45 to exon 50 region can present with weakness localized to the quadriceps (Sunohara *et al.*, 1990), dilating cardiomyopathy (Palmucci *et al.*, 1992), myoglobinuria, myalgia, cramps, fatigue, and a host of other symptoms (Beggs *et al.*, 1991). Thus, it seems that it is difficult to associate specific presentations and progressions with specific deletions. It remains valid, however, to make the generalization that the amino- and carboxyl-terminal regions of dystrophin seem more important for correct dystrophin function than the central rod domain (Figure 2.2). This tentative map will need to be bolstered by additional patients, the correlation of mRNA levels with the protein and gene data, and the finding of point mutations.

2.5 CONCLUSION

In all dystrophinopathies (Duchenne dystrophy, Becker dystrophy, female carriers) there is a stronger biochemical/phenotype correlation than a genotype/phenotype relationship. However, this may be for want of complete data on non-deletion mutations, and mRNA expression and stability.

The study of boys with Duchenne dystrophy has shown us that dystrophin-deficiency causes a remarkably uniform disease. The few clinical variables, such as mental retardation, cannot be correlated with specific genotypes, and thus must be caused by unknown ecogenetic components of the disease process. Perhaps the central nervous system is somewhat compromised by dystrophin-deficiency in neurones or the vasculature, but an environmental insult is required to express the central nervous system pathology.

Becker dystrophy has provided us with the unparalleled opportunity to study the effects of alterations of protein structure on membrane cytoskeleton structure and function. Most studies to date have been done *in vivo* by correlating gene mutations with dystrophin protein expression and clinical phenotype. These studies have permitted certain generalizations: the integrity of the amino- and carboxyl-terminal domains of dystrophin seem more

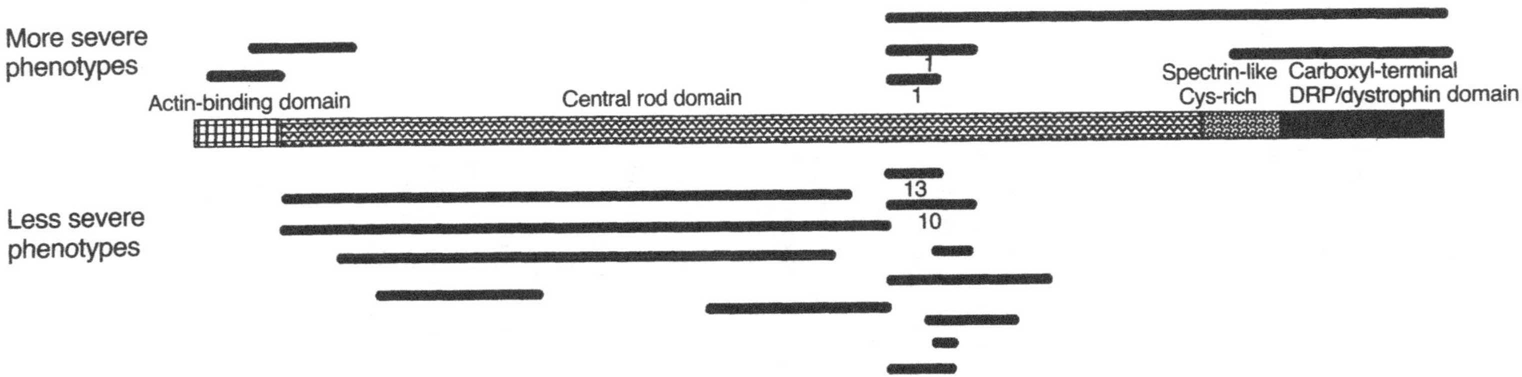

Figure 2.2 Genotype/phenotype correlations in Becker dystrophy: nested deletion map of the dystrophin protein. Shown is a schematic diagram of the dystrophin protein with the four domains indicated, and a summary of gene deletions found in Duchenne/Becker patients (deleted region indicated by a black bar). The 'more severe phenotypes' indicate gene deletions in severe-Becker (wheelchair-bound before age 20) or Duchenne (wheelchair-bound age 11) patients. 'Less severe phenotypes' indicate mild/moderate Becker dystrophy patients (wheelchair-bound after age 20). The majority of patients are from Beggs *et al.* (1991). The numbers under four specific common deletions of the central rod domain indicate the number of patients found with that deletion in Beggs *et al.* While most patients with these common deletions have relatively mild phenotypes (bottom), single patients can be found with the same exon deletions and relatively severe phenotypes (top). This shows the clinical variability in Becker dystrophy which may be due to extragenic factors, or the unknown effect of specific deletions on mRNA splicing or stability.

important for dystrophin structure and function that does the integrity of the central rod domain. However, substantial clinical variability exists in Becker patients with the same deletion. Do patients with the same deletion actually have comparable mutations, i.e. could there be alterations of mRNA expression and stability due to different alterations of intronic sequences? Can studies of dystrophin structure and function in Becker patients be extended to biochemistry and cell biology? For example, what effect do different amino-acid deletions of dystrophin have on the binding to the other proteins with which dystrophin interacts?

The women have shown us that muscle requires nearly its full complement of dystrophin if it is to escape leakage and susceptibility to necrosis: women with 60–70% of normal dystrophin can still show clinical symptoms. Clearly this is a discouraging observation for those pursuing biochemical or genetic rescue of dystrophin-deficiency: nearly every muscle fibre must be supplied with near-normal levels of dystrophin if a true 'cure' is sought, although the disease progression could certainly be slowed with lesser amounts. On the other hand, the findings in both human and animal female carriers have agreed that there is biochemical normalization within the syncytial muscle fibre, although this process is poorly understood. Given adequate time, the dystrophin-positive nuclei will compensate for neighbouring dystrophin-negative nuclei. This is a promising finding for those interested in pursuing biochemical or genetic rescue of dystrophin-deficiency: isolated sites of dystrophin overproduction in a muscle fibre may be adequate to rescue large areas of the myofibre through this biochemical normalization.

REFERENCES

Arahata, K., Ishiura, S., Ishiguro, T. *et al.* (1988) Immunostaining of skeletal and cardiac muscle surface membrane with antibody against Duchenne muscular dystrophy peptide. *Nature*, **333**, 861–3.

Arahata, K., Ishihara, T., Kamakura, K. *et al.* (1989a) Mosaic expression of dystrophin in symptomatic carriers of Duchenne's muscular dystrophy. *New Eng. J. Med.*, **320**, 138–42.

Arahata, K., Hoffman, E.P., Kunkel, L.M. *et al.* (1989b) Dystrophin diagnosis: comparison of dystrophin abnormalities by immunofluorescent and immunoblot analyses. *Proc. Natl. Acad. Sci. U.S.A.*, **86**, 7154–8.

Arahata, K., Beggs, A.H., Honda, H. *et al.* (1991) Preservation of the C-terminus of dystrophin molecule in the skeletal muscle from Becker muscular dystrophy. *J. Neurol. Sci.*, **101**, 148–56.

Arikawa, E., Hoffman, E.P., Kaido, M. *et al.* (1991) The frequency of patients having dystrophin abnormalities in a limb-girdle patient population. *Neurology*, **41**, 1491–6.

Asano, J., Tomatsu, S., Sukegawa, K. *et al.* (1991) Gene deletions in Japanese patients with Duchenne and Becker muscular dystrophies: deletion study and carrier detection. *Clin. Genet.*, **39**, 419–24.

Ballo, R., Hitzeroth, H.W. and Beighton, P.H. (1991) Duchenne muscular dystrophy – a molecular service. *S. Afr. Med. J.*, **179**, 209–12.

Baumbach, L.L., Chamberlain, J.S., Ward, P.A. *et al.* (1989) Molecular and clinical correlations of deletions leading to Duchenne and Becker muscular dystrophies. *Neurology*, **39**, 465–74.

Becker, P.E. and Liener, F. (1955) Eine neue X-chromosomale muskeldystrophie. *Arch. Psych. Z. Neurol.*, **193**, 427–48.

Beggs, A.H., Hoffman, E.P., Snyder, J.R. *et al.* (1991) Exploring the molecular basis for variability among patients with Becker muscular dystrophy: dystrophin gene and protein studies. *Am. J. Hum. Genet.*, **49**, 54–67.

Bieber, F.R., Hoffman, E.P. and Amos, J. (1989) Dystrophin analysis in Duchenne muscular dystrophy: Use in fetal diagnosis and in genetic counseling. *Am. J. Hum. Genet.*, **45**, 362–7.

Bonilla, E., Schmidt, B., Samitt, C.E. *et al.* (1988) Normal and dystrophin-deficient muscle fibers in carriers of the gene for Duchenne muscular dystrophy. *Am. J. Path.* **133**, 440–5.

Bonilla E., Younger D.S., Chang H.W. *et al.* (1990) Partial dystrophin-deficiency in monozygous twin carriers of the Duchenne gene discordant for clinical myopathy. *Neurology*, **40**, 1267–70.

Boyd, Y., Buckle, V., Holt, S. *et al.* (1986) Muscular dystrophy in girls with X;autosome translocations. *J. Med. Genet.*, **23**, 484–90.

Bradley, W.G., Jones, M.Z., Mussini, J.M. and Fawcet, P.R.W. (1978) Becker-type muscular dystrophy. *Muscle Nerve*, **1**, 111–32.

Bulman, D.E., Gangopadhyay, S.B., Bebchuck, K.G. *et al.* (1991a) Point mutation in the human dystrophin gene: identification through western blot analysis. *Genomics*, **10**, 457–60.

Bulman, D.E., Murphy, E.G., Zubrzycka-Gaarn, E.E. *et al.* (1991b) Differentiation of Duchenne and Becker muscular dystrophy phenotypes with amino- and carboxy-terminal antisera specific for dystrophin. *Am. J. Hum. Genet.*, **48**, 295–304.

Burrow, K.L., Coovert, D.D., Klein, C.J. *et al.* (1991) Dystrophin expression and somatic reversion in prednisone-treated and untreated Duchenne dystrophy. *Neurology*, **41**, 661–6.

Byers, T.J., Neumann, P.E., Beggs, A.H. and Kunkel, L.M. (1992) ELISA quantitation of dystrophin for the diagnosis of Duchenne and Becker muscular dystrophies. *Neurology*, in press.

Chelly, J., Gilgenkrantz, H., Lambert, M. *et al.* (1990) Effect of dystrophin gene deletions on mRNA levels and processing in Duchenne and Becker muscular dystrophies. *Cell*, **63**, 1239–48.

Clerk, A., Rodillo, E., Heckmatt, J.Z. *et al.* (1991) Characterisation of dystrophin in carriers of Duchenne muscular dystrophy. *J. Neurol. Sci.*, **102**, 197–205.

Clerk, A., Sewry, C.A., Dubowitz, V. and Strong, P.N. (1992a) Characterization of dystrophin in foetuses at risk for Duchenne muscular dystrophy. *J. Neurol. Sci.*, in press.

Clerk, A., Strong, P.N., and Sewry, C.A. (1992b). Characterisation of dystrophin in human foetal skeletal muscle. *Development*, **114**, 395–402.

Cooke, A., Lanyon, W.G., Wilcox, D.E. *et al.* (1990) Analysis of Scottish Duchenne and Becker muscular dystrophy families with dystrophin cDNA probes. *J. Med. Genet.*, **27**, 292–7.

Cooper, B.J., Gallagher, E.A., Smith, C.A. *et al.* (1990) Mosaic expression of dystrophin in carriers of canine X-linked muscular dystrophy. *Lab. Invest.*, **62**, 171–8.

Covone, A.E., Lerone, M. and Romeo, G. (1991) Genotype-phenotype correlation and germline mosaicism in DMD/BMD patients with deletions of the dystrophin gene. *Hum. Genet.*, **87**, 353–60.

Darras, B.T., Blattner, P., Harper, J.F. *et al.* (1988) Intragenic deletions in 21 Duchenne muscular dystrophy (DMD)/Becker muscular dystrophy (BMD) families studied with the dystrophin cDNA: Location of breakpoints on HindIII and BglII exon-containing fragment maps, meiotic and mitotic origin of the mutations. *Am. J. Hum. Genet.*, **43**, 620–9.

Den Dunnen, J.T., Grootscholten, P.M., Bakker, E. *et al.* (1989) Topography of the Duchenne muscular dystrophy (DMD) gene: FIGE and cDNA analysis of 194 cases reveals 115 deletions and 13 duplications. *Am. J. Hum. Genet.*, **45**, 835–47.

Emery, A.E.H. and Skinner, R. (1976) Clinical studies in benign (Becker type) X-linked muscular dystrophy. *Clin. Genet.*, **10**, 189–201.

Ervasti, J.M., Ohlendieck, K., Kahl, S.D. *et al.* (1990) Deficiency of a glycoprotein component of the dystrophin complex in dystrophic muscle. *Nature*, **345**, 315–19.

Evans, M.I., Greb, A., Kunkel, L.M. *et al.* (1991) In utero fetal muscle biopsy for the diagnosis of Duchenne muscular dystrophy. *Am. J. Ob. Gyn*, **165**, 728–32.

Fanin, M., Danieli, G.A., Vitiello, L. *et al.* (1992) Prevalence of dystrophin-positive fibers in 85 Duchenne muscular dystrophy patients. *Neuromuscular Disorders*, **2**, 41–5.

Gospe, S.M., Lazaro, R.P., Lava, N.S. *et al.* (1989) Familial X-linked myalgia and cramps. *Neurology*, **39**, 1277–80.

Gowers, W.R. (1879). *Pseudo-hypertrophic muscular paralysis. A clinical lecture*, J & A Churchill.

Greenberg, C.R., Jacobs, H.K., Halliday, W. and Wrogemann, K. (1991) Three years' experience with neonatal screening for Duchenne/Becker muscular dystrophy: Gene analysis, gene expression and phenotype prediction. *Am. J. Med. Genet.*, **39**, 68–75.

Hodgson, S., Hart, K., Abbs, S. *et al.* (1989) Correlation of clinical and deletion data in Duchenne and Becker muscular dystrophy. *J. Med. Genet.*, **26**, 682–93.

Hoffman, E.P. and Kunkel, L.M. (1989) Dystrophin abnormalities in Duchenne/Becker muscular dystrophy. *Neuron*, **2**, 1019–29.

Hoffman, E.P. and Gorospe, J.R. (1991) The animal models of Duchenne dystrophy: Windows on the pathophysiological consequences of dystrophin deficiency. In *Ordering the Membrane-Cytoskeleton Trilayer, Topics in Membranes, Vol. 38*, (eds M.S. Mooseker and J.S. Morrow), Academic Press, New York, pp. 113–54.

Hoffman, E.P., Fischbeck, K.H., Brown, R.H. *et al.* (1988) Dystrophin characterization in muscle biopsies from Duchenne and Becker muscular dystrophy patients. *New Eng. J. Med.*, **318**, 1363–8.

Hoffman, E.P., Beggs, A.H., Koenig, M. *et al.* (1989a) Cross-reactive protein in Duchenne muscle. *Lancet*, **ii**, 1211–12.

Hoffman, E.P., Kunkel, L.M., Angelini, C. *et al.* (1989b) Improved diagnosis of Becker muscular dystrophy by dystrophin testing. *Neurology*, **39**, 1011–17.

Hoffman, E.P., Morgan, J.E., Watkins, S.C. *et al.* (1990) Somatic reversion/suppression of the mouse mdx phenotype *in vivo*. *J. Neurol. Sci.*, **99**, 9–25.

Hoffman, E.P., Garcia, C.A., Chamberlain, J.S. *et al.* (1991) Is the carboxyl-terminus of dystrophin required for membrane association? A novel case of Duchenne muscular dystrophy. *Ann. Neurol.*, **30**, 605–10.

Hoffman, E.P., Arahata, K., Minetti, C. *et al.* (1992) Dystrophinopathy in isolated cases of myopathy in females. *Neurology*, **42**, 967–75.

Hu, X., Ray, P.N., Murphy, E.G. *et al.* (1990) Duplicational mutation at the Duchenne muscular dystrophy locus: Its frequency, distribution, origin, and phenotype/genotype correlation. *Am. J. Hum. Genet.*, **46**, 682–95.

Khurana, T.S., Hoffman, E.P. and Kunkel, L.M. (1990) Identification of a chromosome 6 encoded dystrophin related protein. *J. Biol. Chem.*, **265**, 16717–20.

Khurana, T.S., Watkins, S.C., Chafey, P. *et al.* (1991) Immunolocalization and developmental expression of DRP in skeletal muscle. *Neuromuscular Disorders*, **1**, 185–94.

Klein, C.J., Coovert, D.D., Bulman, D.E. *et al.* (1992) Somatic reversion/suppression in Duchenne muscular dystrophy (DMD): evidence supporting a frame restoring mechanism in rare dystrophin positive fibers. *Am J. Hum. Genet.*, **50**, 950–9.

Koenig, M., Hoffman, E.P., Bertelson, C.J. *et al.* (1987). Complete cloning of the Duchenne muscular dystrophy (DMD) cDNA and preliminary genomic organization of the DMD gene in normal and affected individuals. *Cell*, **50**, 509–17.

Koenig, M., Beggs, A.H., Moyer, M. *et al.* (1989). The molecular basis for Duchenne versus Becker muscular dystrophy: Correlation of severity with type of deletion. *Am. J. Hum. Genet.*, **45**, 498–506.

Kuller, J.A., Hoffman, E.P., Fries, M.H. and Golbus, M.S. (1992) Prenatal diagnosis of Duchenne muscular dystrophy by fetal muscle biopsy. *Hum. Genet.*, **90**, 34–40.

Laing, N.G., Mears, M.E., Chandler, D.C. *et al.* (1991) The diagnosis of Duchenne and Becker muscular dystrophies: two years' experience in a comprehensive carrier screening and prenatal diagnostic laboratory. *Med. J. Austr.*, **154**, 14–18.

Love, D.R., Hill, D.F., Dickson, G. *et al.* (1989) An autosomal transcript in skeletal muscle with homology to dystrophin. *Nature*, **339**, 55–8.

Love, D.R., Morris, G.E., Ellis, J.M. *et al.* (1991) Tissue distribution of the dystrophin-related gene product and expression in the mdx and dy mouse. *Proc. Natl. Acad. Sci. U.S.A.*, **88**, 3243–7.

Lupski, J.R., Garcia, C.A., Zoghbi, H.Y. *et al.* (1991) Discordance of muscular dystrophy in monozygotic female twins: evidence supporting asymmetric splitting of the inner cell mass in a manifesting carrier of Duchenne dystrophy. *Am. J. Med. Genet.*, **40**, 354–64.

Malhotra, S.B., Hart, K.A., Klamut, H.J. *et al.* (1988) Frame-shift deletions in patients with Duchenne and Becker muscular dystrophy. *Science*, **242**, 755–9.

Matsuo, M., Masumura, T., Nishio, H. *et al.* (1991) Exon skipping during splicing of dystrophin mRNA precursor due to an intraexon deletion in the dystrophin gene of Duchenne muscular dystrophy Kobe. *J. Clin. Invest.*, **87**, 2127–31.

Medori, R., Brooke, M.H. and Waterston, R.H. (1989) Genetic abnormalities in Duchenne and Becker dystrophies: Clinical correlations. *Neurology*, **39**, 461–5.

Minetti, C., Chang, H.W., Medori, R. *et al.* (1991) Dystrophin deficiency in young girls with sporadic myopathy and normal karyotype. *Neurology*, **41**, 1288–91.

Monaco, A.P., Bertelson, C.J., Liechti-Gallati, S. *et al.* (1988) An explanation for the phenotypic differences between patients bearing partial deletions of the DMD locus. *Genomics*, **2**, 90–5.

Morandi, L., Mora, M., Gussoni, E. *et al.* (1990) Dystrophin analysis in Duchenne and Becker muscular dystrophy carriers: correlation with intracellular calcium and albumin. *Ann. Neurol.*, **28**, 674–9.

Moser, H. and Emery, A.E.H. (1974) The manifesting carrier in Duchenne muscular dystrophy. *Clin. Genet.*, **5**, 271–84.

Nance, W.E. (1990) Invited editorial: Do twin lyons have larger spots? *Am. J. Hum. Genet.*, **46**, 646–8.

Naylor, E.W., Hoffman, E.P., Paulus-Thomas, J. *et al.* (1992) Reconsideration of neonatal screening for Duchenne muscular dystrophy based on molecular diagnosis and potential therapeutics. *Screening*, **1**, 99–113.

Nicholson, L.V.B., Davison, K., Johnson, M.A. *et al.* (1989) Dystrophin in skeletal muscle. II. Immunoreactivity in patients with Xp21 muscular dystrophy. *J. Neurol. Sci.*, **94**, 137–46.

Nicholson, L.V.B., Johnson, M.A., Gardner-Medwin, D. *et al.* (1990) Heterogeneity of dystrophin expression in patients with Duchenne and Becker muscular dystrophy. *Acta Neuropath.*, **80**, 239–50.

Palmucci, L., Doriguzzi, C., Mongini, T. *et al.* (1992) Dilating cardiomyopathy as the expression of Xp21 Becker type muscular dystrophy. *J. Neurol. Sci.*, **111**, 218–21.

Patel, K., Voit, T., Dunn, M.J. *et al.* (1988) Dystrophin and nebulin in the muscular dystrophies. *J. Neurol. Sci.*, **87**, 315–26.

Richards, C.S., Watkins, S.C., Hoffman, E.P. *et al.* (1990) Skewed X inactivation in a female MZ twin results in Duchenne muscular dystrophy. *Am. J. Hum. Genet.*, **46**, 672–81.

Roberts, R.G., Bobrow, M. and Bentley, D.R. (1992) Point mutations in the dystrophin gene. *Proc. Natl. Acad. Sci. U.S.A.*, **89**, 2331–5.

Rojas, C. and Hoffman, E.P. (1991) Recent advances in dystrophin research. *Current Opinion Neurobiology*, **1**, 420–9.

Shimizu, T.K., Matsumura, K., Hashimoto, K. *et al.* (1988). A monoclonal antibody against a synthetic polypeptide fragment of dystrophin. *Proc. Jap. Acad.*, **64**, 205–8.

Soong, B.W., Tsai, T.F., Su, C.H. *et al.* (1991) DNA polymorphisms and deletion analysis of the Duchenne-Becker muscular dystrophy gene in the Chinese. *Am. J. Med. Genet.*, **38**, 593–600.

Sunohara, N., Arahata, K., Hoffman, E.P. *et al.* (1990) Quadriceps myopathy: forme fruste of Becker muscular dystrophy. *Ann. Neurol.*, **28**, 634–9.

Takemitsu, M., Ishiura, S., Koga, R. *et al.* (1991) Dystrophin-related protein in the fetal and denervated skeletal muscles of normal and mdx mice. *Biochem. Biophys. Res. Commun.*, **180**, 1179–86.

Tanaka, H., Ikeya, K. and Ozawa, E. (1990) Difference in the expression pattern of dystrophin on the surface membrane between the skeletal and cardiac muscles of mdx carrier mice. *Histochemistry*, **93**, 447–52.

Uchino, M., Araki, S., Miike, T. *et al.* (1989) Localization and characterization of dystrophin in muscle biopsy specimens from Duchenne muscular dystrophy and various neuromuscular disorders. *Muscle Nerve*, **12**, 1009–16.

Upadhyaya, M., Smith, R.A., Thomas, N.S.T. *et al.* (1990) Intragenic deletions in 164 boys with Duchenne muscular dystrophy (DMD) studied with dystrophin cDNA. *Clin. Genet.*, **37**, 456–62.

Voit, T., Haas, K., Leger, J.O. *et al.* (1991a) Xp21 dystrophin and 6q dystrophin-related protein. Comparative immunolocalization using multiple antibodies. *Am. J. Pathol.*, **139**, 969–76.

Voit, T., Stuettgen, P., Cremer, M. and Goebel, H.H. (1991b) Dystrophin as a diagnostic marker in Duchenne and Becker muscular dystrophy. Correlation of immunofluorescence and western blot. *Neuropediatrics*, **22**, 152–62.

Watkins, S.C., Hoffman, E.P., Slayter, H.S. and Kunkel, L.M. (1989) Dystrophin distribution in heterozygote MDX mice. *Muscle Nerve*, **12**, 861–8.

Weller, B., Karpati, G., Lehnert, S. *et al.* (1991) Inhibition of myosatellite cell proliferation by Gamma irradiation does not prevent the age-related increase of the number of dystrophin-positive fibers in soleus muscles of mdx female heterozygote mice. *Am. J. Pathol.*, **138**, 1497–1502.

3

Molecular genetics and genetic counselling for Duchenne/Becker muscular dystrophy

NIGEL G. LAING

3.1 INTRODUCTION

Individuals from families in which inherited disorders have been identified frequently wish to know whether they have their family's disease, the risk of passing the disease to their children, and the genetic status of their children including their unborn children. The guilt associated with passing the family disease to offspring, and the stress and worry for parents wondering if the child they are watching grow will later have a family disorder is incomprehensible to those who have not experienced it. The vast majority of couples at risk of passing a serious inherited disorder to their children do not wish to do so. They do not wish to subject their child to being handicapped by, or slowly or rapidly dying from, the family disease. However, one of the things frequently associated with being a couple is a desire to have one's own children. Genetic counselling is 'a communication process, the intent of which is to provide individuals and families having a genetic disease or at risk of such a disease with information about their condition and to provide information that would allow couples at risk to make informed reproductive decisions' (Gelehrter and Collins, 1990). Successful genetic counselling may perhaps be defined as facilitating couples to have the pregnancy outcomes they wish including not having affected children if they do not wish to do so. The accuracy of the information provided to those seeking counselling is crucial. Successful genetic counselling is only possible after correct diagnosis of the family disease and subsequent correct determination of the genetic status (genotyping) of individuals in the family.

Molecular and Cell Biology of Muscular Dystrophy
Edited by Terence Partridge
Published in 1993 by Chapman & Hall, London. ISBN 0 412 43440 7

The impact of molecular genetics on counselling for Duchenne and Becker muscular dystrophies (DMD and BMD) can be summarized as increased accuracy of information, including more accurate diagnosis of the family disease and increased accuracy of genotyping, including genotyping of offspring *in utero*. Molecular diagnosis is allowing the identification of DMD and BMD sufferers amongst patients previously diagnosed with a variety of other neuromuscular disorders. It is also providing much more accurate identification of individuals at risk of having affected children and facilitating the birth of unaffected children. Molecular genetics applied to DMD/BMD is saving the unaffected male fetuses which would have been terminated on the basis of sex. Molecular genetic analysis of Duchenne and Becker muscular dystrophy is acting as a prototype for other inherited diseases. Many of the techniques which have been pioneered in DMD/BMD are already being applied to other disorders such as cystic fibrosis and neurofibromatosis whose disease genes have now been localized and identified.

3.2 DUCHENNE AND BECKER MUSCULAR DYSTROPHY

Duchenne and Becker muscular dystrophy are X-linked recessive disorders. This means that generally only males are affected; that the diseases are passed through 'carrier' women, who are generally unaffected since the mutated gene is compensated for by the normal gene on their other X-chromosome; that all daughters of affected men are carriers, and that the disease is never transmitted from father to son (Gelehrter and Collins, 1990).

Duchenne muscular dystrophy is the more severe disorder, resulting in affected individuals becoming wheelchair bound before 12 years of age and death usually by the end of the third decade. It is commonly estimated to affect 1:3000 to 1:3500 live male births (Emery, 1980; McKusick, 1990). Approximately one-third of DMD patients are educationally subnormal (Rowland, 1988).

Becker muscular dystrophy is a less severe disorder with, on average, later onset and longer lifespan. However the Becker phenotype may vary from close to the phenotype of Duchenne muscular dystrophy, when the disease may be termed 'intermediate', through to a patient using walking aids in old age. Becker muscular dystrophy has an incidence of approximately 1:30 000 live male births (McKusick, 1990). Debate continued over many years as to whether Duchenne and Becker muscular dystrophies were separate entities or allelic. However, linkage data (Murray *et al.*, 1982; Kingston *et al.*, 1983; Kingston *et al.*, 1984) and later identification of mutations in the same gene provided categorical proof that the two disorders are allelic (Kunkel *et al.*, 1986).

3.2.1 Inheritance of DMD/BMD

The X-linked nature of DMD and BMD means that on average, half of the sons of a carrier will be affected, while half of her daughters will be carriers (Figure 3.1). Thus the risk of a carrier having an affected child is one in four: an unacceptably high risk to most.

Duchenne muscular dystrophy is an X-linked recessive lethal disorder since, although at least two cases of DMD patients having children have been reported (Walton, 1955; Thompson, 1978), affected males do not usually father children. From theoretical considerations therefore (Haldane, 1935), one-third of Duchenne muscular dystrophy cases must have 'new mutations' and their mothers will not be carriers, so long as the mutation rates in males and females are equal (Figure 3.2). With an incidence of 1 in 3000 to 1 in 3500 live male births, the mutation rate for DMD must be of the order of 1:10 000, which is very high in comparison with other disorders (Cummings, 1991).

For the one-third of mothers of DMD boys who are not carriers (Figure 3.2a), there is very little risk of any other woman in the pedigree having an affected child: only the small risk of an affected son by another new mutation. At each generation back through the female line, the chance that the mother is a carrier is reduced by 50%. Thus, two-thirds of mothers with affected boys are carriers and half of this two-thirds will carry new mutations (Figure 3.2b). In this case also, where the mother herself carries a new mutation, there will again only be the mutation rate chance that any other woman in the pedigree will have an affected boy. However, half of the daughters of such women will be carriers and half of any subsequent sons will be affected. Only when at least the grandmother is a carrier (one-third of pedigrees, Figure 3.2b) is the chance of affected offspring for other women in the pedigree greater than the chance of another new mutation. Thus, for more than two-thirds of DMD pedigrees (in our series 75%), there

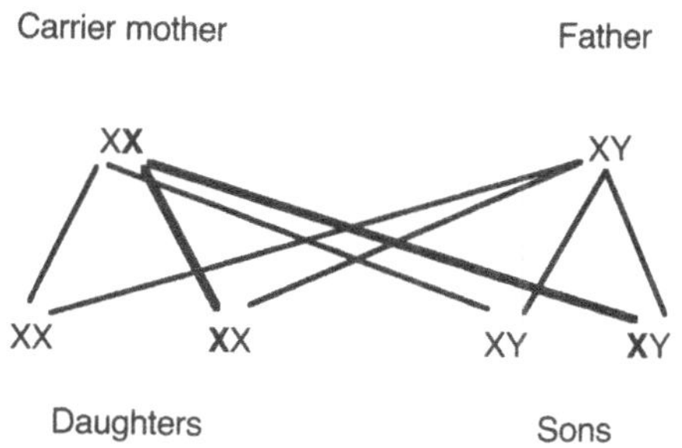

Figure 3.1 Inheritance of X-linked disorders.

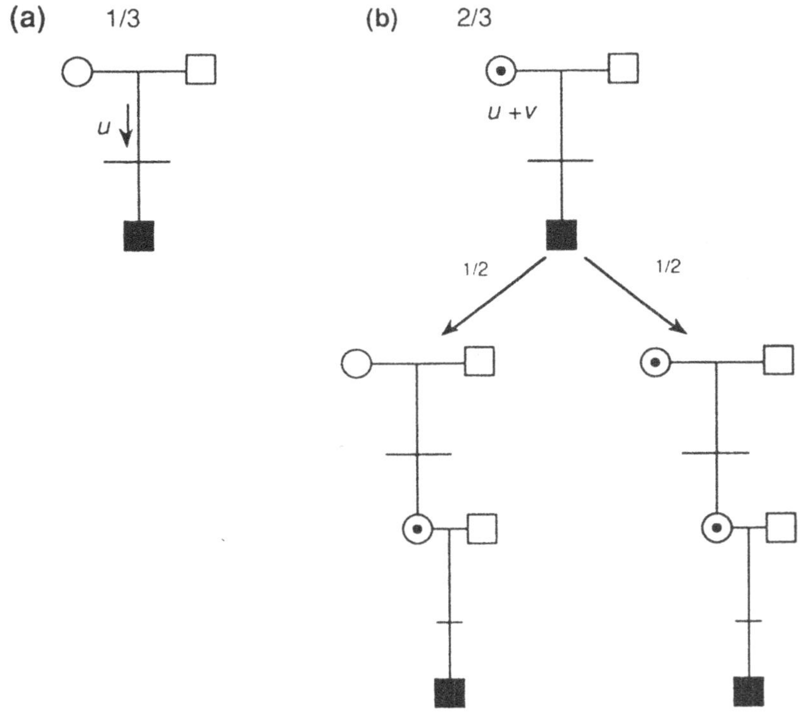

Figure 3.2 Explanation of the proportion of DMD patients who are new mutations and the proportion of mothers and grandmothers of DMD patients who are carriers.

is no previous family history. The patients are so-called sporadic or simplex cases, because the affected boy or his mother are new mutations, or there were by chance no previous affected males in the pedigree. The likelihood of a carrier being detected through an affected child will depend on the number of offspring she has: the larger the number of offspring, the greater the chance that one will be affected. The trend to smaller-sized families in industrialized nations means that fewer families will be detected early through affected boys.

The percentage of mothers of BMD cases who are carriers is higher. BMD is not a lethal disorder: most BMD patients can have children and all daughters of BMD patients will be carriers. Thus a smaller percentage of BMD cases, estimated to be 10% (Emery and Rimoin, 1990), have new mutations.

3.2.2 The importance of accurate diagnosis for counselling

Accurate diagnosis of affected individuals and precise carrier-status determination are the foundations of sound genetic counselling for DMD/BMD.

The importance of accurate diagnosis for counselling is that the risks of a carrier of DMD/BMD having an affected child is orders of magnitude higher than for most of the other inherited neuromuscular disorders with which DMD/BMD can be confused. These disorders include autosomal recessive Duchenne-like dystrophy (ARDMD), spinal muscular atrophy (SMA), limb girdle muscular dystrophy (LGMD), quadriceps myopathy, and metabolic and mitochondrial disorders. Thus an unaffected sister of a DMD/BMD patient has a 1/4 chance of an affected child if she is a carrier, a 1/2 chance of being a carrier if her mother is a carrier and a 2/3 chance that her mother is a carrier. Multiplying all the risks together ($2/3 \times 1/2 \times 1/4$) gives the sister of a sporadic DMD/BMD case a combined risk of 1/12 of an affected child. An unaffected sister of a patient with an autosomal recessive disorder has a 2/3 chance of being a carrier. In order to have an affected child, a carrier of an autosomal recessive disorder must have children by another carrier and then one in four of their children will be affected. The carrier frequency for spinal muscular atrophy, the second most common human autosomal recessive disorder, is estimated to be 1/40. Therefore, the combined risk of an unaffected sister of an SMA patient having an affected child is $2/3 \times 1/40 \times 1/4$ or 1/240. This is considerably less than the 1/12 chance if the family disease is DMD/BMD and emphasizes the importance to counselling of correct diagnosis.

3.2.3 Women affected by DMD/BMD

Frequently the diagnosis of DMD/BMD is not made because there are affected women in the pedigree. However, although X-linked disorders usually only strike males, women can, under certain specific circumstances, also be affected. During early female embryonic development, one or other of the two X chromosomes in each cell is inactivated in a process termed lyonization (Lyon, 1988). If in a carrier lyonization inactivates a sufficiently high proportion of the X chromosomes carrying the normal DMD/BMD gene, then the woman concerned will be the equivalent of an affected male, with an inactivated normal X and an active mutated X in each cell. Thus, depending on the proportion of mutated X chromosomes inactivated, a carrier may suffer from DMD/BMD, may be partially affected (a so-called manifesting carrier), or, if a sufficiently high proportion of the mutated X chromosomes are inactivated, may be indistinguishable from a non-carrier by all tests relying on the phenotypic expression of the disease. There are apparently two ways in which the normal X chromosomes are always inactivated. One is for a translocation of the X chromosome to have occurred (Verellen-Dumoulin *et al.*, 1984), and the other is during monozygous twinning (Richards *et al.*, 1990). In the situation where a woman has one translocated and one normal X, the normal X is always inactivated.

Then, if the translocation cuts through the DMD gene, all the woman's cells will contain an inactivated normal X and an active X with a mutated and inoperable DMD gene. Such women will suffer from DMD. Translocations of the X chromosome in women with DMD were always found to involve the same sub-region of the short arm of the X chromosome (Xp21) and were instrumental in the localization and one approach to the initial cloning of the DMD/BMD gene (Ray *et al.*, 1985; Thompson *et al.*, 1986). Women may also be affected through suffering from Turner's syndrome, having only one X, and the X having a DMD/BMD mutation (reviewed in Moser, 1984; see also Chelly *et al.*, 1986). Theoretically, women may also be affected (a) in uniparental disomy where the two X chromosomes are a duplication of one parental X and (b) if there are separate mutations in the DMD/BMD genes on both X chromosomes. These latter causes of DMD/BMD in women should occur very rarely, e.g. with the carrier frequency of DMD being approximately 1 in 2250 women, one woman in 5 million should have separate DMD mutations on both X chromosomes. Rare events do happen and tend to be reported, e.g. a male XXY carrier of DMD (Hennekam *et al.*, 1989). It can be anticipated therefore that a woman with separate mutations on each X will be reported now that molecular-based diagnosis is available.

3.3 GENETIC COUNSELLING FOR DUCHENNE AND BECKER MUSCULAR DYSTROPHY: PRIOR TO MOLECULAR GENETICS

Unless a genetic counselling programme is instigated, the incidence of DMD and BMD will remain at 1:3000 and 1:30 000 male births respectively. The aim of genetic counselling is to reduce the distress and burden on the patients, their families and society. Prevention is the desire of the majority of families. The X-linked recessive nature of DMD/BMD, with phenotypically normal women being carriers of the disease, and the high mutation rate add particular difficulties to providing accurate information to families.

In order to appreciate the impact of molecular genetics on counselling, it is useful to review counselling practices for DMD/BMD prior to the introduction of molecularly-based techniques. Patient and carrier diagnosis are the essential elements of successful DMD/BMD counselling and will be discussed separately. Two of our families, Families A and B (Figures 3.3 and 3.4) will be used as examples.

3.3.1 Diagnosis of DMD/BMD prior to molecular genetics – what disease does this patient have?

Prior to the application of molecularly-based techniques, the diagnosis of DMD/BMD was made on the basis of: (a) family history; (b) clinical

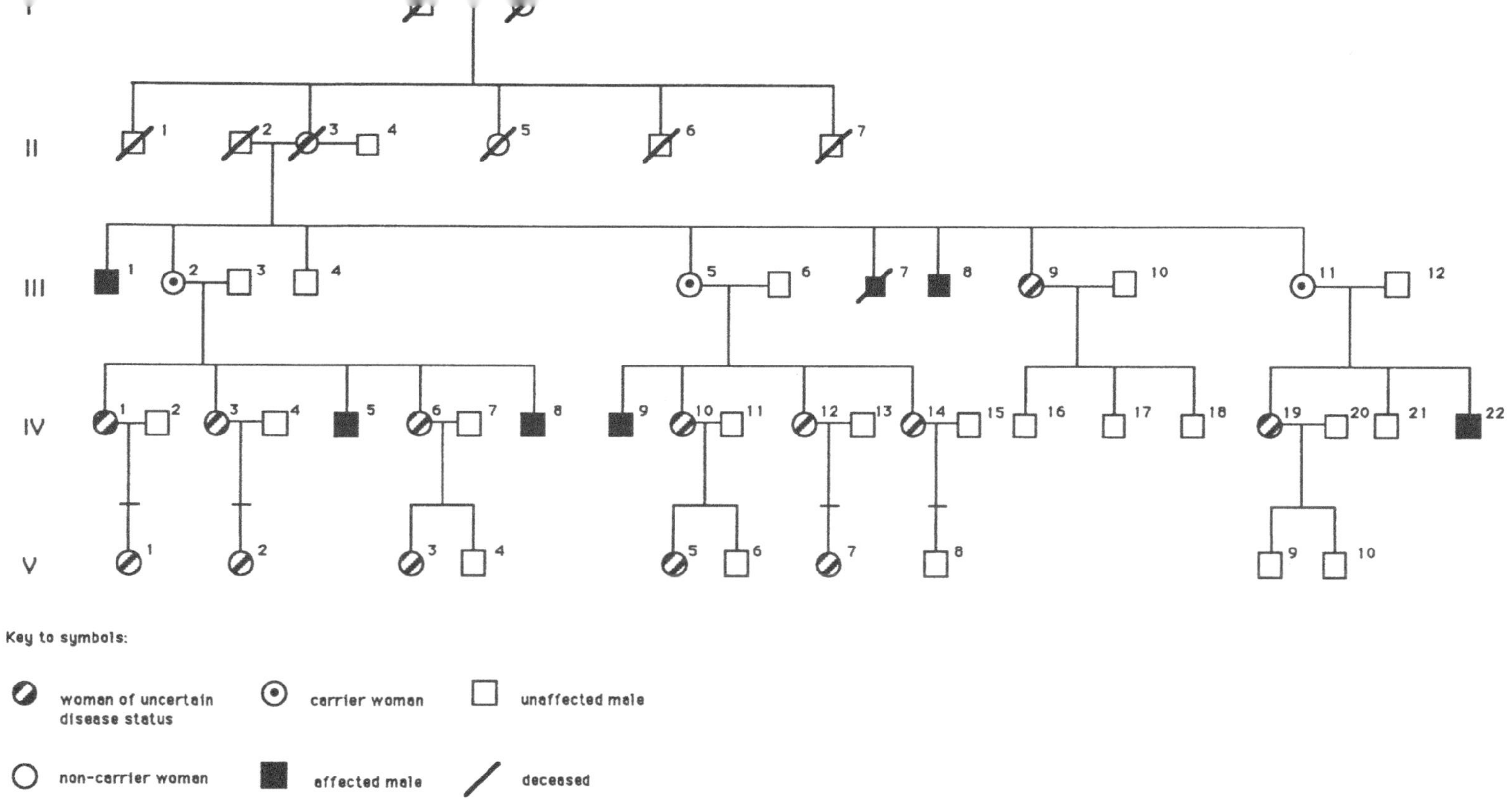

Figure 3.3 Pedigree of family A segregating BMD.

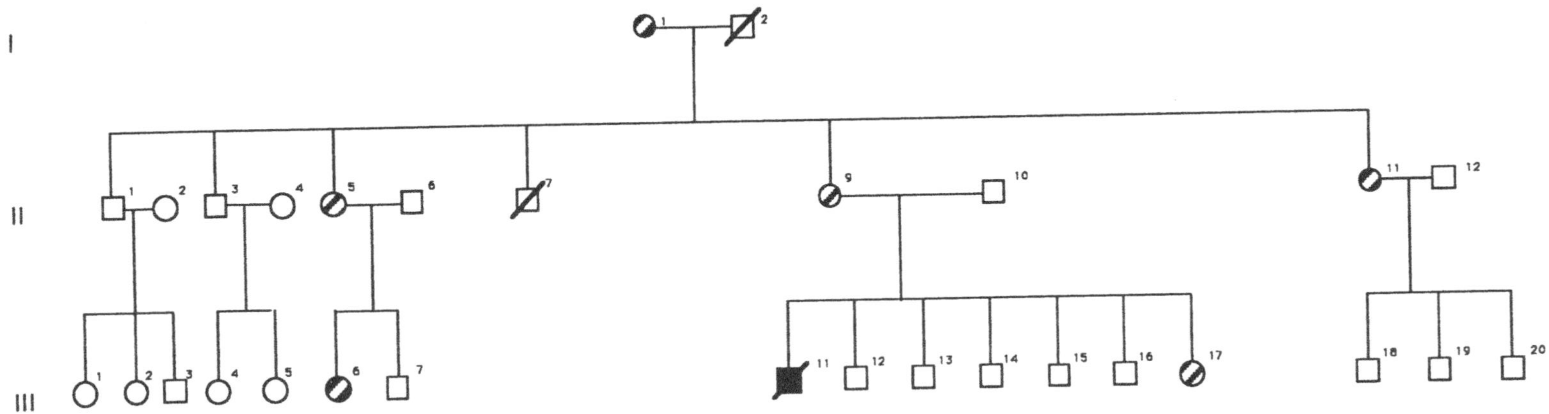

Figure 3.4 Pedigree of family B – a sporadic case of DMD.

findings; (c) serum muscle enzyme levels; (d) muscle biopsy (MB) and (e) electromyography (EMG) and electrocardiography (ECG) (Kakulas and Adams, 1985). In our laboratory, this battery of tests is called 'conventional testing'. Each of the elements of conventional testing has its weaknesses, leading to difficulties in making an accurate diagnosis in some cases. The main weakness is that all the elements rely on phenotypic tests of genotypic status and the phenotype can be highly variable.

3.3.1.1 Mode of inheritance

If a family demonstrates obvious X-linked inheritance, then the diagnosis of DMD or BMD is facilitated. One is more likely to see an obvious X-linked pattern of inheritance in BMD families (Family A, Figure 3.3), where affected men do have children, than in DMD families where new mutations comprise a larger proportion of cases (Family B, Figure 3.4). Patients without a family history of the disorder could be very difficult to categorize as DMD/BMD and ran the risk of being diagnosed as having other diseases with perhaps different modes of inheritance. A family with both affected males and females in a sibship is also difficult to categorize, and would likely be labelled as having an autosomal recessive neuromuscular disorder. But, as has been described, women can be affected by DMD/BMD, and thus a sibship containing both affected males and females may have DMD/BMD, autosomal recessive Duchenne-like dystrophy, limb-girdle muscular dystrophy, etc.

3.3.1.2 Clinical findings

The clinical features of DMD/BMD include among others: proximal distribution of muscle weakness, winging of scapulae, pseudohypertrophic calves and positive Gowers' manoeuvre (rising from the floor by using the arms to walk up the legs). None of these clinical signs are unique to DMD/BMD. It has been claimed that autosomal recessive Duchenne-like dystrophy shows clinical differences to DMD such as slower progression, but there is little to provide a definitive differential diagnosis. It is also difficult to differentiate between DMD and BMD in boys prior to the more rapid deterioration of DMD becoming clinically obvious. The clinical definition of DMD includes becoming wheelchair bound by age 12, and therefore prior to age twelve discrimination between DMD and BMD can be problematical.

3.3.1.3 Serum muscle enzyme levels

The focal necrosis of muscle fibres characteristic of DMD and BMD releases into the bloodstream enzymes normally restricted to the inside of muscle fibres. This leads to elevated serum levels of such muscle enzymes and a simple test for muscle disease through examination of levels of muscle enzymes in serum. The most commonly used such enzyme is creatine phosphokinase (CK). The normal range of serum creatine kinase levels is 25–200 units/l in our laboratory. In boys with DMD, the level can be in the order of 10 000 to 20 000 units/l. The CK levels in BMD can be lower in early years concomitant with the lesser degree of muscle fibre damage, but as ambulation continues, CK levels in BMD can rise as high as those in DMD. Though it is generally true that only DMD/BMD can produce such high levels of creatine kinase consistently, any disorder or event that causes muscle damage, e.g. an accident or sports injury, will raise serum CK levels. The use of CK in diagnosing DMD/BMD is reviewed by Moser (1984).

3.3.1.4 Muscle biopsy, EMG and ECG

Analysis of muscle biopsy and EMG both show myopathic changes in DMD and BMD (Kakulas and Adams, 1985). The degeneration and regeneration of muscle fibres is visible by both light and electron microscopy. Despite the generally myopathic changes however, it is accepted that a mixed picture of neurogenic and myopathic features can be seen in muscle biopsies and EMG from BMD patients (Bradley *et al.*, 1978). This mixed picture in BMD can make differential diagnosis of BMD from the benign form of spinal muscular atrophy (Kugelberg-Welander disease) difficult and can lead to many isolated male patients being diagnosed as showing 'mixed' neuropathic/myopathic changes and therefore to be Kugelberg-Welander disease, especially if CK level is low.

3.3.2 Patient diagnosis prior to molecular genetics – summary

None of the elements of conventional diagnosis provided a 'definitive' diagnosis of DMD or BMD. It was less easy to mistake the diagnosis of DMD than that of BMD once the patients were old enough. However, it was the experience of this group, and others, that when sporadic cases first presented, at perhaps around five years of age, it was difficult to distinguish between DMD and BMD. BMD was difficult to distinguish from other conditions such as limb-girdle muscular dystrophy, quadriceps myopathy, and spinal muscular atrophy. The difficulties were compounded for female cases, and sibships containing affected males and females.

3.3.3 Carrier detection for DMD/BMD prior to molecular genetics – is this woman a carrier or not?

As previously discussed, accurate carrier status determination is crucial to successful genetic counselling for X-linked disorders. A son of a carrier has a 50% chance of being affected (Figure 3.1). Sons of women who are not carriers have only a very small chance of being affected with DMD/BMD equal to the new mutation rate for DMD or BMD. The risk of a woman being a carrier was frequently calculated by Bayesian analysis where both prior and conditional probabilities are used to calculate a posterior probability of carrier status (Bundey, 1978; Emery, 1980). Prior probability is based on the family history of the individual, conditional probability on the individual's phenotype, including clinical signs and test results, and disease status of children. Families A and B (Figures 3.3 and 3.4) will be used as examples of the effectiveness of various carrier tests.

Prior to input from molecular genetics, the same 'conventional testing' as was used to diagnose DMD/BMD was also used to determine carrier status. The difficulty of the conventional tests being phenotypic tests of genotypic status was even more pronounced in the diagnosis of carriers, because only minor sub-clinical alterations might be expected.

3.3.3.1 Pedigree analysis

The first test for carrier status was the relationship of the woman to the affected patient. The closer the relationship to a patient, the greater the carrier risk.

Obligate carriers are women who have one or more affected sons and another affected relative in their maternal lineage (e.g. Family A III-2, -5, -11, Figure 3.3). They are called obligate carriers since they almost certainly are carriers of the family mutation. The only way in which they would not be carriers of the family mutation would be through the very small chance that their affected son had a new mutation. Women in the pedigree who do not have affected sons but are the daughters of carriers (Figure 3.3: III-9, IV-1, -3, -6, -10, -12, -14, -19) have 50% prior probability of being carriers. This prior probability is modified in Bayesian analysis by the number of unaffected sons that that woman has. Let us take as an example III-9 who has three unaffected sons. (The calculations are summarized in Table 3.1). Her prior probabilities of being a carrier or of not being a carrier are equal

Table 3.1 Calculation of posterior probability of III-9 in Figure 3.3 being a carrier

	Carrier	*Non-carrier*
Prior probability	1/2	1/2
Conditional probability		
(3 unaffected sons)	1/8 $\{(1/2)^3\}$	1
Joint probability	1/16	1/2
Relative probability	1 (1/16)	8 (8/16)
Posterior probability	1/9	8/9

at 1/2. If she is not a carrier the conditional probability of her having three unaffected sons is virtually 1. If she is a carrier, the conditional probability of her three sons being unaffected is $(1/2)^3$, or 1/8. The joint probability (prior multiplied by conditional) is 1/16 that she is a carrier, 1/2 that she is not. The relative probabilities of the two events are 1 that she is a carrier to 8 that she is not a carrier. So, the posterior probabilities are 1/9 that she is a carrier, 8/9 that she is not a carrier.

The probability of II-3 in Figure 3.3 being a carrier with 3 affected sons and 3 obligate carrier daughters out of eight children is very high and she is termed a probable carrier. She may possibly not be a carrier if she has germline mosaicism. When a woman has germline mosaicism for DMD/BMD she herself is not a carrier but her ovaries contain ova with a mutation in the DMD/BMD gene. Germline mosaicism is more important in sporadic cases, but is a question which must be considered when trying to determine the point at which the mutation arose in any multi-generational family such as Family A. Germline mosaicism may also occur in males so that an unaffected male may have more than one carrier daughter (Darras and Francke, 1987). The possibility of germline mosaicism means that any mother of an affected child with DMD/BMD should be treated as a carrier and counselled accordingly.

In families with no previous history (Family B, Figure 3.4) the perspective is different. The prior probability that the sister of a sporadic case (e.g. Figure 3.4 III-17) is a carrier is half of the probability that their mother is a carrier or 1/3. This is modified in Bayesian analysis for III-17 in Family B (Figure 3.4) by the fact that she has five normal brothers. This gives her a posterior probability of being a carrier of 1/34. The probability of the daughter of an unaffected maternal uncle of a DMD/BMD patient (e.g. Figure 3.4: III-1, 2, 4 and 5) being a carrier is very low. Such women would have only the background probability of being a carrier of any woman in the population which is 4 times the mutation rate ($2u + 2v$, or $4u$ when $u = v$) (Figure 3.2).

3.3.3.2 Clinical findings

The majority of carriers do not exhibit clinical signs of disease. Only approximately 8% of carriers of DMD show clinical signs including pseudohypertrophic calves and winging of the scapulae (Emery, 1980). Frequently there is familial concordance of signs in carriers (Emery, 1980).

3.3.3.3 Serum muscle enzyme levels

An elevated serum CK is defined as being greater than 3 standard deviations above the mean or a similar threshold. The CK levels in carriers are discussed extensively by Moser (1984) amongst others. In summary, the CK level in two-thirds of carriers is above the 95th percentile, while one third have levels below the 95th percentile, i.e. within the 'normal' range. Since the measure of 'elevated' is statistical, some unaffected women will have 'elevated' CK especially after physical exercise or any other event which has caused muscle damage. Thus both carrier and non-carrier women can have either elevated or normal CK levels. With these caveats in mind, if a woman from a family segregating DMD/BMD has greatly elevated CK on several separate occasions this would strongly indicate that she is a carrier. However, women with normal CKs or borderline elevated CKs are problematical. In addition, CK tends to be more elevated in younger carriers than older carriers and CK level falls during pregnancy, making it an even less reliable measure just when one may wish it to be at its most reliable. (The frequency with which aunts of newly diagnosed sporadic cases are pregnant is astonishing!).

In Family A (Figure 3.3), IV-3 and IV-6 had significantly elevated CK on several occasions and therefore must be considered carriers. Other women, e.g. IV-1, -10, -12 and -14, -19 did not or had borderline elevated CKs and their carrier status remained questionable (Figure 3.5). In Family B (Figure 3.4), the mother of the sporadic case had a significantly elevated CK on separate occasions and must be considered a carrier (Figure 3.6). This alters the carrier probability of her daughter III-17 to 1/2 from 1/34 on pedigree analysis alone, showing the power of positive CK results. III-17 herself did not have elevated CK making her posterior probability 26%. The fact that II-9 is a definite carrier also increases the risk that her sisters and niece (II-5, -11 and III-6) are carriers. This is not a real problem for II-11 who has completed her family. However, it is a problem for III-6, who is of reproductive age and does not have elevated CK levels. She would like to know her carrier status categorically.

3.3.3.4 *Muscle biopsy, EMG and ECG*

Abnormalities in the muscle biopsy specimens from carrier women, including signs of degeneration and regeneration, are sometimes detected (Kakulas and Adams, 1985), as are abnormalities in the EMG (Kakulas and Adams, 1985) or ECG (Emery, 1972). If abnormalities are detected, they can be used as conditional modifiers in Bayesian analysis.

3.3.4 Carrier diagnosis prior to molecular genetics – summary

All the conventional tests described for carriers rely on phenotypic indicators of genotypic status. Therefore, women with disproportionately high inactivation of the X chromosome containing the mutated DMD/BMD gene

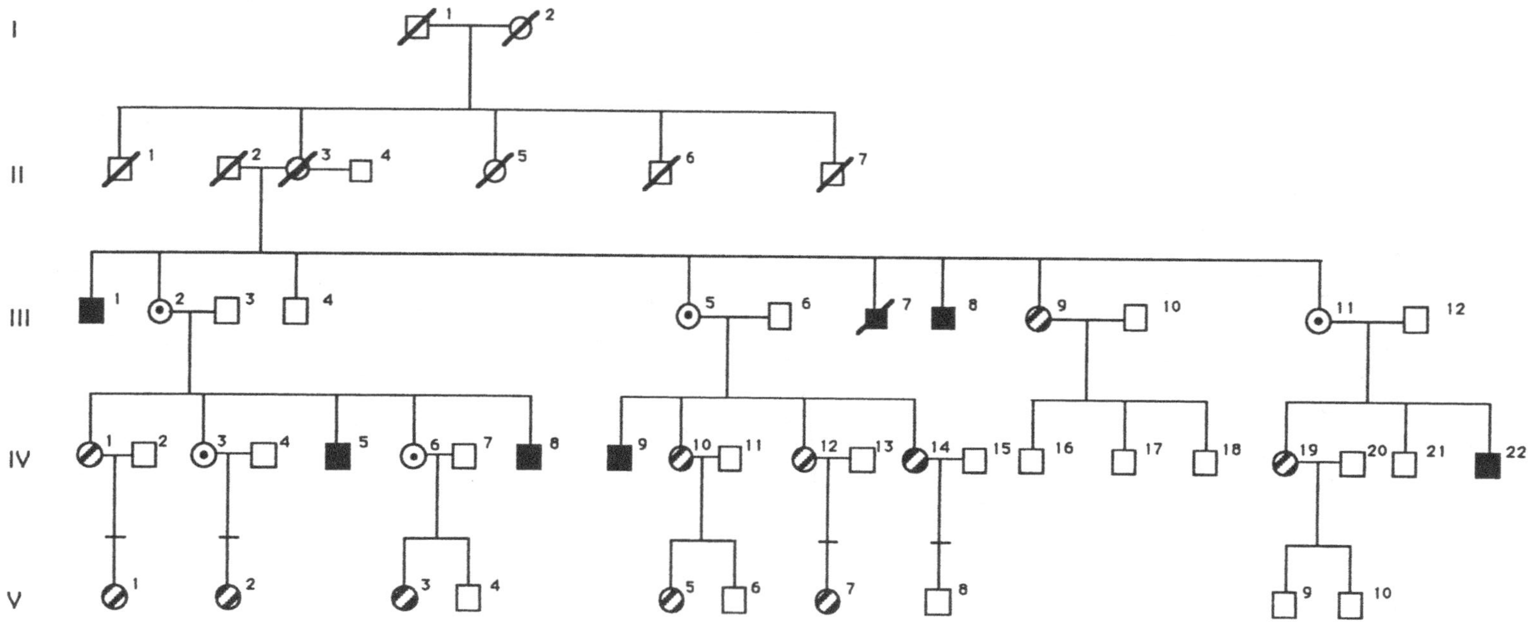

Figure 3.5 Family A: carrier status following CK testing.

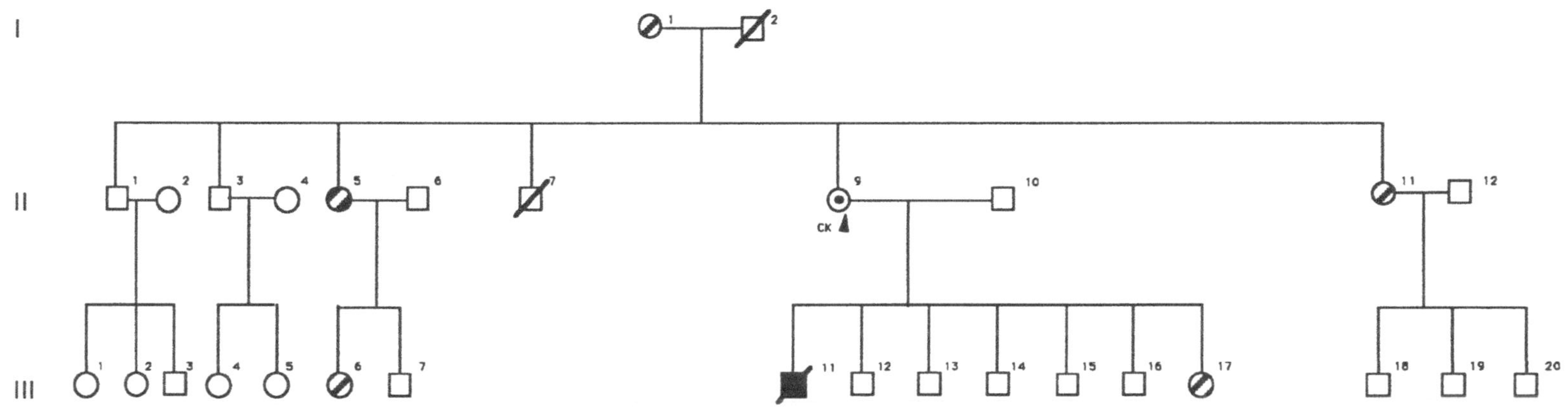

Figure 3.6 Family B: carrier status following CK testing.

were extremely vulnerable to error in conventional carrier-testing programmes because they would be normal by all tests of phenotypic status. Conventioinal carrier diagnosis cannot escape the shadow of lyonization.

In our laboratory (Hurse and Kakulas, 1974) and others (Hutton and Thompson, 1976) schemes for carrier detection using conventional tests and including enforced rest followed by exercise stress (running up stairs) to raise serum CK were initiated in an effort to detect a larger proportion of carriers. However, carrier status remained uncertain for many women.

As we have seen, women seeking genetic counselling regarding their carrier status were generally given a posterior statistical probability of being a carrier based on the prior and modifying probabilities. Many medical geneticists have found these probability estimates difficult (Bundey, 1978), therefore compounding the problems involved. Luckily there are now computer programmes available which will calculate the probabilities. The single posterior probability estimate calculated would lead to a woman being told that she had a 90% chance of not being a carrier, an 80% chance of being a carrier, etc. A woman given a 90% probability of not being a carrier has a 10% chance of being a carrier. If she were in fact that 1 in 10 chance, any sons she had would have a 50% chance of being affected. For example III-9 in Family A (Figure 3.3) with a posterior probability of 8/9 of not being a carrier, would perhaps rather like to know categorically whether she is one of the 8 chances of not being a carrier or the 1 chance of being a carrier. She would like to know whether, if she had another son, the son would have a 50% probability of being affected or only the probability of a new mutation. Given an 80% probability of being a carrier, a woman has a 20% chance of not being a carrier and therefore a one in five chance of not needing to take any precautions.

Prior to molecularly-based techniques, the options facing women told that they were carriers were:

1. To have no children at all. Women who chose this option frequently had themselves sterilized by tubal ligation.
2. To become pregnant but terminate all male fetuses, knowing that by doing so they would have a 50% chance of terminating a boy who would have been totally unaffected by the known family disease.
3. To have children, not interfere in any way and accept fate.

The stress associated with being given an 80% probability of being a carrier is perhaps greater than being told categorically that you are a carrier. If you are told you are a carrier, then the options may be clearer. Counsellors were well aware of the limitations and difficulties of percentage diagnosis. Counsellors were generally very conservative. Despite the conservatism, affected boys were born to women given low estimated carrier risks, which was bound to happen, with tests relying upon statistical analysis. Nevertheless, carrier detection programmes based on conventional tests did reduce the incidence of DMD/BMD (Hurse and Kakulas, 1974; Hutton and Thompson, 1976) when undertaken comprehensively in a confined community.

3.4 GENETIC COUNSELLING OF DUCHENNE AND BECKER MUSCULAR DYSTROPHY POST MOLECULAR GENETICS

3.4.1 Introduction

The most significant inputs by molecular genetics have been more accurate patient and carrier diagnosis through the identification of the gene mutated in DMD/BMD and its product, and the establishment of prenatal diagnosis. These have resulted in major advances for known families. The increased accuracy of diagnosis and carrier testing decreases social tension and relieves feelings of guilt. The prenatal diagnosis, by saving the lives of the unaffected male fetuses, also reduces social tensions and because it is acceptable to more couples is allowing more couples to have their own children.

The first improvements in diagnosis through molecular genetics arose when the position of the DMD/BMD gene on the X-chromosome was determined through linkage to genetic markers (Murray *et al.*, 1982; Kingston *et al.*, 1983; Kingston *et al.*, 1984) (Figure 3.7). This introduced a number of new possibilities: tracing of the mutated DMD/BMD gene through families, pinpointing where the mutation arose in some families and prenatal diagnosis with a degree of accuracy dependent upon the closeness of linkage of the informative marker to the disease location. Identification of the DMD/BMD gene (Kunkel *et al.*, 1986) followed by complete cloning of the DMD/BMD messenger RNA (Koenig *et al.*, 1987) allowed the detection of the disease-causing mutations in the majority of families, leading to more accurate carrier detection and prenatal diagnosis. The DMD/BMD gene at a size of 2.5 million base pairs, is close to one-thousandth of the human genome, and the largest human gene so far identified. The huge size of the gene is probably one of the main reasons for the high mutation rate in DMD.

The previously unidentified protein product of the gene was identified by the raising of antibodies to parts of the molecule synthesized as fusion proteins in expression vectors. This protein was named dystrophin (Hoffman *et al.*, 1987) and led to the renaming of the DMD/BMD gene as the dystrophin gene. The amino-acid sequence of dystrophin predicted that it would be a large (427 kilodaltons) rod-shaped protein similar to the known cytoskeletal proteins α-actinin and spectrin (Koenig *et al.*, 1988). By immunohistochemistry dystrophin is seen to lie just interior to the sarco-

Recombination is the exchange of material between homologous chromosomes. It is the basis of the imprecision in linkage analysis.

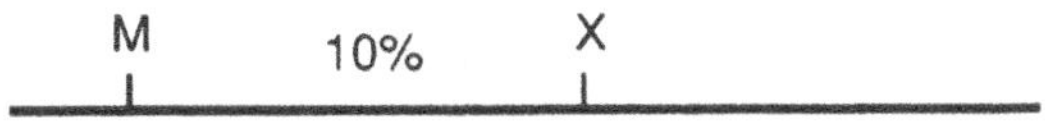

If there is a polymorphic marker at a certain distance θ from the disease-causing mutation X then the greater the distance θ, the greater is the chance of recombination occurring between the marker and the disease mutation. The distance θ is measured as percentage recombination. Ten percent recombination means that for every 100 people examined, the disease will be inherited in 90 along with the allele of the marker linked to the disease in the affected or carrier parent. In the other 10 people, there will be recombination, and the other parental allele will now be linked to the disease. In those 10 people, diagnosis using the linked marker will be wrong. A marker at 10% recombination from a disease gene can be used with 90% accuracy to diagnose the disease.

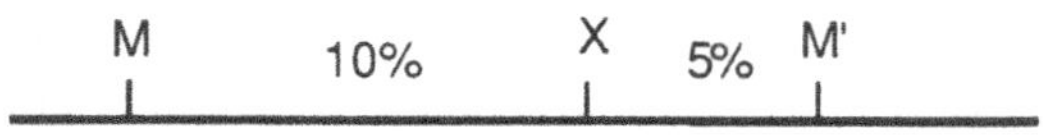

If there are two polymorphic markers M and M′ flanking the disease mutation at 10% and 5% recombination, then, if both markers are informative, the linkage results from both can be combined. The chance of error in diagnosis becomes the chance of recombination occurring between both markers and the disease mutation which is 0.10×0.05 which is 0.005. Therefore linkage analysis with two flanking markers at 10% and 5% recombination can be 99.5% accurate.

Figure 3.7 Linkage analysis.

lemma (Arahata *et al.*, 1988; Zuberzycka-Gaarn *et al.*, 1988) and has been hypothesized to attach the internal cytoskeleton of the muscle fibre to the muscle cell membrane (Bonilla *et al.*, 1988; Zuberzycka-Gaarn *et al.*, 1988; Campbell and Kahl, 1989). The availability of antibodies to dystrophin has permitted diagnosis of patients and carriers by immunohistochemical and immunoblotting methods. In DMD there is a total lack of dystrophin, i.e. the mutations are apparently null mutations, whereas in BMD there is usually an alteration in size or quantity of the protein (Hoffman *et al.*, 1987, 1988; Arkahata *et al.*, 1988).

The specific diagnostic tests which have been developed for DMD patients through the application of molecularly-based techniques are close to 100% accurate. The accuracy of those for BMD is uncertain but they are also perhaps close to 100%. These tests allow differential diagnosis in boys before the age when the distinction between DMD and BMD becomes clinically obvious, and are identifying DMD or BMD patients who were previously diagnosed as having other neuromuscular disorders. Carrier diagnosis is also much more accurate. Women who were diagnosed by conventional tests as carriers are being rediagnosed as non-carriers and vice versa. Prenatal diagnosis of affected and unaffected boys is reducing the incidence of DMD/BMD in the community and is facilitating the birth of unaffected sons to known carrier women.

Most of the techniques which have arisen through molecular genetics can be illustrated in Families A and B.

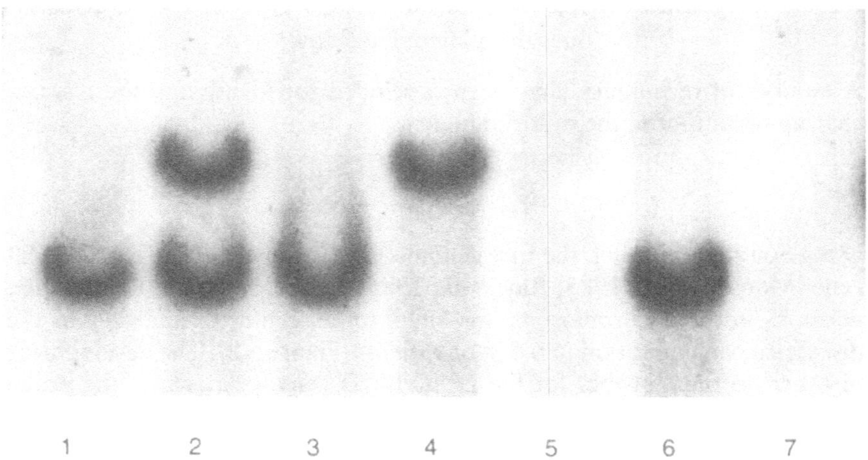

Figure 3.8 Southern blot for members of the multiplex family A (Figure 3.5) The disease-causing mutation in this family is a deletion which includes the polymorphic genomic marker XJ1.1 illustrated here. Lane 1: IV-1, Lane 2: V-1, Lane 3: IV-3, Lane 4: V-2, Lane 5: IV-5, Lane 6: V-3, Lane 7: prenatal diagnosis of a male fetus (after Laing *et al.*, 1991).

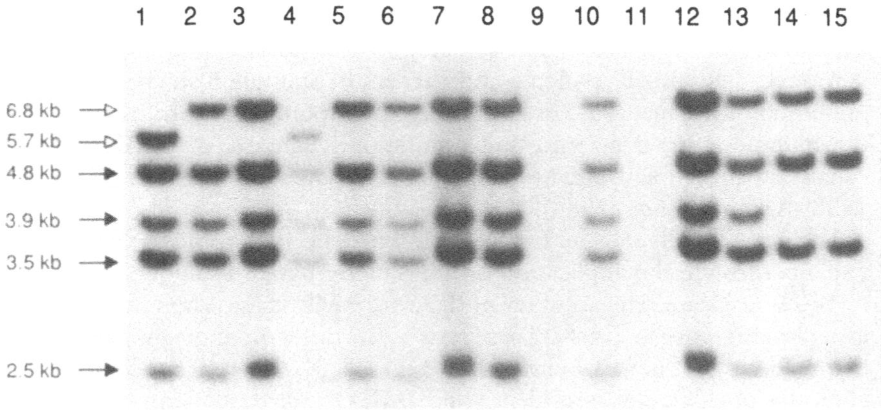

Figure 3.9 DMD cDNA 8 hybridized with *Taq*I digested DNA from West Australians with various diagnoses following conventional testing. The DNA of the two patients in lanes 14 and 15 show the same deletion of the 3.9 kb fragment. The patient in lane 14 had a conventional diagnosis of BMD, the patient in lane 15 a conventional diagnosis of Kugelberg-Welander disease. The 5.7 kb band in lanes 1 and 4 is the less common allele of the 6.8 kb/5.7 kb *Taq*I polymorphism of cDNA8 (Laing *et al.*, 1988).

3.4.2 Diagnosis of DMD/BMD post molecular genetics – what disease does this patient have now?

A number of techniques have been developed for identifying the disease-causing mutations in the dystrophin gene.

3.4.2.1 Genomic probes

Using Southern blotting, the first genomic clones of part of the DMD/BMD gene (Monaco *et al.*, 1985; Ray *et al.*, 1985; Kunkel *et al.*, 1986) identified deletions in the X chromosome too small to be seen cytogenetically as the disease-causing mutation in 6.5% of patients (Figure 3.8). The development of other genomic probes for the DMD/BMD gene, particularly the probe P20 (Wapenaar *et al.*, 1988) allowed the detection of the disease-causing mutation in around 40% of families (Bartlett *et al.*, 1988).

3.4.2.2 cDNA probes

Using the complete cDNA clone in Southern blotting, deletions were identified in 60–70% of families (Koenig *et al.*, 1987, 1989; Forrest *et al.*, 1987; Laing *et al.*, 1991) (Figure 3.9) and duplications were identified in

6% of families (Hu *et al.*, 1990). The complete cDNA clone demonstrated unique-sized junction fragments associated with 17% of the demonstrable mutations (den Dunnen *et al.*, 1989; Prior *et al.*, 1990). These junction fragments occur when the deletion or duplication includes part of a restriction fragment detected by the probe being used. This alters the size of the fragment to produce one unique to that mutation.

The advent of the complete cDNA clone revolutionized DMD/BMD diagnosis, since for the first time the disease-causing mutation could be identified in the majority of families. This led to a multiplication in the number of laboratories which offered genotype-based diagnostic services for DMD/BMD and therefore increased availability of services to families.

3.4.2.3 *Pulsed-field gel electrophoresis (PFGE)*

PFGE can detect most major deletions or duplications in the dystrophin gene (den Dunnen *et al.*, 1987, 1989; Chen *et al.*, 1988a). Deletions/duplications are only missed when they are smaller than approximately 30 kilobases (kb) (den Dunnen *et al.*, 1987). The rare-cutting enzyme most frequently used, *Sfil*, cuts the dystrophin gene into just five fragments and a major disease-causing deletion or duplication virtually always produces a junction fragment with one or other of the four genomic probes commonly used (den Dunnen *et al.*, 1987, 1989). PFGE was the first technique to suggest that the majority of DMD/BMD mutations were large deletions/duplications (den Dunnen *et al.*, 1987).

3.4.2.4 *Polymerase chain reaction (PCR)*

Multiplex PCR in which more than one region of DNA is amplified in the one reaction was first described by Chamberlain and colleagues for detecting deletions in the dystrophin gene (Chamberlain *et al.*, 1988) and subsequently refined both by that group and others (Chamberlain *et al.*, 1990; Beggs *et al.*, 1990; Abbs *et al.*, 1991). It offers a huge advantage in speed in identifying DMD/BMD deletions: detecting in 24 h or less 98% of all deletions found by laborious screening of Southerns with the complete cDNA clone; a process which can take weeks. PCR, including multiplex PCR, can also be performed from buccal cells obtained by mouth scrape or mouth wash (Lench *et al.*, 1988) which is especially useful for any young or old patients or family members who are disinclined to give blood.

3.4.2.5 *mRNA screening*

Screening the 14 kb mRNA for mutations by first synthesizing cDNA using reverse transcriptase and then amplifying defined regions by PCR (RT–PCR), perhaps offers the possibility of detecting all mutations more simply than by screening the 2.5 million base pairs of the gene. Major deletions or duplications are easily detected and point mutations will be more readily identified in the mRNA than in the genomic DNA. In addition, examination of the transcript will easily detect splice-site mutations, such as that which causes golden retriever muscular dystrophy (Sharp *et al.*, 1992) whereas they would be inefficient to detect in genomic DNA.

It is not necessary to carry out a muscle biopsy in order to examine the mRNA since the illegitimate transcription of dystrophin mRNA can be detected in circulating lymphocytes (Roberts *et al.*, 1990, 1991; Schlosser *et al.*, 1990; Chelly *et al.*, 1991).

Intuitively, it would seem that point mutations altering single amino acids in such a large protein would be unlikely to affect protein function unless they are in crucial positions. Therefore point mutations might seldom lead to BMD. On the other hand, point mutations could easily create a false stop codon (a nonsense mutation) leading to a truncated protein and DMD. Thus point mutations may more frequently lead to DMD rather than BMD. This is perhaps being born out by recent data on a large series of patients demonstrating gross cDNA deletions or duplications more frequently in BMD (88%) than in DMD (70%) (Beggs *et al.*, 1991).

3.4.2.6 Immunohistochemistry and immunoblotting

The absence of dystrophin in DMD muscle and the alteration in quality and quantity of dystrophin in BMD muscle detectable by either immunohistochemistry or immunoblotting can be used to diagnose DMD/BMD efficiently in muscle biopsy specimens (Hoffman *et al.*, 1987, 1988; Arahata *et al.*, 1988). The diagnosis can even be made in fetuses of 9 weeks and older (Patel *et al.*, 1988). By immunohistochemistry, the lack of dystrophin is less subjective than the patchy, variably negative or reduced staining seen in BMD. Thus immunoblotting, with its ability to detect the alteration in dystrophin size seen in 86% of BMD cases (Beggs *et al.*, 1991), as well as documenting the reduction in quantity in normal-sized dystrophin seen in 9% of BMD cases (Beggs *et al.*, 1991) may be more useful for diagnosing BMD.

3.4.3 Diagnosis of DMD/BMD post molecular genetics – discussion

3.4.3.1 Accurate diagnosis

Prior to the introduction of molecularly-based techniques, the difficulty with patient diagnosis meant there was uncertainty that counselling for an X-linked disorder was appropriate. The new techniques of mutation and linkage analysis and immunohistochemistry and immunoblotting have solved this problem to a large extent. The techniques created by the application of molecular genetics to DMD/BMD can all be used to provide a differential diagnosis between DMD and BMD and between DMD/BMD and the other neuromuscular disorders with which they can be confused. Identifying mutations in the dystrophin gene allows accurate diagnosis of DMD/BMD. Some individuals who were diagnosed on the basis of conventional tests as having spinal muscular atrophy (Lunt *et al.*, 1989; Laing *et al.*, 1991), (Figure 3.9); limb girdle muscular dystrophy (Norman *et al.*, 1989a; Laing *et al.*, 1990), quadriceps myopathy (Beggs *et al.*, 1991), myalgia with cramps (Gospe *et al.*, 1989) or other unusual variants (Bettecken and Muller, 1989; de Visser *et al.*, 1990) are now being identified as BMD patients. It has become obvious that BMD in particular can very successfully masquerade clinically as a benign spinal muscular atrophy such as Kugelberg-Welander disease or as limb-girdle muscular dystrophy.

Re-examination of pedigrees with affected male and female sibs has revealed some probably to have autosomal recessive Duchenne-like dystrophy since no dystrophin abnormality has been found (Norman *et al.*, 1989b; Francke *et al.*, 1989), while other such families have been shown to be true DMD pedigrees with the affected females being manifesting carriers (Francke *et al.*, 1989). The families can be counselled according to their now more certain disease diagnosis. Recessive Duchenne-like dystrophy may involve mutation in the genes coding for the glycoproteins which apparently are the site of attachment of the dystrophin molecule in the membrane (Campbell and Kahl, 1989; Ervasti *et al.*, 1990) or may result from mutations in the genes for entirely different protein systems.

Definitive diagnosis of the minority of cases in which DMD/BMD can still not be discriminated from other disorders may be helped by the identification of the disease genes for these other disorders.

3.4.3.2 Incidence of BMD

The incidence of BMD will have to be reviewed in light of findings that many patients who were thought previously to have other disorders have in fact phenotypic variants of BMD. The estimates of the incidences of the other disorders will have to be revised downwards, and the estimate of the incidence of BMD will have to be revised upwards.

3.4.3.3 Families with no living affected patients

For families with no living affected patients, mutation detection can now be performed easily by multiplex PCR on any frozen muscle biopsy material. It also may be possible to use formalin fixed tissue. In our laboratory we have successfully examined stored frozen material from male fetuses terminated on the basis of sex (Laing *et al.*, 1992).

3.4.3.4 *Prognostic value*

Molecularly-based diagnostic procedures are better prognostic tests than those previously available. Determination of dystrophin levels in muscle biopsy specimens when the patient is too young for definitive clinical differentiation between DMD and BMD can allow accurate diagnosis of the family disease. If the muscle biopsy is dystrophin negative by immunohistochemistry or immunoblotting then the family disease is almost undoubtedly DMD. If the immunohistochemistry shows patchy or variable staining, or if there is either an altered size or abundance of dystrophin detected following immunoblotting, then the family disease is likely to be BMD.

Precise analysis of the mutations detected in patients may also give accurate prognosis. The reading frame theory (Monaco *et al.*, 1988) states that the major deletions or duplications which cause DMD produce a shift in the translational reading frame while those which cause BMD do not alter the reading frame. If a deletion or duplication alters the translational reading frame, a false stop codon will be produced which will truncate the protein. The N-terminal end of the dystrophin molecule will be synthesized, but the C-terminal end will not. This will lead to a non-functional dystrophin molecule and DMD. If on the other hand the mutation maintains the translational reading frame, a dystrophin molecule of altered size but partial function will be produced and BMD will result. The severity of the BMD will be controlled by the degree of alteration in function. Using cDNA probes in Southern blot analysis, 92% of deletions obey the reading frame rule (Koenig *et al.*, 1989). But investigation of dystrophin mRNA using RT–PCR has demonstrated that at least some of the deletions which appear upon Southern analysis to break the rule may be explained by alternate splicing (Chelly *et al.*, 1990).

3.4.3.5 *Clarification of varying phenotype in the one family*

The basis of variable phenotype seen in some families can now be approached at the molecular level. In at least one family, the answer has been identified as separate mutations (Laing *et al.*, 1992). A dystrophin-negative fetus which was terminated on the basis of sex in a family segregating BMD raised the possibility that the fetus would later have had DMD if it had not been terminated. Subsequent investigation revealed two separate mutations in the same dystrophin gene in the family. These were an unidentified (non-major) mutation giving rise to a BMD phenotype and a frame-shift deletion correlating with the dystrophin negativity of the fetus (Laing *et al.*, 1992). In the future it may be shown that variable phenotype in other families is due to alternate splicing around the mutation in different individuals (Chelly *et al.*, 1990; Roberts *et al.*, 1990, 1991) especially, perhaps, where the family mutation produces DMD in some families and BMD in others, e.g. exon 3–7 deletions (Malhotra *et al.*, 1988).

3.4.3.6 Confirmation of neonatal/infant screening

In a number of regions of the world, neonatal or infant screening by raised CK has been carried out to detect the first DMD/BMD case in a family with no previous history of the disease (Plauchu *et al.*, 1980, 1989; Scheurbrandt *et al.*, 1986; Greenberg *et al.*, 1991). Neonatal or infant screening (the relative merits of both have been discussed at length (e.g. Moser, 1984; Scheurbrandt *et al.*, 1986)) can prevent the tragedy of multiple affected boys in a family with no previous history of DMD/BMD. 15%–30% of DMD cases are born into families with a previously affected child (Plauchu *et al.*, 1980; Moser, 1984; Greenberg *et al.*, 1991). Neonatal screening has the potential to prevent this, perhaps reducing the incidence of DMD by as much as 20% (Plauchu *et al.*, 1987). One of our own families, which had four affected boys but no previous history, could have been spared some of the subsequent tragedy if screening had been in place. Neonatal or infant screening also allows detection of carrier aunts, cousins and other family members who might not otherwise have been detected before they themselves had had affected children.

There were always difficulties with neonatal and infant screening prior to definitive tests for mutations and dystrophin abnormalities, since one could never be absolutely certain that the infant or neonate with an elevated CK had DMD or BMD or another disorder entirely. Now, with the accurate tests available, the infants with elevated CK can be screened for DMD/BMD disease-causing mutations using multiplex PCR or for dystrophin abnormalities (Greenberg *et al.*, 1988, 1991; Naylor, 1991).

There are severe ethical and moral dilemmas associated with neonatal or infant screening. For example, when some parents have been told that their apparently normal boy has DMD/BMD they have felt robbed of five years of relatively normal life with their boys. On the other hand, early identification of the first affected child in a family prevents multiple cases. These dilemmas will need to be confronted if health services are serious about reducing the incidence of DMD/BMD.

3.4.3.7 Diagnostic algorithms

Diagnostic algorithms for DMD/BMD, whether to first search for mutations in the gene or dystrophin abnormalities, are established in different laboratories, it would seem, according to the biases of those involved. As far as the patient is concerned, it is perhaps better to test for DMD/BMD by multiplex PCR or RT–PCR first rather than the relatively more invasive muscle biopsy to examine dystrophin immunohistochemistry or immunoblotting. The only difficulty which may arise would be in neutral mutations of the dystrophin gene. (A deletion of an intronic probe in affected and unaffected brothers (Koh *et al.*, 1987) has been resolved by the identification of a second disease-causing deletion in the affected boy (Bartlett *et al.*, 1989)). However, deletion of an exon-containing fragment has been reported in an apparently unaffected male (Nordenskjold *et al.*, 1990). If a mutation is identified in the dystrophin gene of a patient previously diagnosed as having facioscapulohumeral muscular dystrophy (FSHMD), and there is no confirmation by immunohistochemical or immunoblot evidence of dystrophin abnormality, does this mean that the FSHMD diagnosis is wrong?

3.4.4 Prenatal diagnosis

Prenatal diagnosis allows known carriers the choice of terminating the 50% of male fetuses which will later in life be affected and of saving the 50% of male fetuses not carrying the known family mutation. The necessary fetal DNA can be obtained either by amniocentesis at around 16 weeks of pregnancy or preferably by chorionic villus sampling (CVS) at around 9–11 weeks of pregnancy. Any mutation that can be detected by genomic probes or cDNA probes in Southern analysis, by genomic or cDNA probes in PFGE, or by multiplex PCR can be used for extremely accurate prenatal diagnosis. This is most often used to diagnose the disease state of male fetuses (Figure 3.8, lane 7) but it can also diagnose the carrier status of female fetuses (Bartlett *et al.*, 1987).

In the 30% of DMD families and 14% of BMD families which do not show easily detectable mutations, linked polymorphisms can be used for prenatal diagnosis with a lesser degree of accuracy than direct analysis of mutations. The accuracy of linked polymorphisms is determined by their distance from the disease-causing mutation (Figure 3.7). The further the marker is from the mutation, the less accurate the polymorphism is for diagnosis because of the possibility of recombination between the mutation and the polymorphic marker. The huge size of the dystrophin gene, plus other undefined factors, leads to the significant 11–12% recombination across the gene (Abbs *et al.*, 1990; Oudet *et al.*, 1991). Unless tightly linked flanking markers or several markers along the dystrophin gene are used, there may be significant error in prenatal diagnosis for DMD/BMD by linkage analysis.

Prenatal diagnosis is best performed using PCR technology either in multiplex detection of deletions or in determining alleles of linked polymorphisms, especially the highly informative microsatellites (Weber and May, 1989) flanking and within the dystrophin gene. Results using these methods are obtained much more quickly, in less than 24 h, as compared to Southern analysis which can take weeks when culturing of CVS samples is necessary.

Fetal muscle biopsy followed by dystrophin immunoblotting may also be used for prenatal diagnosis if there is no other alternative (Evans *et al.*, 1991).

3.4.5 DMD/BMD diagnosis summary

Diagnosis of DMD/BMD patients is much more accurate in the post-molecular age and prenatal diagnosis is now possible. The phenotype of BMD especially is expanding. The benefit to known families from these changes is almost immeasurable and the potential of the techniques now available is facing health services and society with difficult decisions.

3.4.6 Carrier detection for DMD/BMD post molecular genetics – is this woman a carrier or not?

To accurately diagnose the carrier status of a woman in a DMD/BMD family one ideally wishes to detect whether she carries the family mutation. The same post-molecular genetic techniques which are so successful in accurately diagnosing cases of DMD and BMD can also be applied to determine carrier status, but with certain difficulties.

The deletions or duplications which cause the majority of both DMD and BMD cases, and which are so easy to detect in affected males, are not so easy to detect in women. This is because the presence of a normal dystrophin gene on the woman's other X chromosome masks the deletion or duplication in the carrier. A carrier of a deletion should have a half dose of the region of deletion from the normal dystrophin gene, while a carrier of a duplication should have a 150% dose. These alterations should be detectable. The use of dosage in Southern blot analysis to determine carrier status has been described (Hejtmancik *et al.*, 1986; Laing *et al.*, 1989; Mao and Cremer, 1989). It should be a simple technique to use but in practice it is not (den Dunnen *et al.*, 1989). Some clinical geneticists do not trust the results that dosage analysis gives as a basis for advice to a woman about her carrier status. However, detection of carriers by dosage PCR has now been described (Abbs and Bobrow, 1992) and may, if it proves reliable, become a standard technique.

In many ways, it is far better to have more tangible evidence of the presence of the family mutation. A number of techniques have been used to provide this: (a) segregation of polymorphic markers within the family mutation, (b) presence of junction fragments, (c) analysis of cDNA.

3.4.6.1 Segregation of polymorphisms

If a possible carrier does not inherit her mother's allele for a polymorphic marker lying within the family deletion then she is highly likely to have inherited the deletion from her mother and therefore to be a carrier (Bartlett *et al.*, 1987). However, there are some very unlikely scenarios in which she might not be a carrier. For example, if the allele inherited from her mother had spontaneously mutated to the same allele as her father has, or she was a Turner's syndrome patient with only one X. On the other hand, if a possible carrier in a family segregating a disease-causing deletion is heterozygous for a polymorphic marker lying within the deletion then she is unlikely to be a carrier. This is because she has one allele of the polymorphism on one of her X chromosomes, another allele on her other X chromosome and therefore cannot carry the deletion. Again there are some highly unlikely scenarios in which she could still carry the family mutation. For example if she were XXX or carrying a duplication with two different alleles of the marker on one chromosome and the deletion on the other.

The effectiveness of such results in determining carrier status can be illustrated in Family A which segregates a deletion including the intragenic genomic polymorphic marker XJ1.1 (Figure 3.8). In lane 2, DNA from V-1 shows both alleles for the XJ1.1 polymorphism. Therefore V-1 is unlikely to carry the family deletion. However, V-2 having only the larger 3.8 kb allele (Figure 3.8, lane 4), has not inherited her mother's smaller 3.1 kb allele (Figure 3.8, lane 3: IV-3). V-2 is highly likely to be a carrier. The carrier status of women in Family A following DNA analysis is summarized in Figure 3.10. Since the family mutation has been identified, the carrier status of many women in the pedigree is known with a high degree of accuracy and all those determined to be carriers can be offered prenatal diagnosis.

It should be noted that there are no affected family members in generation V of Family A. The disease has effectively been stopped in this family. There need never again be an affected person in the family unless a couple do not wish to terminate a male fetus which has inherited the family mutation.

3.4.6.2 Presence of junction fragments

The unique-sized junction fragments seen in 12% of all DMD/BMD families using the complete cDNA in Southern analysis (Prior *et al.*, 1990), in PFGE (den Dunnen *et al.*, 1989), or with cosmids for deletion-prone regions (Blonden *et al.*, 1991) can be used to follow the disease-causing mutation

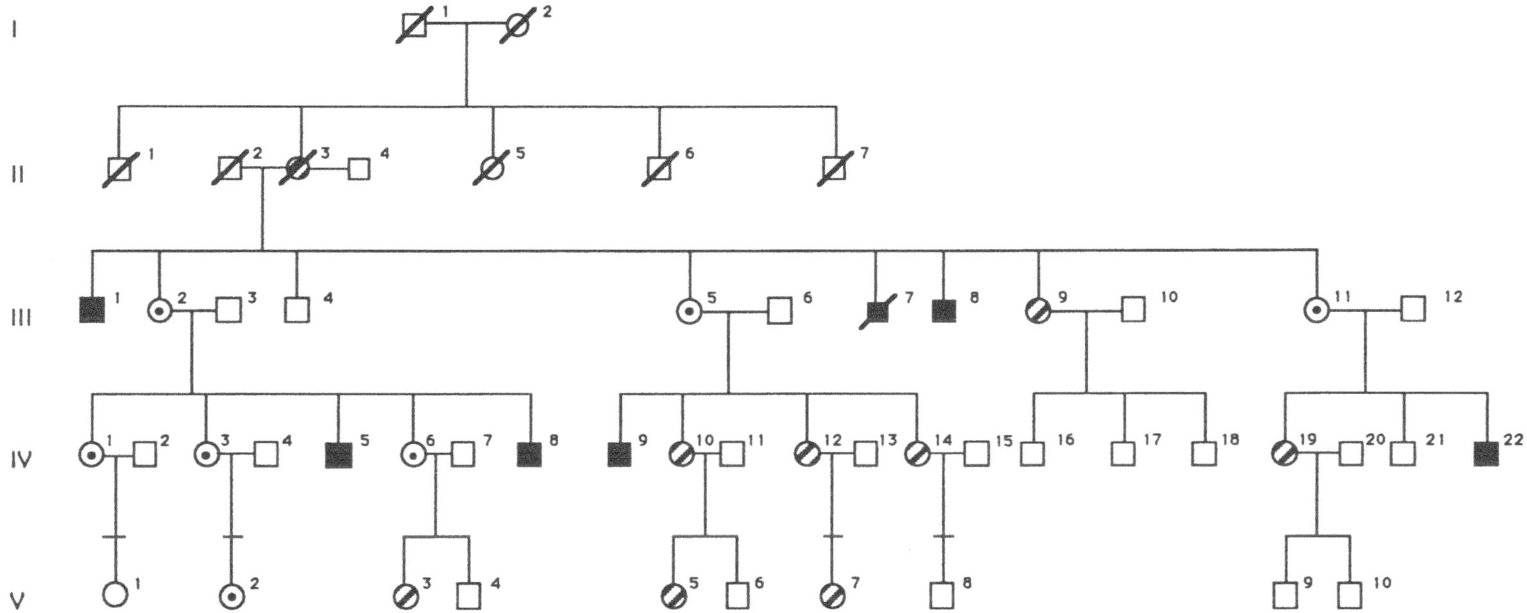

Figure 3.10 Family A: carrier status following DNA analysis.

through the family and thus detect carriers and non-carriers. This is a very useful method. However, one risk is that the unique fragment is an allele of a rare polymorphism which then may or may not (following recombination) segregate with the disease-causing mutation. Another risk is that when performing diagnosis of carrier status using junction fragments, one is relying on the mutation staying the same in the family. If the junction fragment is present, one can be fairly certain that the woman is a carrier. However, should the junction fragment be absent there must always be a very slight doubt that the mutation has changed from one generation to the next causing the loss of the junction fragment. This would be more of a problem in Southern blot analysis with cDNA probes, where a small alteration in the size of the deletion could theoretically prevent the formation of the junction fragment, than in analysis using PFGE, which should not be affected to the same extent. The risk of a mutation altering between generations must be very low, and will have to be determined empirically as examples appear.

3.4.6.3 Analysis of cDNA

RT–PCR may be used to detect carriers by showing the two different-sized dystrophin mRNA transcripts in women segregating major deletions or duplications in RNA isolated from a muscle biopsy (Chelly *et al.*, 1990) or from circulating lymphocytes (Roberts *et al.*, 1990, 1991; Schlosser *et al.*, 1990; Chelly *et al.*, 1991). The disease-causing deletion may remove one of the primer sites for the PCR, and then a product of altered size would not be seen with that primer. However, if multiple primers are used, RT–PCR should always produce an altered fragment size (Roberts *et al.*, 1991).

The proportions of the two different messages present in RNA from muscle and circulating lymphocytes of carriers will depend not only on the randomness of inactivation in lyonization (Lyon, 1988) but will also depend on the proportion of nuclei still expressing the mutated gene in muscle. This proportion decreases with age (see Section 3.4.6.5) and therefore it may be that a larger proportion of carriers may be detected by RT–PCR of lymphocyte RNA than by RT–PCR of muscle RNA. How serious a problem lyonization will be to the usefulness of lymphocyte RT–PCR will also have to be determined empirically by testing a large number of known carriers of deletions or duplications.

3.4.6.4 Families with minor mutations

In the approximately 30% of DMD families, or 12% of BMD families which do not show a gross alteration of gene or mRNA structure, mutational analysis cannot be performed to detect carriers or perform prenatal diagnosis unless the minor mutation can be identified. Identification of a disease-causing point mutation in one DMD family was achieved by examining the size of the dystrophin transcript in the patient, predicting the approximate site of the mutation, then sequencing that region of the cDNA (Bulman *et al.*, 1991). Roberts *et al.* (1992) have identified point mutations in the dystrophin gene using RT–PCR followed by chemical cleavage of mis-match and then sequencing. This latter technique will probably prove more suitable for use in routine diagnosis of minor mutations.

However, in most of the families without readily detectable mutations, linkage analysis must still be used to determine carrier status and perform prenatal diagnosis. Linkage analysis can identify the DMD gene carrying the mutation and, in many cases, where the mutation arose in the family (providing sufficient family members can be tested). Analysis of unaffected male relatives is especially important. For example, if one can show that an unaffected brother of a sporadic case has inherited the same dystrophin gene as the affected person then one knows, within the limits of double recombination and germline mosaicism, that the sporadic case has a new mutation.

The effectiveness of linkage analysis in Family B is illustrated in Figure 3.11. The mother, II-9, of the sporadic case, III-11, was diagnosed as a carrier by elevated CK and was informative for the restriction fragment length polymorphisms OTC, 754, P20 and C7 which lie both within and flanking the dystrophin gene. Analysis of the three unaffected brothers of the deceased affected boy who were willing to participate in the analysis, revealed, as would be expected, that they had all inherited the same (normal) dystrophin gene region from their carrier mother. Their sister (III-17) had inherited the same dystrophin gene region as the unaffected brothers, and could be given a very high probability of not being a carrier. The only risk of III-17 being a carrier is the risk of double recombination between the intragenic probe P20 and the flanking markers 754 and C7, with one of the recombinational events lying within the dystrophin gene. This reduces her posterior probability of being a carrier to a very low level since double recombination is unlikely to occur within this distance (Edwards, 1986).

The dystrophin gene inferred to carry the DMD mutation in Figure 3.11, was inherited by the mother (II-9) of the affected boy from her unaffected father (I-2). This implies that the mother must have a new mutation unless her father is a germline mosaic. If II-9 has a new mutation, no other woman in the pedigree, including III-6, who wishes to know her carrier status, can be carriers of the known family mutation. The fact that the unaffected brother (III-7) of III-6 has inherited the same dystrophin gene region as the affected boy is inferred to have had (Figure 3.11) means that III-6, without

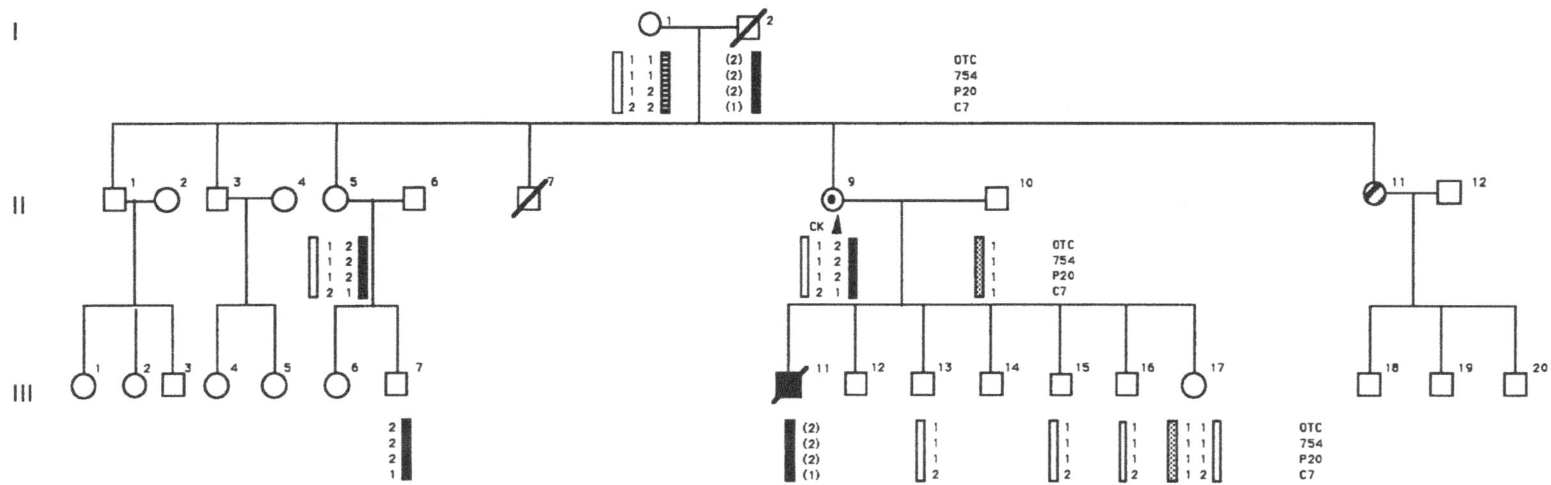

Figure 3.11 Family B: carrier status following DNA analysis and showing dystrophin gene region haplotypes for markers OTC, 754, P20 and C7.

even analysing her DNA, can be said not to be a carrier. Only negligible risk remains of her grandfather being a germline mosaic combined with the possibility of a double recombination in her brother.

Figure 3.12 illustrates confirmatory evidence that the dystrophin region carrying the mutation in this family was as hypothesized above. An apparently sporadic case (IV-2) was brought to our attention, whose mother (III-9) turned out to be the half-sister of the affected boy (adopted by another family); her affected son, IV-2, had the same haplotype for the dystrophin gene region as had been inferred for the deceased affected boy.

It is interesting to summarize the carrier probabilities estimated for III-17 in Family B following the various methods of carrier status determination. On pedigree analysis alone (Figure 3.4) her risk was 1/34. When the CK level of her mother was included (Figure 3.6) her risk was 50%. Including her own CK level her risk was 26%. By DNA linkage her risk is conservatively 2% (Figure 3.11).

Linkage analysis should now be performed using PCR techniques to analyse the microsatellites at either end of the huge dystrophin gene (Abbs *et al.*, 1990; Beggs and Kunkel, 1990; Feener *et al.*, 1991; Oudet *et al.*, 1991; Powell *et al.*, 1991) and other intragenic and linked markers. PCR techniques should be used since they are so much faster than the tedious and time-consuming search for informative polymorphisms using Southern analysis of the many now-known polymorphisms within and flanking the dystrophin gene. Microsatellites also tend to be more informative than markers used in Southern analysis.

It can be seen from Family B that screening for elevated CK in the molecular age is still invaluable for families with minor mutations since it can help to elucidate where the mutation arose in those families where linkage analysis is the only recourse (see also Hyser *et al.*, 1987).

3.4.6.5 Accurate carrier diagnosis – immunohistochemistry

Muscle biopsies from carriers of DMD can show a patchwork of dystrophin-positive and negative muscle fibres (Arahata *et al.*, 1989; Bieber and Hoffman, 1990). This has allowed women previously diagnosed as limb girdle muscular dystrophy patients to be rediagnosed as manifesting carriers of DMD (Arikawa *et al.*, 1991). However, non-manifesting carriers are more easily detected by this method when younger since in them the mosaic pattern is gradually lost (Bieber and Hoffman, 1990). It is hypothesized that the focal necrosis of muscle fibres seen in carriers occurs in the area of control of nuclei whose active X chromosome is expressing the mutated dystrophin gene. The focal necrosis destroys not only the myofibrillar structure but also the nucleus expressing DMD. Gradually the muscle nuclei expressing DMD are replaced by offspring of satellite cells whose active

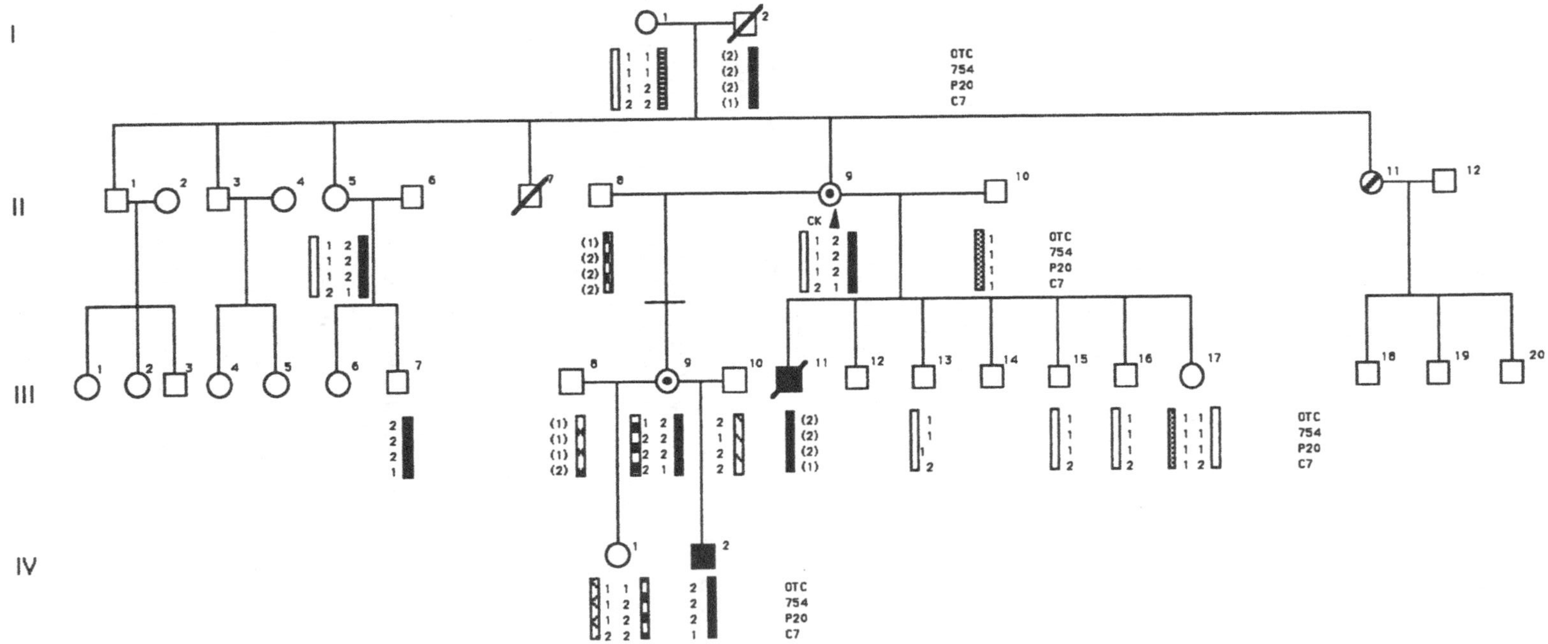

Figure 3.12 Family B: carrier status of the complete family including the previously unknown branch.

X chromosome is expressing a normal dystrophin gene. Thus, it can be impossible to determine carrier status in mature non-manifesting carriers by immunohistochemistry. The gradual loss of the mosaic pattern correlates with the gradual loss of elevated CK in carriers.

Immunohistochemical and immunoblot analysis of dystrophin in clonal muscle culture has been used to overcome the difficulty caused by the loss of the mosaic pattern (Bieber *et al.*, 1989; Hurko *et al.*, 1989; Miranda *et al.*, 1989). But it is an extremely time-consuming and expensive technique only to be used as a last resort.

3.4.6.6 Carrier-status determination from fetuses

If all other methods of ascertaining the carrier status of a woman fail, analysis of fetuses may be used to determine her status.

If the carrier status of a woman remains uncertain in a family with a known mutation for any reason, e.g. lack of concrete evidence that she carries the family mutation, fetal DNA analysis may clarify the woman's carrier status (Chamberlain *et al.*, 1988; Boelter *et al.*, 1990; Miciak *et al.*, 1992). In a variation on this theme, we recently performed DNA analysis on a blighted ovum from a woman of uncertain carrier status. The blighted ovum was found to have the family mutation and thus proved that the mother, who had an 80% carrier probability, was in fact a carrier (unpublished observations).

If the carrier status of a woman seeking counselling remains uncertain in a family where linkage analysis has to be used, e.g. through recombination with flanking markers or through the affected family member being a sporadic case, immunohistochemistry or immunoblotting of muscle from a terminated male fetus can be used to diagnose the carrier status of the woman (Boelter *et al.*, 1990). Immunoblot analysis following fetal muscle biopsy has also been used as a last resort (Evans *et al.*, 1991).

3.4.6.7 Families with no living affected patient

In families where there is no source of DNA from an affected person, determination of carrier status is more difficult. Carrier diagnosis can be made in such families by segregation analysis for polymorphisms (Chen *et al.*, 1988b). A search can also be made, especially in obligate carriers, for junction fragments since these will be found in one in eight of all DMD/BMD families using the complete cDNA clone, or in the majority of families using PFGE. The risk that a junction fragment detected by Southern blotting was a rare RFLP would be increased in such families.

However, junction fragments, detected by screening Southern blots with cDNA probes, could be cross-checked using RT–PCR for alterations in the transcript or with PFGE for alterations in genomic structure. Analysis of fetal material as above (Section 3.4.6.6) may clarify carrier status. Linkage analysis concentrating upon unaffected males in the pedigree may also be performed to clarify the point at which the family mutation arose, if no mutation can be identified in at-risk female members of the family.

3.4.6.8 Germline mosaicism

The current best estimates of the proportion of mothers of sporadic cases who are germline mosaics is 14% (Bakker *et al.*, 1987, 1989; Lanman *et al.*, 1987; Bartlett *et al.*, 1988). Therefore it is obviously an important mechanism in the generation of mutations. Germinal mosaicism can severely affect the accuracy of prenatal diagnosis (Boileau and Junien, 1989; Gutmann and Fischbeck, 1989; van Ommen, *et al.*, 1990). The theoretical impact of germline mosaicism on the percentage of affected patients who have new mutations has also been investigated (Grimm *et al.*, 1990).

3.4.7 Carrier detection for DMD/BMD post molecular genetics – discussion

Counselling for DMD/BMD carriers has been vastly simplified now that the disease-causing mutation can be traced in so many families. The vast majority of women can be told categorically whether or not they are carriers. Women who are told categorically that they are carriers now have the option through prenatal diagnosis of terminating affected male fetuses and saving unaffected boys. This is more acceptable to many couples than running the risk of terminating unaffected male fetuses, and, in our experience, many women are reversing tubal ligations. The advent of prenatal diagnosis of DMD/BMD is saving lives. In one of our early cases, the family mutation was identified on the morning that a male fetus was to be terminated on the basis of sex. The termination was cancelled, and subsequent investigation showed that the fetus did not have the family mutation. The fetus has now become a healthy young boy.

The ability to tell mothers of sporadic cases definitively, years afterwards, that they are not carriers, by either demonstrating heterozygosity for markers within the deletion found in their affected son or by other methods, and that they could have done nothing to prevent the birth of their affected child, lifts in many cases a very heavy burden of guilt.

Accurate detection of carriers allows the prevention of cases of DMD/

BMD. Given the opportunity, the majority of couples will terminate fetuses which have a high probability of suffering later from DMD/BMD. Whether a couple will terminate a mild BMD is their personal decision, and is based on their perception of the burden of the disease. Each couple is different. Some will accept nothing more than the background risk, others will not terminate a Duchenne boy. It is a very individual decision. However, some couples still have to be given a statistical probability of having an affected child and have to make decisions based on unacceptably high levels of uncertainty.

3.5 GENETIC COUNSELLING OF DUCHENNE AND BECKER MUSCULAR DYSTROPHY: FUTURE

The future for many laboratories is the present for some. Molecular biological and other techniques being applied to DMD/BMD are changing rapidly and a standard algorithm for laboratory investigation is yet to emerge because of the state of flux in techniques.

3.5.1 Future laboratory procedures

3.5.1.1 PCR techniques

On the technical side, laboratories should move as much as possible towards all testing being performed by PCR techniques. These techniques will include multiplex analysis for detecting deletions in genomic DNA, linkage analysis using microsatellites rather than polymorphisms detected using Southern analysis and amplification of mRNA from lymphocytes to detect carriers of major gene alterations. Methods of sufficiently accurate dosage analysis using multiplex PCR may also be developed for detecting carriers.

3.5.1.2 Detection of minor mutations

It can be anticipated that one area of technical development will be in the identification of minor mutations since minor mutations are the major remaining area of doubtful prenatal and carrier diagnosis. One can expect greater use in diagnostic laboratories of methods for detecting minor mutations (Roberts *et al.*, 1992), including single-strand conformation polymorphism analysis (Orita *et al.*, 1989), and denaturing gradient gel electrophoresis (Attree, 1989).

The distribution of disease-causing minor mutations in the dystrophin gene is not yet known. It will be interesting to see if the distribution of disease-causing minor mutations shows the same peaks around intron 7 and the P20 intron as the major deletions, or whether minor mutations are more evenly spread. This is not simply of academic interest, as the answer may allow more accurate estimation of risk in families segregating minor mutations.

3.5.1.3 Analysis of dystrophin transcripts

One can also expect greater emphasis on analysis of dystrophin gene transcripts. This does not require muscle biopsy, since it can be effected from lymphocyte RNA. The power of RT–PCR to detect carriers and its acceptability in a diagnostic setting will be determined in time.

3.5.2 DMD and mental retardation

In the future, it can be anticipated that the relationship between dystrophin gene mutations and mental retardation will be clarified. The reason for one-third of DMD patients being educationally subnormal (Rowland, 1988), is not yet known. It may be that some patients with X-linked mental retardation but no myopathy may also have mutations of the dystrophin gene.

3.5.3 Male and female mutation rates

The question of whether the male and female mutation rates are equal should also be answered through molecular analysis. This is possible as molecular diagnosis can identify whether the mothers and grandmothers of sporadic cases are carriers. Concentrating solely on major deletions or duplications may not however give a true answer to this question, since there are theoretical reasons (Moser, 1984), and the beginnings of experimental evidence (Baumbach *et al.*, 1989) for a larger percentage of major deletions or duplications occurring in oogenesis than in spermatogenesis. A statistical approach as suggested by Muller and Grimm (1986) may provide a more accurate answer.

3.5.4 Early identification of manifesting carriers

It will be interesting to see whether a means of identifying female fetuses who will become manifesting carriers will be developed. At present, no female carriers are terminated. At some time in the future, a couple participating in a prenatal screening programme will have a daughter who becomes either a manifesting carrier, or worse, affected with DMD/BMD.

3.5.5 DMD/BMD screening

3.5.5.1 Patient screening

The aim of prevention in DMD/BMD is to have as few patients as possible who require therapy, which could take the form of myoblast transfer (Partridge *et al.*, 1989; Karpati *et al.*, 1989) or injection or transfection with gene constructs (Wolff *et al.*, 1990; Acsadi *et al.*, 1991) or something as yet undreamt of. More countries or states should perhaps adopt schemes for DMD/BMD screening. Screening identifies affected boys at a very early age, alerts genetic counsellors and the family and has the potential to prevent the birth of other affected siblings or relatives. Moser (1984) argued that the emphasis should be on prevention of DMD/BMD since there is no cure. However, the dominant view is that since there is no cure at present or perhaps for the foreseeable future, one should not screen for DMD/BMD.

Moser argued against screening of neonates or infants and instead for screening of boys with delayed motor milestones at around two years of age since early diagnosis may interfere with the proper bonding between parent and child. These considerations are addressed by Plauchu *et al.* (1980) and Scheurbrandt *et al.* (1986) and influenced the decision to make screening voluntary in Germany (Scheubrandt *et al.*, 1986). Adopting a voluntary screening model, rather than any compulsory screening, may be the preferable route.

3.5.5.2 Carrier screening

Prenatal diagnosis for known carriers of DMD does not have the potential to prevent the majority of DMD/BMD cases. The majority of carriers are not recognized until they have an affected son because most pedigrees show no previous family history. If the emphasis was changed to identification of carriers, as has been argued by others (Hutton and Thompson, 1976; Roses, 1988; Scheurbrandt *et al.*, 1986), the mothers or grandmothers of affected boys still to be born might be detected. If carrier detection through screening were a practical proposition, it would have the potential to prevent most DMD, since the mothers of two-thirds of patients are carriers. It must be preferable to identify a family before DMD patients are born, rather than waiting for diagnosis through neonatal or infant screening or even later clinical presentation. In addition, detecting carriers by screening would not have the ethical and moral dilemmas associated with early detection of DMD/BMD patients, since there are no phenotypic consequences in the vast majority of carriers.

The effectiveness of screening CK in infant girls to detect carrier status is uncertain. Since the mosaic pattern of dystrophin-positive and dystrophin-negative fibres is greatest in younger carriers, a better differential than the commonly quoted two-thirds may be expected in much younger girls than are normally tested. Plauchu *et al.* (1980) tested CK levels in female infants, but the results were 'disappointing'. On the other hand it is known that CK activities can be high in young carrier girls (Beckmann *et al.*, 1978).

Carrier status can be determined *in utero* for the daughters of known carriers participating in prenatal screening programmes. Carrier and non-carrier women diagnosed as fetuses, could then be followed prospectively in order to determine the CK ranges of carrier and non-carrier neonates and infants for possible initiation of screening. The results of general screening of female infants for elevated CK could be cross-checked using molecular techniques including accurate multiplex PCR-based dosage analysis (Abbs and Bobrow, 1992), or perhaps RT–PCR on stored umbilical cord blood if large quantities of RNA remain essential for this technique.

3.5.5.3 Fetal screening

If through prevention the incidence of DMD could be reduced to zero, there would be no need for a cure. Zero incidence, as pointed out by Edwards (1986), could only be achieved however by fetal screening because one-third of cases have new mutations. Edwards (1986) also argues that since a small reduction in a common disease is as beneficial as a major reduction in a rare disease, when one is screening by amniocentesis or CVS for other reasons one should perhaps screen for DMD/BMD. In other words, if one is screening for cystic fibrosis or Down's, it may be a pity if the child later turns out to have an easily detectable DMD mutation. General fetal screening would seem to be unthinkable by currently available techniques. However, if the isolation of fetal cells from maternal blood (Mueller *et al.*, 1990; Bianchi *et al.*, 1990; Bruch *et al.*, 1991), becomes more feasible, even though the technique is fraught with difficulties due mainly to maternal contamination (Adinolfi, 1991), fetal screening may become possible in the future.

3.6 CONCLUSION

Molecular genetics has increased the options open to DMD/BMD families and through the application of new technology greater aid will become available. As the genes for other muscular dystrophies are linked and identified, the techniques pioneered for DMD/BMD will be applied. The aim of those involved in genetic services should be to see that as many families as possible are made aware of their options and have the options available.

ACKNOWLEDGEMENTS

The diagnostic and laboratory work used in examples in this chapter was supported by the Neuromuscular Foundation and Muscular Dystrophy Research Association of Western Australia. I thank Sister Margaret Mears for helping compile the family information, Sue Laing and members of the West Australian Diagnostic Molecular Neurogenetics Laboratory: Professor Byron Kakulas, Professor Frank Mastaglia, Dr Jack Goldblatt, Dr Steve Wilton, Patrick Akkari, Dave Chandler, Phillip Jacobsen, Bernadette Majda, Margaret Mears and Jo-Ann Stanton for critical reading of the manuscript.

REFERENCES

Abbs, S., Roberts, R.G., Mathew, C.G. *et al.* (1990) Accurate assessment of intragenic recombination frequency within the Duchenne muscular dystrophy gene. *Genomics*, 7, 602–6.

Abbs, S., Yau, S.C., Clark, S. *et al.* (1991) A convenient multiplex PCR system for the detection of dystrophin gene deletions: a comparative analysis with cDNA hybridisation shows mistypings by both methods. *J. Med. Genet.*, **28**, 304–11.

Abbs, S. and Bobrow, M. (1992) Analysis of quantitative PCR for the diagnosis of deletion and duplication carriers in the dystrophin gene. *J. Med. Genet.*, **29**, 191–6.

Acsadi, G., Dickson, G., Love, D.R. *et al.* (1991) Human dystrophin expression in mdx mice after intramuscular injection of DNA constructs. *Nature*, **352**, 816–18.

Adinolfi, M. (1991) On a non-invasive approach to prenatal diagnosis based on the detection of fetal cells in maternal blood samples. *Prenatal Diagnosis*, **11**, 799–804.

Arahata, K., Ishiura, S., Ishiguro, T. *et al.* (1988) Immunostaining of skeletal and cardiac muscle surface membrane with antibody against Duchenne muscular dystrophy peptide. *Nature*, 333, 861–3.

Arahata, K., Ishihara, T., Kamakura, K. *et al.* (1989) Mosaic expression of dystrophin in symptomatic carriers of Duchenne's muscular dystrophy. *N. Eng. J. Med.*, 320, 138–42.

Arikawa, E., Hoffman, E.P., Kaido, M. *et al.* (1991) The frequency of patients with dystrophin abnormalities in a limb-girdle patient population. *Neurology*, **41**, 1491–6.

Attree, O. (1989) Mutations in the catalytic domain of human coagulation factor IX: rapid characterization by direct sequencing of DNA fragments displaying an altered melting behaviour. *Genomics*, 4, 266–72.

Bakker, E., van Broekhoven, C., Bonten, E.J. *et al.* (1987) Germline mosaicism and Duchenne muscular dystrophy mutations. *Nature*, **329**, 554–6.

Bakker, E., Veenema, H., den Dunnen, J.T. *et al.* (1989) Germinal mosaicism increases the recurrence risk for 'new' Duchenne muscular dystrophy mutations. *J. Med. Genet.*, **26**, 553–9.

Bartlett, R.J., Pericak-Vance, M.A., Lanman, J.T. *et al.* (1987) Prenatal detection of an inherited Duchenne muscular dystrophy deletion allele. *Neurology*, **37**, 355–6.

Bartlett, R.J., Pericak-Vance, M.A., Koh, J. *et al.* (1988) Duchenne muscular dystrophy: high frequency of deletions. *Neurology*, **38**, 1–4.

Bartlett, R.J., Walker, A.P., Laing, N.G. *et al.* (1989) Inherited deletion at Duchenne dystrophy locus in normal male. *Lancet*, **i**, 496–7.

Baumbach, L.L., Ward, P.A., Fenwick, R. *et al.* (1989) Analysis of mutations at the Duchenne muscular dystrophy locus provides no evidence for illegitimate recombination in deletion formation. *Am. J. Hum. Genet.*, **45** (suppl.), A173.

Beckmann, R., Sauer, M., Ketelsen, U.-W. *et al.* (1978) Early diagnosis of Duchenne muscular dystrophy. *Lancet*, **ii**, 105.

Beggs, A.H. and Kunkel, L.M. (1990) A polymorphic CACA repeat in the 3′ untranslated region of dystrophin. *Nuc. Acids Res.*, **18**, 1931.

Beggs, A.H., Koenig, M., Boyce, F.M. *et al.* (1990) Detection of 98% of DMD/BMD gene deletions by polymerase chain reaction. *Hum. Genet.*, **86**, 45–8.

Beggs, A.H., Hoffman, E.P., Snyder, J.R. *et al.* (1991) Exploring the molecular basis for variability among patients with Becker muscular dystrophy: dystrophin gene and protein studies. *Am. J. Hum. Genet.*, **49**, 54–67.

Bettecken, T. and Muller, C.R. (1989) Identification of a 220kb insertion into the Duchenne gene in a family with an atypical course of muscular dystrophy. *Genomics*, **4**, 592–6.

Bianchi, D.W., Flint, A.F., Pizzimenti, M.F. *et al.* (1990) Isolation of fetal DNA from nucleated erythrocytes in maternal blood. *Proc. Nat. Acad. Sci. USA*, **87**, 3279–83.

Bieber, F.R. and Hoffman, E.P. (1990) Duchenne and Becker muscular dystrophies: genetics, prenatal diagnosis and future prospects. *Clin. Perinatol.*, **17**(4), 845–65.

Bieber, F.R., Hoffman, E.P. and Amos, J.A. (1989) Dystrophin analysis in Duchenne muscular dystrophy: use in fetal diagnosis and in genetic counselling. *Am. J. Hum. Genet.*, **45**, 362–7.

Blonden, L.A.J., Grootscholten, P.M., den Dunnen, J.T. *et al.* (1991) 242 breakpoints in the 200-kb deletion-prone region of the DMD gene are widely spread. *Genomics*, **10**, 631–9.

Boelter, W.D., Burt, B.A., Spector, E.B. *et al.* (1990) Dystrophin protein and RFLP analysis for fetal diagnosis and carrier confirmation of Duchenne muscular dystrophy. *Prenatal Diagnosis*, **10**, 703–15.

Boileau, C. and Junien, C. (1989) Misdiagnosis of normal fetus owing to undetected germinal mosaicism for DMD deletion. *J. Med. Genet.*, **26**, 790–2.

Bonilla, E., Samitt, C.E., Miranda, A.F. *et al.* (1988) Duchenne muscular dystrophy: deficiency of dystrophin at the muscle cell surface. *Cell*, **54**, 447–52.

Bradley, W.G., Jones, M.Z., Mussini, J.-M. *et al.* (1978) Becker-type muscular dystrophy. *Muscle Nerve*, **1**, 111–32.

Bruch, J.F., Metezeau, P., Garcia-Fonknechten, N. *et al.* (1991) Trophoblast-like cells sorted from peripheral maternal blood using flow cytometry: a multiparametric study involving transmission electron microscopy and fetal DNA amplification. *Prenatal Diagnosis*, **11**, 787–98.

Bulman, D.E., Gangopadhyay, S.B., Bebchuk, K.G. *et al.* (1991) Point mutation in the human dystrophin gene: identification through western blot analysis. *Genomics*, **10**, 457–60.

Bundey, S. (1978) Calculation of genetic risks in Duchenne muscular dystrophy by geneticists in the United Kingdom. *J. Med. Genet.*, **15**, 249–53.

Campbell, K.P. and Kahl, S.D. (1989) Association of dystrophin and an integral membrane glycoprotein. *Nature*, **338**, 259–62.

Chamberlain, J.S., Gibbs, R.A., Ranier, J.E. *et al.* (1988) Deletion screening of the Duchenne muscular dystrophy locus via multiplex DNA amplification. *Nucl. Acids Res.*, **16**, 11141–56.

Chamberlain, J.S., Gibbs, R.A., Rainier, J.E. *et al.* (1990) Multiplex PCR for the diagnosis of Duchenne muscular dystrophy, in *PCR Protocols: a Guide to Methods and Applications*, (eds M. Innis, D. Gelfland, J. Sninsky and T.S. White), Academic Press, San Diego, pp. 272–81.

Chelly, J., Marlhens, F., Marec, B.L. *et al.* (1986) De novo microdeletion in a girl with Turner syndrome and Duchenne muscular dystrophy. *Hum. Genet.*, **74**, 193–6.

Chelly, J., Gilgenkrantz, H., Lambert, M. *et al.* (1990) Effect of dystrophin gene deletions on mRNA levels and processing in Duchenne and Becker muscular dystrophies. *Cell*, **63**, 1239–48.

Chelly, J., Gilgenkrantz, H., Hugnot, J.P. *et al.* (1991) Illegitimate transcription. Application to the analysis of truncated transcripts of the dystrophin gene in nonmuscle cultured cells from Duchenne and Becker patients. *J. Clin. Invest.*, **88**, 1161–6.

Chen, J.D., Denton, M.J., Morgan, G. *et al.* (1988a) The use of field inversion gel electrophoresis for deletion detection in Duchenne muscular dystrophy. *Am. J. Hum. Genet.*, **42**, 777–80.

Chen, J.D., Denton, M.J., Serravale, S. *et al.* (1988b) Prenatal diagnosis and carrier detection by DNA studies in a Duchenne muscular dystrophy family with no living affected male. *Austr. Paed. J.*, **24**, 351–3.

Cummings, M.R. (1991) *Human Heredity, Principles and Issues*, West Publishing Company, St Paul, New York, Los Angeles, San Francisco.

Darras, B.T. and Francke, U. (1987) A partial deletion of the muscular dystrophy gene transmitted twice by an unaffected male. *Nature*, **329**, 556–8.

de Visser, M., Bakker, E., Defesche, J.C. *et al.* (1990) An unusual variant of Becker muscular dystrophy. *Ann. Neurol.*, **27**, 578–81.

den Dunnen, J.T., Bakker, E., Klein-Breteler, E.G. *et al.* (1987) Direct detection of more than 50% of the Duchenne muscular dystrophy mutations by field inversion gels. *Nature*, **329**, 640–2.

den Dunnen, J.T., Grootscholten, P.M., Bakker, E. *et al.* (1989) Topography of the Duchenne muscular dystrophy (DMD) gene: FIGE and cDNA analysis of 194 cases reveals 115 deletions and 13 duplications. *Am. J. Hum. Genet.*, **445**, 835–47.

Edwards, J.H. (1986) The population genetics of Duchenne: natural and artificial selection in Duchenne muscular dystrophy. *J. Med. Genet.*, **23**, 521–30.

Emery, A.E.H. (1972) Abnormalities of the electrocardiogram in hereditary myopathies. *J. Med. Genet.*, **9**, 8–12.

Emery, A.E.H. (1980) Duchenne muscular dystrophy genetic aspects, carrier detection and antenatal diagnosis. *Br. Med. Bull.*, **36**, 117–22.

Emery, A.E.H. and Rimoin, D.L. (1990) *Principles and Practice of Medical Genetics*, Churchill Livingstone, Edinburgh, London, Melbourne, New York.

Ervasti, J.M., Ohlendieck, K., Kahl, S.D. *et al.* (1990) Deficiency of a glycoprotein component of the dystrophin complex in dystrophic muscle. *Nature*, **345**, 315–9.

Evans, M.I., Greb, A., Kazazian, H. *et al.* (1991) In utero fetal muscle biopsy for the diagnosis of Duchenne muscular dystrophy. *Am. J. Obst. Gyn.*, **165**, 728–32.

Feener, C.A., Boyce, F.M. and Kunkel, L.M. (1991) Rapid detection of CA polymorphisms in cloned DNA: application to the 5′ region of the dystrophin gene. *Am. J. Hum. Genet.*, **48**, 621–7.

Forrest, S.M., Smith, T.J., Cross, G.S. *et al.* (1987) Effective strategy for prenatal prediction of Duchenne and Becker muscular dystrophy. *Lancet*, **ii**, 1294–7.

Francke, U., Darras, B.T., Hersh, J.H. *et al.* (1989) Brother/sister pairs affected with early-onset, progressive muscular dystrophy: molecular studies reveal etiologic heterogeneity. *Am. J. Hum. Genet.*, **45**, 63–72.

Gelehrter, T.D. and Collins, F.S. (1990) *Principles of Medical Genetics*, Williams and Wilkins, Baltimore.

Gospe, S.M., Lazaro, R.P., Lava, N.S. *et al.* (1989) Familial X-linked myalgia and cramps: a non-progressive myopathy associated with a deletion in the dystrophin gene. *Neurology*, **39**, 1277–80.

Greenberg, C.R., Rohringer, M. and Jacobs, H.K. (1988) Gene studies in newborn males with Duchenne muscular dystrophy detected by neonatal screening. *Lancet*, **ii**, 425–7.

Greenberg, C.R., Jacobs, H.K., Halliday, W. *et al.* (1991) Three years' experience with neonatal screening for Duchenne/Becker muscular dystrophy: gene analysis, gene expression and phenotype prediction. *Am. J. Med. Genet.*, **39**, 68–75.

Grimm, T., Muller, B., Muller, C.R. *et al.* (1990) Theoretical considerations on germline mosaicism in Duchenne muscular dystrophy. *J. Med. Genet.*, **27**, 683–7.

Gutmann, D.H. and Fischbeck, K.H. (1989) Molecular biology of Duchenne and Becker's muscular dystrophy: clinical applications. *Ann. Neurol.*, **26**, 189–94.

Haldane, J.B.S. (1935) The rate of spontaneous mutation of a human gene. *J. Genet.*, **31**, 317–26.

Hejtmancik, J.F., Harris, S.G., Tsao, C.C. *et al.* (1986) Carrier diagnosis of Duchenne muscular dystrophy using restriction fragment length polymorphisms. *Neurology*, **36**, 1553–62.

Hennekam, R.C.M., Veenema, H., Bakker, E. *et al.* (1989) A male carrier for Duchenne muscular dystrophy. *Am. J. Hum. Genet.*, **44**, 591–2.

Hoffman, E.P., Brown, R.H. and Kunkel, L.M. (1987) Dystrophin: the protein product of the Duchenne muscular dystrophy locus. *Cell*, **51**, 919–28.

Hoffman, E.P., Fischbeck, K.H., Brown, R.H. *et al.* (1988) Characterization of dystrophin in muscle-biopsy specimens from patients with Duchenne's or

Becker's muscular dystrophy. *N. Eng. J. Med.*, **318**, 1663–8.

Hu, X., Ray, P.N., Murphy, E.G. *et al.* (1990) Duplicational mutation at the Duchenne muscular dystrophy locus: its frequency, distribution, origin and phenotype/genotype correlation. *Am. J. Hum. Genet.*, **46**, 682–5.

Hurko, O., Hoffman, E.P., McKee, L. *et al.* (1989) Dystrophin analysis in clonal myoblasts derived from a Duchenne muscular dystrophy carrier. *Am. J. Hum. Genet.*, **44**, 820–6.

Hurse, P.V. and Kakulas, B.A. (1974) Genetic counselling in neuromuscular diseases in Western Australia. *Proc. Austr. Ass. Neurol.*, **11**, 145–53.

Hutton, E.M. and Thompson, M.W. (1976) Carrier detection and genetic counselling in Duchenne muscular dystrophy. *Can. Med. Ass. J.*, **115**, 749–52.

Hyser, C.L., Doherty, R.A., Griggs, R.C. *et al.* (1987) Carrier assessment for mothers and sisters of isolated Duchenne dystrophy cases: the importance of serum enzyme determinations. *Neurology*, **37**, 1476–80.

Kakulas, B.A. and Adams, R.D. (1985) *Diseases of Muscle, Pathological Foundations of Clinical Myology*, Harper and Row, Philadelphia.

Karpati, G., Pouliot, Y., Zuberzycka-Gaarn, E. *et al.* (1989) Dystrophin is expressed in mdx skeletal muscle fibres after normal myoblast implantation. *Am. J. Pathol.*, **135**, 27–32.

Kingston, H.M., Thomas, N.S.T., Pearson, P.L. *et al.* (1983) Genetic linkage analysis between Becker muscular dystrophy and a polymorphic DNA sequence on the short arm of the X chromosome. *J. Med. Genet.*, **20**, 255–8.

Kingston, H.M., Sarfarazi, M., Thomas, N.S.T. *et al.* (1984) Localization of the Becker muscular dystrophy gene on the short arm of the X chromosome by linkage to cloned DNA sequences. *Hum. Genet.*, **67**, 6–17.

Koenig, M., Hoffman, E.P., Bertelson, C.J. *et al.* (1987) Complete cloning of the Duchenne muscular dystrophy (DMD) cDNA and preliminary genomic organization of the DMD gene in normal and affected individuals. *Cell*, **50**, 509–517.

Koenig, M., Monaco, A.P. and Kunkel, L.M. (1988) The complete sequence of dystrophin predicts a rod-shaped cytoskeletal protein. *Cell*, **53**, 219–28.

Koenig, M., Beggs, A.H., Moyer, M. *et al.* (1989) The molecular basis for Duchenne versus Becker muscular dystrophy: correlation of severity with type of deletion. *Am. J. Hum. Genet.*, **45**, 498–506.

Koh, J., Bartlett, R.J., Pericak-Vance, M.A. *et al.* (1987) Inherited deletion at Duchenne dystrophy locus in normal male. *Lancet*, **ii**, 1154–5.

Kunkel, L.M., Hejtmancik, J.F., Caskey, C.T. *et al.* (1986) Analysis of deletions in DNA from patients with Becker and Duchenne muscular dystrophy. *Nature*, **322**, 73–7.

Laing, N.G., Siddique, T., Bartlett, R.J. *et al.* (1988) RFLP for Duchenne muscular dystrophy cDNA clone 44-1. *Nucl. Acids Res.*, **16**, 7209.

Laing, N.G., Siddique, T., Bartlett, R.J. *et al.* (1989) Duchenne muscular dystrophy: detection of deletion carriers by spectrophotometric densitometry. *Clin. Genet.*, **35**, 393–8.

Laing, N.G., Mears, M.E., Thomas, H.E. *et al.* (1990) Differentiation of Becker muscular dystrophy from limb-girdle muscular dystrophy and Kugelberg-Welander disease using a cDNA probe. *Med. J. Austr.*, **152**, 270–71.

Laing, N.G., Mears, M.E., Chandler, D.C. *et al.* (1991) The diagnosis of Duchenne and Becker muscular dystrophies: two year's experience in a comprehensive carrier-screening and prenatal diagnosis laboratory. *Med. J. Austr.*, **154**, 14–18.

Laing, N.G., Layton, M.G., Johnsen, R.D. *et al.* (1992) Two distinct mutations in a single dystrophin gene: chance occurrence or pre-mutation. *Am. J. Med. Genet.*, **42**, 688–92.

Lanman, J.T., Pericak-Vance, M.A. and Bartlett, R.J. (1987) Familial inheritance of a DXS164 deletion mutation from a heterozygous female. *Am. J. Hum. Genet.*, **41**, 138–44.

Lench, N., Stanier, P. and Williamson, R. (1988) Simple non-invasive method to obtain DNA for gene analysis. *Lancet*, **i**, 1356–8.

Lunt, P.W., Cumming, W.J.K., Kingston, H. *et al.* (1989) DNA probes in differential diagnosis of Becker muscular dystrophy and spinal muscular atrophy. *Lancet*, **i**, 46–7.

Lyon, M.F. (1988) The William Allen memorial address: X-chromosome inactivation and the location and expression of X-linked genes. *Am. J. Hum. Genet.*, **42**, 8–16.

Malhotra, S.B., Hart, K.A., Klamut, H.J. *et al.* (1988) Frame-shift deletions in patients with Duchenne and Becker muscular dystrophy. *Science*, **242**, 755–9.

Mao, Y. and Cremer, M. (1989) Detection of Duchenne muscular dystrophy carriers by dosage analysis using the DMD cDNA clone 8. *Hum. Genet.*, **81**, 193–5.

McKusick, V.A. (1990) *Mendelian Inheritance in Man. Catalogues of Autosomal Dominant, Autosomal Recessive and X-Linked Phenotypes*, John Hopkins University Press, Baltimore and London.

Miciak, A., Keen, A., Jadayel, D. and Bundey, S. (1992) Multiple mutation in an extended Duchenne muscular dystrophy family. *J. Med. Genet.*, **29**, 123–6.

Miranda, A.F., Francke, U., Bonilla, E. *et al.* (1989) Dystrophin immunocytochemistry in muscle culture: detection of a carrier of Duchenne muscular dystrophy. *Am. J. Med. Genet.*, **32**, 268–73.

Monaco, A.P., Bertelson, C., Middlesworth, W. *et al.* (1985) Detection of deletions spanning the Duchenne muscular dystrophy locus using a tightly linked DNA segment. *Nature*, **316**, 842–5.

Monaco, A.P., Bertelson, C.J., Liechti-Gallati, S. *et al.* (1988) An explanation for the phenotypic differences between patients bearing partial deletions of the DMD locus. *Genomics*, **2**, 90–5.

Moser, H. (1984) Duchenne muscular dystrophy: pathogenetic aspects and genetic prevention. *Hum. Genet.*, **66**, 17–40.

Mueller, U.W., Hawes, C.S., Wright, A.E. *et al.* (1990) Isolation of fetal trophoblast cells from peripheral blood of pregnant women. *Lancet*, **336**, 197–200.

Muller, C.R. and Grimm, T. (1986) Estimation of the male to female ratio of mutation rates from the segregation of X-chromosomal DNA haplotypes in Duchenne muscular dystrophy. *Hum. Genet.*, **74**, 181–3.

Murray, J.M., Davies, K.E., Harper, P.S. *et al.* (1982) Linkage relationship of a cloned DNA sequence on the short arm of the X chromosome to Duchenne muscular dystrophy. *Nature*, **300**, 69–71.

Naylor, E.W. (1991) New technologies in newborn screening. *Yale J. Biol. Med.*, **64**, 21–4.

Nordenskjold, M., Nicholson, L., Edstrom, M. *et al.* (1990) A normal male with an inherited deletion of one exon within the DMD gene. *Hum. Genet.*, **84**, 207–9.

Norman, A., Thomas, N., Coakley, J. *et al.* (1989a) Distinction of Becker from limb-girdle muscular dystrophy by means of dystrophin cDNA probes. *Lancet*, i, 466–8.

Norman, A.M., Hughes, H.E., Gardner-Medwin, D. *et al.* (1989b) Dystrophin analysis in the diagnosis of muscular dystrophy. *Arch. Dis. Child.*, **64**, 1501–3.

Orita, M., Iwahana, H., Kanazawa, H. *et al.* (1989) Detection of polymorphisms of human DNA by gel electrophoresis as single strand conformation polymorphisms. *Proc. Nat. Acad. Sci. USA*, **86**, 2766–70.

Oudet, C., Heilig, R., Hanauer, A. *et al.* (1991) Non-radioactive assay for new microsatellite polymorphisms at the 5′ end of the dystrophin gene, and estimation of intragenic recombination. *Am. J. Hum. Genet.*, **49**, 311–19.

Partridge, T.A., Morgan, J.E., Coulton, G.R. *et al.* (1989) Conversion of mdx myofibres from dystrophin-negative to -positive by injection of normal myoblasts. *Nature*, 337, 176–9.

Patel, K., Voit, T., Dunn, M.J. *et al.* (1988) Dystrophin and nebulin in the muscular dystrophies. *J. Neurol. Sci.*, **87**, 315–26.

Plauchu, H., Dellamonica, C., Cotte, J. *et al.* (1980) Duchenne muscular dystrophy: systematic neonatal screening and earlier detection of carriers. *Journal de Genetique Humaine*, **28**, 65–82.

Plauchu, H., Cordier, M.P., Carrier, H.N. *et al.* (1987) Depistage neonatal systematique de la dystrophie musculaire de Duchenne: Bilan apres dix ans d'experience dans la region de Lyon (France). *Journal de Genetique Humaine*, **35**, 217–30.

Plauchu, H., Dorche, C., Cordier, M.P. *et al.* (1989) Duchenne muscular dystrophy: neonatal screening and prenatal diagnosis. *Lancet*, **i**, 669.

Powell, J.F., Fodor, F.H., Cockburn, D.J. *et al.* (1991) A dinucleotide repeat polymorphism at the DMD locus. *Nucl. Acids Res.*, **19**, 1159.

Prior, T.W., Friedman, K.J., Highsmith, W.E. *et al.* (1990) Molecular probe protocol for determining carrier status in Duchenne and Becker muscular dystrophies. *Clin. Chem.*, **36**, 441–5.

Ray, P.N., Belfall, B., Duff, C. *et al.* (1985) Cloning of the breakpoint of an X;21 translocation associated with Duchenne muscular dystrophy. *Nature*, **318**, 672–5.

Richards, C.S., Watkins, S.C., Hoffman, E.P. *et al.* (1990) Skewed X inactivation in a female MZ twin results in Duchenne muscular dystrophy. *Am. J. Hum. Genet.*, **46**, 672–81.

Roberts, R.G., Bentley, D.R., Barby, T.F.M. *et al.* (1990) Direct diagnosis of carriers of Duchenne and Becker muscular dystrophy by amplification of lymphocyte RNA. *Lancet*, **336**, 1523–6.

Roberts, R.G., Barby, T.F.M., Manners, E. *et al.* (1991) Direct detection of dystrophin gene rearrangements by analysis of dystrophin mRNA in peripheral blood lymphocytes. *Am. J. Hum. Genet.*, **49**, 298–310.

Roberts, R.G., Bobrow, R. and Bentley, D.R. (1992) Point mutations in the dystrophin gene. *Proc. Nat. Acad. Sci. USA*, **89**, 2331–5.

Roses, A.D. (1988) Mutants in Duchenne muscular dystrophy. *Arch. Neurol.*, **45**, 84–5.

Rowland, L.P. (1988) Clinical concepts of Duchenne muscular dystrophy: the impact of molecular genetics. *Brain*, **111**, 479–95.

Scheurbrandt, G., Lundin, A., Lovgren, T. *et al.* (1986) Screening for Duchenne muscular dystrophy: an improved screening test for creatine kinase and its application in an infant screening program. *Muscle Nerve*, **9**, 11–23.

Schlosser, M., Slomski, R., Wagner, M. *et al.* (1990) Characterization of pathological dystrophin transcripts from the lymphocytes of a muscular dystrophy carrier. *Mol. Biol. Med.*, 7, 519–23.

Sharp, N.J.H., Kornegay, J.N., Van Camp, S.D. *et al.* (1992) An error in dystrophin mRNA processing in golden retriever muscular dystrophy, an animal homologue of Duchenne muscular dystrophy. *Genomics*, **13**, 115–21.

Thompson, C.E. (1978) Reproduction in Duchenne dystrophy. *Neurology*, **28**, 1045–7.

Thompson, M.W., Ray, P.N., Belfall, B. *et al.* (1986) Linkage analysis of polymorphisms within the DNA fragment XJ cloned from the breakpoint of an X;21 translocation associated with X linked muscular dystrophy. *J. Med. Genet.*, **23**, 548–55.

van Ommen, G.J.B., Bakker, E., Blonden, L. *et al.* (1990) Possibilities, pitfalls and prospects of the diagnosis of Duchenne and Becker muscular dystrophy, in *Molecular Probes: Technology and Medical Applications*, (eds A. Albertini, R. Paoletti and R.A.S. Reisfeld), Raven Press, New York, pp. 25–32.

Verellen-Dumoulin, C., Freund, M., Meyer, R.D. *et al.* (1984) Expression of an X-linked muscular dystrophy in a female due to translocation involving Xp21 and non-random inactivation of the normal X chromosome. *Hum. Genet.*, **67**, 115–19.

Walton, J.N. (1955) On the inheritance of muscular dystrophy. *Ann. Hum. Genet., London*, **20**, 1–38.

Wapenaar, M.C., Kievits, T., Hart, K.A. *et al.* (1988) A deletion hotspot in the Duchenne muscular dystrophy gene. *Genomics*, **2**, 101–8.

Weber, J.L. and May, P.E. (1989) Abundant class of human DNA polymorphisms which can be typed by the polymerase chain reaction. *Am. J. Hum. Genet.*, **44**, 388–96.

Wolff, J.A., Malone, R.W., Williams, P.*et al.* (1990) Direct gene transfer into mouse muscle in vivo. *Science*, **247**, 1465–8.

Zuberzycka-Gaarn, E., Bulman, D., Karpati, G. *et al.* (1988) The Duchenne muscular dystrophy gene product is localized in sarcolemma of human skeletal muscle. *Nature*, **333**, 466–9.

4

Inheritance and pathogenicity of myotonic dystrophy

KEITH JOHNSON

4.1 INTRODUCTION

Myotonic dystrophy (DM) is a fascinating disorder at both the clinical and genetic level. Tremendous variation is seen in the range, severity and age at onset of symptoms, both between families and within a single sibship. Its classification as a form of muscular dystrophy is slightly misleading given that many, if not all, organs and tissues of the body are affected. At the clinical level it is possible to distinguish a homogeneous group of patients by various phenotypic symptoms. The criteria used to do this are borne out by the genetic findings, which clearly show that a single locus on chromosome 19 is responsible for the inheritance of this genetically determined condition. These findings make DM one of the most interesting human diseases to study because any postulated mechanism by which the disease may be caused must explain so many puzzling biochemical, histopathological and physiological phenomena. For this reason most of the current research into this disease has focused on a genetic approach to finding the responsible gene. This chapter will therefore give only a cursory summary of the phenotypic findings and, like the research, focus on the search for the gene itself, a search that at the time of writing is close to attaining its target. It is also historically appropriate to focus on the genetics of DM, because many of the first useful results from new approaches, such as linkage analysis and the application of computer analysis to genetic linkage data, were developed by researchers investigating the genetics of DM.

4.2 THE CLINICAL PICTURE

Myotonic dystrophy (DM) is inherited as an autosomal dominant trait with an incidence of between 1 in 10 000 and 1 in 100 000 (Emery, 1991), with

Molecular and Cell Biology of Muscular Dystrophy
Edited by Terence Partridge
Published in 1993 by Chapman & Hall, London. ISBN 0 412 43440 7

an overall figure of 1 in 20 000 (Mathieu *et al.*, 1990). This makes it the commonest of the adult muscular dystrophies, although its exact incidence is difficult to determine accurately because it exhibits variable expressivity (see Figure 4.1). The frequency of the disease is virtually identical in all racial groups in which this disease has been reported, with the exception of populations where there is a founder effect, or where genetic isolation has occurred, as in Quebec (Mathieu et al., 1990), Guam (Chen *et al.*, 1968) and Northern Transvaal (Lotz and van der Meyden, 1985). There is a recent report which suggests that the prevalence of the disease is much lower in central and southern Africa and southern China (Ashizawa and Epstein, 1991). This finding suggests that the distribution of the disease worldwide reflects a major mutation causing DM which occurred in the North Eurasian groups migrating out of Africa to Europe.

DM is characterized clinically by myotonia, both clinical and electrical, weakness of the distal muscles and atrophy of the facial muscles and the

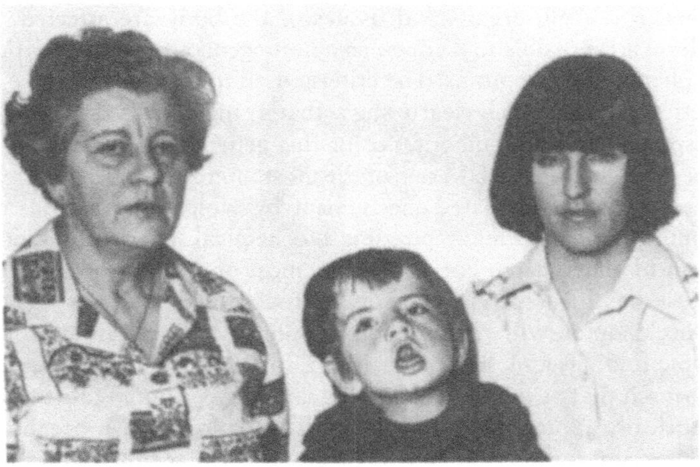

Figure 4.1 Facial appearance in myotonic dystrophy. This picture shows three generations of DM individuals from the same family and graphically illustrates the variability of expression seen for many of the clinical features of this disease. Note especially the variation in facial appearance. The grandmother (left) had bilateral cataracts and had percussion myotonia on examination. She has no muscle symptoms or facial weakness. Her daughter (right) had no symptoms until after the birth of her son (centre). She has facial weakness, ptosis, cataracts and both active and percussion myotonia. The congenital child shows facial diplegia and jaw weakness, present from birth. (Reproduced with permission, Harper, P. (1989) Saunders, Philadelphia).

sternomastoids. Additionally, frontal baldness, lens opacities, gonadal atrophy, cardiomyopathy, mild dementia and mild endocrine changes have all been seen in varying combinations. Because of the variable expressivity, even within the same family, a set of diagnostic criteria for the inclusion of family members in genetic studies of DM has been drawn up (Griggs *et al.*, 1989). As the age at onset is also extremely variable, only affected individuals whose diagnosis fits these criteria are included in genetic studies. This is because it is not uncommon to have parents of affected individuals who remain asymptomatic even in their seventies and eighties. This situation appears as a non-penetrant individual in the pedigree, however, the designation of these individuals as heterozygotes stems from the fact that the new mutation rate for DM is one of the lowest observed for any Mendelian trait (Harper, 1989). Virtually every case of DM is found to be familial on taking a history from the patient; this is one of the commonest means of diagnosis for this disorder (Harper, 1989).

DM is usually asymptomatic until adulthood, although some cases show onset in adolescence. The phenomenon of genetic anticipation has been reported since early studies of DM, but has gained greater credence following the appearance of recent critical reviews of the data (Howeler, 1986; Howeler *et al.*, 1989). Anticipation describes the situation where the onset of the disease appears to be at an earlier age in successive generations of families, usually associated with increasing severity. The reason for this late acceptance of the existence of anticipation as a real phenomenon by geneticists, is probably due to the paper by Penrose (1948), who pointed out all of the possible biases that might lead to the apparent observation of anticipation.

There is a much more severe, congenital form of this disorder, which is almost exclusively maternally transmitted (Vanier, 1960). The mechanism underlying this mode of transmission has not been resolved, but several theories have been proposed to explain it. One invokes an intra-uterine factor that leads to the clinical condition (Harper and Dyken, 1972). A second postulates that genomic imprinting of the region of the genome containing the DM gene is involved (Erickson, 1985). Genomic imprinting describes the phenomenon of differential expression of the two copies of a gene in a diploid cell according to their parental origin, and was first recognized in the mouse (Cattenach and Kirk, 1986). A third suggests that there are mitochondrial mutations (Merril and Harrington, 1985) and a fourth that more than one locus is involved (Bundey and Carter, 1972).

A recent study of 101 DM families has shown that the incidence of congenital DM closely follows the clinical course of multisystemic involvement in the mother (Koch *et al.*, 1991). Congenital DM children are born only to mothers who are clearly manifesting disease symptoms themselves during the pregnancy. This strongly argues against imprinting and mitochondrial mutations as the explanation for the maternal mode of inheritance, and supports the possible existence of an intra-uterine factor. The probability of another locus being involved is very small, as the congenital disease co-segregates with the same haplotype of tightly linked markers on chromosome 19 as the later-onset form of the disease.

4.3 BIOCHEMICAL, ELECTROPHYSIOLOGICAL AND HISTOLOGICAL FINDINGS

Although DM patients usually present with symptoms of muscle weakness, other symptoms are often present, and it is clear from biochemical studies that many systems are involved in the pathogenesis of the disorder (reviewed by Harper, 1989). The electrophysiological and biochemical changes seen in patients point towards a defective cell membrane component which causes both structural and functional changes. The heart and the lens of the eye are commonly affected; the slit lamp is routinely used to examine at-risk individuals for lens opacities, which are an early sign of DM in some patients (Harper, 1989). However, it is likely that the striated muscles are most susceptible to these perturbations, and show the earliest pathological changes in most patients.

Histologically, many changes of muscle tissues have been described in DM patients (summarized in Harper, 1989). Increased numbers of central nuclei are seen in the fibres. This is probably a reflection of the regeneration occurring in those fibres as a result of the prior selective loss of Type I fibres. Nuclear chains are observed as are sarcoplasmic masses. These are fibre structure anomalies, with increased fibre splitting and the appearance of ringed fibres, accompanied by changes in innervation. However, histological changes are not limited to skeletal muscle, there are demonstrable changes in smooth muscle, cardiac muscle and also in many other non-muscle systems.

Electrophysiological studies of cultured myotonic muscle cells, and fibres isolated from patients, have shown a lowered resting potential accompanied by an increased intracellular sodium concentration (Rudel and Lehmann-Horn, 1985). More recently this same group of workers performed an extensive comparison of the electrophysiological properties of normal, myotonic and dystrophic muscle (Franke *et al.*, 1990). Intracellular recordings showed that myotonic fibres exhibited runs of action potentials in response to a single stimulus. The K^+ conductance in DM patient fibres was normal using the three-electrode voltage clamp to determine the total membrane conductance and the ion component conductances. In contrast, Na^+ channels opened at potentials which in normals had no effect and these late-opening channels were found frequently in surface patches of myotonic fibres. These workers concluded that a reduced Cl^- conductance, which had been postulated as a possible mechanism by which the myotonia in DM muscles is produced, was not producing the myotonia, but that the Na^+ channels were reopening abnormally.

Extensive studies performed on DM erythrocyte membranes have detected an altered sodium/potassium exchange rate across them (Hull and Roses, 1976). A more recent study has found a sodium–potassium ATPase with altered electrical properties in patients with myotonia (Fenton *et al.*, 1991). This is almost certainly not the site of the primary defect in DM. Although the alpha3 sub-unit gene (*ATP1A3*) of the sodium–potassium ATPase maps into the region of chromosome 19 containing markers linked to the defect segregating in DM families, this gene has been excluded as a candidate by genetic studies (Harley *et al.*, 1988) (Section 4.4 gives further details).

Several other observations of altered membrane properties indicate that the changes in channel properties are likely to be due to a generalized membrane defect. Studies by electron-spin resonance have detected alterations in the fluidity and polarity of DM patient erythrocyte membranes (Butterfield *et al.*, 1976). These cell membranes have also been shown to have an abnormally low permeability to water (Kuwabara *et al.*, 1991).

A significant decrease in the whole body sensitivity to insulin has been demonstrated in DM patients (Moxley *et al.*, 1984). The locus for the insulin receptor gene (*INSR*), which is located on chromosome 19p, has been excluded as a candidate genetically (Shaw *et al.*, 1986) and no biochemical defects in the insulin receptor protein have been detected in DM patients (Kakehi *et al.*, 1990).

Changes in cell surface markers have also been reported in the muscles of DM patients when compared with normal controls. The bee-venom toxin, apamin, has receptors on the membranes of adult DM fibres which are not found on normal adult fibres (Renaud *et al.*, 1986). Also, the muscle form of nerve cell adhesion molecule (N-CAM), which is usually only found on fetal and embryonic muscle fibres, is expressed in adult fibres of DM patients (Walsh *et al.*, 1988).

The number of altered parameters of normal membrane function suggest that these are secondary events in the aetiology of DM. The site of the defect in DM is likely to be an alteration in the structure of a fundamental membrane component, given the widespread tissue effects and the diverse changes that have been detected in DM muscle. In view of the extreme complexity of biological membranes, this information does not point to any specific protein that could be a candidate for the causative mutation. For this reason a reverse genetic (or as it is now known, positional cloning) approach, tracking the gene's inheritance with DNA markers in DM families, is the only certain way of identifying the site of the mutations that cause DM. Positional cloning techniques have been recently reviewed

(Wicking and Williamson, 1991); they rely on being able to map DNA fragments (probes) by both genetic and physical methods. Probes are mapped genetically by studying the segregation of polymorphisms in families (reviewed by Nimmo and Johnson, 1991) and physically by a variety of methods described later.

4.4 GENETIC MAPPING OF THE MYOTONIC LOCUS

The positional cloning approach has now been successfully applied to the isolation and characterization of several human genetic disease loci, including those for Duchenne muscular dystrophy (Koenig *et al.*, 1987), cystic fibrosis (Kerem *et al.*, 1989; Riordan *et al.*, 1989; Rommens *et al.*, 1989), neurofibromatosis type I (Wallace *et al.*, 1990; Viskochil *et al.*, 1990), colorectal cancer (Fearon *et al.*, 1990) and X-linked mental retardation (Fragile-X) (Oberle *et al.*, 1991; Verkerk *et al.*, 1991; Yu *et al.*, 1991). This approach relies on the stepwise narrowing down of the target region of the genome containing a disease locus using parallel genetic and physical mapping methods (see Wicking and Williamson, 1991 for a review). Briefly, once the gene has been assigned to a specific chromosome by linkage to a marker, a detailed genetic map of the region containing the disease gene is constructed so that flanking markers can be identified.

The first task in a 'reverse genetic' approach to finding a disease gene is to generate a detailed genetic map of the chromosome region into which the mutation maps. In order to achieve this, new genetic markers from the autosome have to be isolated and mapped by linkage analysis in disease families. The genetic map can also be generated in non-disease families because the gene order on the chromosomes in question is the same. However, no information concerning the location of the disease gene can be derived from the latter studies.

The way in which the genetic maps of all chromosomes have developed is illustrated by the dramatic increase in the number of polymorphic markers (probes) between the last two human gene mapping (HGM) workshops (Table 4.1). The major contributing factor to this exponential growth was the realization of the applicability of restriction fragment length polymorphisms (RFLPs) to human genetics (Botstein *et al.*, 1980). The methods by which polymorphisms are detected for any DNA probe and the uses to which they are put are summarized in Nimmo and Johnson (1991). Initially, genetic maps proliferated around loci of interest, with an initial bias towards the X chromosome because of the ease of assigning loci to that chromosome on the basis of sex linkage. The advent of molecular cloning techniques enabled DNA probes from all chromosomes to be isolated and mapped. DNA probes are usually assigned to a specific autosome by a physical method, such as mapping on metaphase spreads by *in situ*

Table 4.1 The growth in gene mapping in one year (1989–1990) as detailed in the differences for chromosome 19 and the whole genome between *HGM10* and *HGM10.5*

	Genes		Anonymous markers	
	Total	New	Total	New
Chromosome 19	88*	12	152	100
Genome	1867	234	10 000	3677

*Excludes gene families, fragile sites and anonymous markers. (There are several large gene families clustered on chromosome 19.) Data extracted from Williamson *et al.* (1991)

hybridization and by hybridization to somatic cell hybrid panels of cell lines carrying different combinations of human chromosomes in a rodent background. Once assigned to a given chromosome more detailed mapping information is generated by performing linkage studies. Many cloned genes and anonymous DNA probes have been assigned to individual chromosomes and incorporated into genetic maps of those autosomes in this way.

DM was one of the first human autosomal disease loci to be linked to genetic markers by family studies. The ABH/secretor blood group was shown to be linked to the DM locus in 1954 (Mohr). The cloning of the complement component 3 (*C3*) gene and its localization to chromosome 19 (Whitehead *et al.*, 1982) following the inclusion of the *C3* locus in the ABH/DM linkage group (Eiberg *et al.*, 1981) assigned DM to this autosome. This assignment predates that of Huntington's disease (HD) to chromosome 4 (Gusella *et al.*, 1983), which is often quoted as the first human autosomal disease gene to be chromosomally assigned by molecular genetic techniques. HD was, more correctly, the first autosomal disease gene to be mapped by a systematic linkage analysis of the human genome with polymorphic DNA markers.

The linkage to DNA markers on chromosome 19 in DM families has been demonstrated in all populations, including the Japanese (Takemoto *et al.*, 1990), ruling out genetic heterogeneity in this disorder. This does not rule out the possibility that different mutations within a single gene will be found in DM patients but makes it unlikely that mutations within different genes will be found.

The genetic map of chromosome 19 initially consisted of the blood group markers that had been shown to be linked to DM and a number of cloned genes that mapped on to it. These included the *LDLR*, *INSR* and *C3*. As more markers became available a two-point linkage map of the chromosome emerged along with a physical map that positioned markers relative to translocation break-points (see next section and HGM 10.5). The aim was to merge these two maps, as the markers within them should be colinear.

Markers which were found to be tightly linked to DM were investigated to determine their usefulness in presymptomatic and antenatal diagnosis and to generate order information. This was essential to the overall objective of generating tightly linked flanking markers that define the DM interval and then to isolate that interval as a series of overlapping genomic clones.

The first genetic marker that was found to be tightly linked to DM was the gene for apolipoprotein C2 (*APOC2*) (Humphries *et al.*, 1984; Shaw *et al.*, 1985). This gene is now known to map at 19q13.2 and is linked to DM at a distance of 1–2 cM (1 centimorgan, cM, is the equivalent of 1% recombination). This tight linkage, coupled with the high polymorphic information content (PIC) at this locus, make *APOC2* a useful marker in prenatal and presymptomatic diagnosis (Meredith *et al.*, 1986). The PIC value is an estimate of the number of heterozygotes in a population from the observed allele frequencies, calculated using the Hardy-Weinberg equilibrium.

Linkage of DM to other markers soon followed, including *D19S19* (*LDR152*) (Bartlett *et al.*, 1987), *D19S50* (*pEFD4.2*) (Korneluk *et al.*, 1989a), *BCL3* a breakpoint associated with neoplasia (Korneluk *et al.*, 1989b) and the gene encoding the muscle-specific form of creatine kinase (*CKMM*) (Brunner *et al.*, 1989). These tightly linked probes were genetically mapped by members of an international collaboration that genotyped thousands of DM patients (MDA/Piton Foundation working party for myotonic dystrophy). This was done in order to find recombinants between these markers on DM chromosomes that would enable them to be ordered relative to each other and to DM (Johnson *et al.*, 1988a, 1989; Brunner *et al.*, 1989; Korneluk *et al.*, 1989a). This order is dependent on a few critical recombination events in affected individuals as defined by a set of clinical criteria adopted by the working group (Griggs *et al.*, 1989). These studies also excluded two candidate genes, the protein kinase C gamma-gene (Johnson *et al.*, 1988b) and the alpha-subunit of the sodium–potassium ATPase (Harley *et al.*, 1988).

The outcome of these efforts was to generate a consistent map that placed DM distal to *CKMM* and proximal to *D19S50*, Figure 4.2. It also identified the major problem facing DM research: the closest linked markers were all on the proximal side of the locus with a relatively large genetic interval, 10 cM, to the nearest probe on the distal side, *D19S50* (Korneluk *et al.*, 1989a). In the human genome 1 cM is equal to 1 Mb of DNA assuming that recombination is spread evenly over the 3×10^9 bp of the haploid human genome. This meant that the flanking markers defined in Figure 4.2 were about 10 Mb apart, a substantial problem in physical mapping terms, as discussed in the next section.

DM research was therefore concentrated on isolating new markers on the

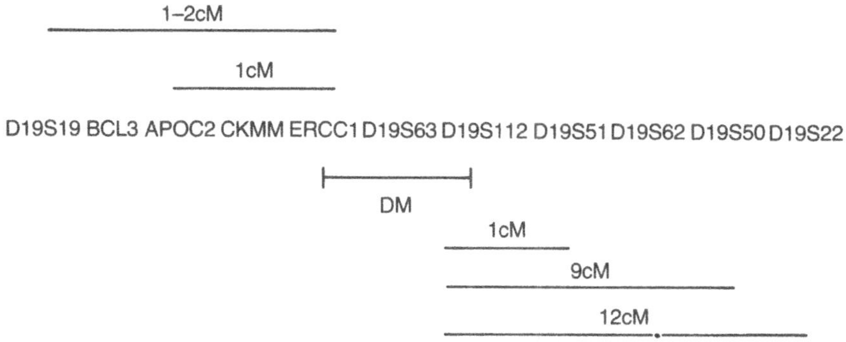

Figure 4.2 Genetic map of DM region. This shows the positions of markers determined from linkage and recombination studies. The maximum likelihood recombination fraction (θ_{max}) and the maximum lod score (Z_{max}) are indicated where they are known. This figure is a composite of data from many groups as summarized in Le Beau *et al.* (1989), updated with the marker information from Harley *et al.* (1991), Jansen *et al.* (1991), Johnson *et al.* (1990), Shutler *et al.* (1991b) and Tsilfidis *et al.* (1991a).

distal side of DM that would increase the accuracy of antenatal and presymptomatic diagnosis and which would link up to the proximal markers in physical maps. In order to achieve this, it was necessary to identify an enriched source of the myotonic region of chromosome 19q13.2-13.3. A human–chinese hamster hybrid cell line was identified that contained all of the closely linked, proximal flanking markers (*BCL3*, *APOC2*, *CKMM* and *ERCC-1*) but not the distal markers (*D19S22* and *D19S50*) (Stallings *et al.*, 1988). This cell line, 20XP3542-1-4, was isolated from another hybrid cell line that had been shown to contain an intact chromosome 19 as its only human genetic material, 5HL9-4 (Thompson *et al.*, 1989). Both of these cell lines were generated using the human *ERCC-1* gene as a dominant selectable marker in a UV repair-deficient hamster host.

ERCC-1 encodes one of the excision repair proteins that removes from DNA nucleotides that have become damaged due to exposure to UV light. There is a mutant hamster cell line which lacks a functional copy of this gene and so cannot survive exposure to UV light unless it receives a copy of *ERCC-1* which can complement this defect. In this case the human gene can complement a defect in the hamster gene. This forms the basis of a selection system by which human/hamster hybrid cell lines containing *ERCC-1* and associated regions of human chromosome 19 can be generated.

Cosmid and bacteriophage libraries were constructed from 20XP3542-1-4 cell line DNA. This cell line contains an intact hamster genome and several chromosomal fragments of human DNA including the DM region markers. The human DNA in this cell line comprises only about 0.1% of the total DNA and a method of identifying clones containing human DNA inserts had to be used to take account of this. Clones were screened by hybridization to human specific repetitive probes (Johnson *et al.*, 1990). The human-positive clones identified in this way were then used as the source of new polymorphic probes to genotype the recombinants already identified with the closely linked proximal and distal markers. In this way a new marker, *D19S51*, was identified which mapped distal to DM at a recombination fraction of < 1 cM (Johnson *et al.*, 1990; Tsilfidis *et al.*, 1991a).

D19S51 identifies RFLPs with the enzymes *Pst*I and *Bgl*I, Figure 4.3, and because of its proximity on the distal side of the gene has greatly increased

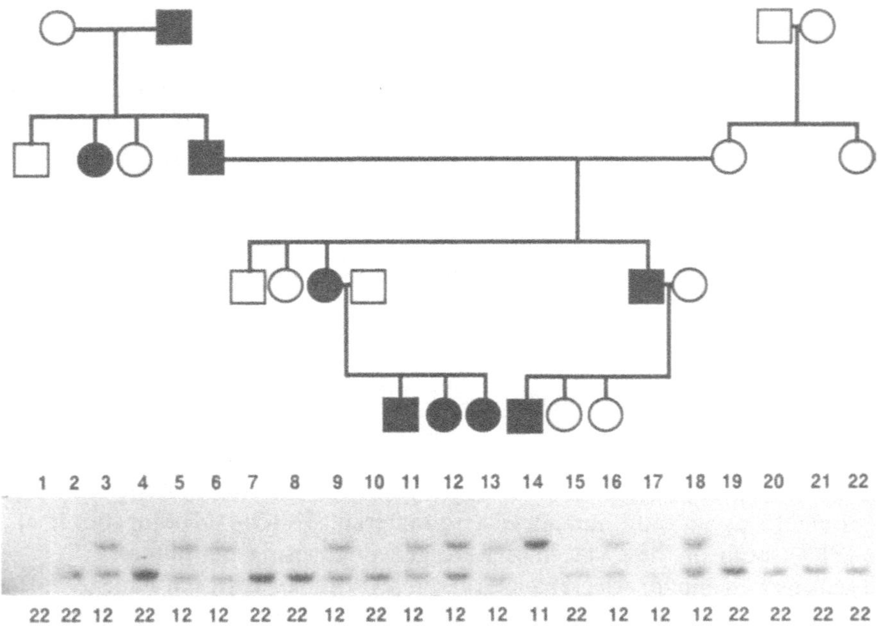

Figure 4.3 RFLP with *Bgl*I detected by D19S51 (p134C). Each lane contains DNA from the individual from the pedigree as illustrated at the top of the figure. The DNA was digested with the restriction enzyme *Bgl*I and electrophoresed in a 1.2% agarose gel. The DNA was denatured by soaking the gel in alkali (0.4 M NaOH) for 30 minutes prior to transfer of the DNA to a nylon membrane by Southern transfer. The immobilized DNA was hybridized with probe 134C which had been radioactively labelled. The filter was washed to 0.1 × SSC and autoradiographed for 48 hours at –70°C. The bands that can be seen correspond to the two alleles detected by this probe. In this family the DM gene is segregating with allele 1 (the upper band). The genotypes are indicated below the autoradiograph. Shaded symbols refer to DM patients.

the accuracy of antenatal diagnoses in families where the affected parent is heterozygous for this marker (Lavedan *et al.*, 1991). More recently other groups have reported genetic markers from the DM region that are either closer on the distal side than *D19S51* or map between it and *ERCC-1*, see Figure 4.4 (Brook *et al.*, 1990a, 1990b; Harley *et al.*, 1991; Hermens *et al.*, 1991; Korneluk *et al.*, 1991; Shutler *et al.*, 1991a, 1991b, 1991c; Tsilfidis *et al.*, 1991b, 1991c). These markers are so close to each other that it is impossible to distinguish them using genetic methods. This is because they lie within 1 cM of each other and there are too few recombinants between them to establish their order with respect to each other and to the DM locus.

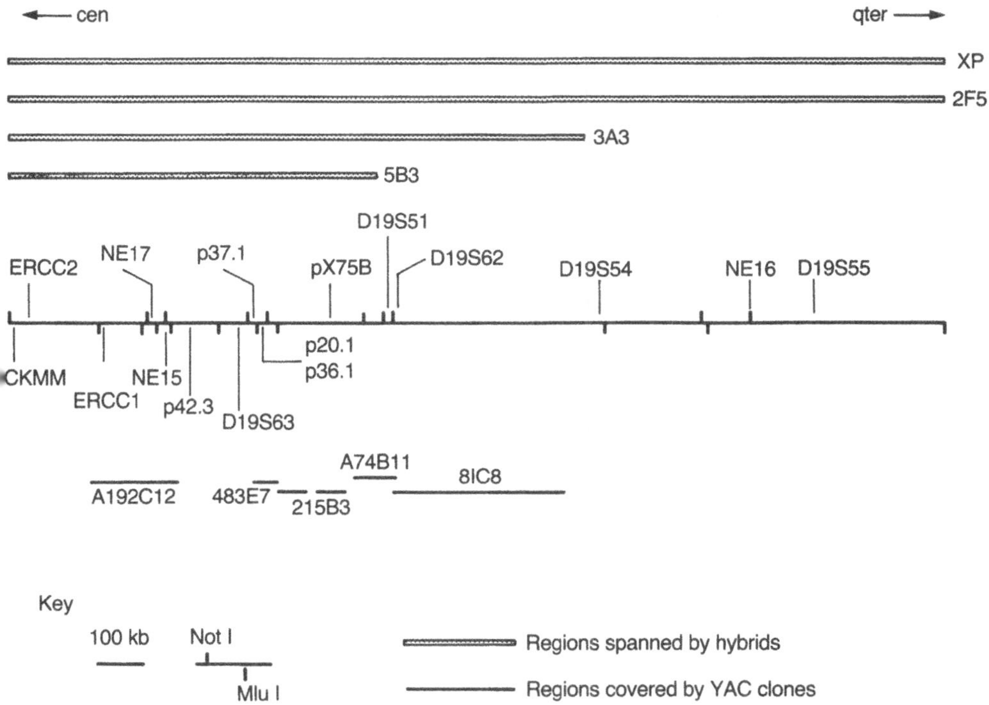

Figure 4.4 Physical map of the DM region on 19q13.3. This map shows the positions and relationships of the markers in the chromosome 19q13.3 interval distal to the creatine kinase muscle-form gene (*CKMM*). The interval contained within the hybrid 20XP3542-1-4 is marked 'XP' and the intervals within the radiation hybrids from Brook *et al.* (1992) are shown. The markers and their positions on the pulsed-field map are shown. For details see references. Below the PF map the positions of the YAC inserts are marked. (Data compiled from Brook *et al.*, 1992; Buxton *et al.*, 1992a, 1992b; Harley *et al.*, 1991; Jansen *et al.*, 1991; Smeets *et al.*, 1990). The DM critical region is between ERCC1 and pX75B.

The availability of this density of markers in the DM region made it possible to begin constructing physical maps and to commence the search for candidate genes.

Genetic analysis of recombination events between markers that are very tightly linked to a disease locus is limited in humans by the finite number of families available for study. However, as the recombination between markers and the disease locus shrinks to effectively zero, the phenomenon of linkage disequilibrium is observed. A common haplotype for a set of tightly linked markers is observed in affected individuals which is rare or absent in the control population. This effect is only seen when the new mutation rate for a genetic disease is low so that most of the affected individuals derive from a common ancestor. Linkage disequilibrium was first widely reported for the CF locus (Estivill *et al.*, 1987) and has subsequently been found in DM populations in Canada (MacKenzie *et al.*, 1989), the UK (Harley *et al.*, 1991) and Finland (Nokelainen *et al.*, 1991). Currently the strongest linkage disequilibrium is detected with markers from the interval between *D19S63* and *D19S62* (Harley *et al.*, 1991).

4.5 PHYSICAL MAPPING

Physical mapping of human chromosomes first became possible when somatic cell genetics enabled translocated chromosomes from patients to be isolated and maintained stably in human/rodent somatic cell hybrids. As the genetic map of chromosome 19 became more comprehensive it became clear that several of the earlier hybrids contained rearranged versions of the chromosome. Eventually, a panel of hybrids was assembled that divided the long arm of the chromosome into distinct regions where the probes mapped in accordance with the genetic map (Schonk *et al.*, 1989; Brook *et al.*, 1991). These hybrids were a tool for the rapid assignment of any new markers to one of these subdivisions of 19q by hybridization to a Southern blot of digested DNA from them.

Radiation hybrids, which were first used for gene mapping some time ago (Goss and Harris, 1975), are currently being widely used to isolate small regions of chromosomes harbouring important human disease loci like HD (Cox *et al.*, 1989; Doucette-Stamm *et al.*, 1991) and the Wilms' tumour gene (Glaser *et al.*, 1990). The 20XP3542-1-4 cell line contains two selectable markers which allow for their segregation in resultant reduced hybrids to be determined. These markers allow for selection in CHO-UV20 cells for retention of the *ERCC1* gene using mitomycin C and negative selection for the loss of the polio virus receptor (PVS) using live virus. The principle of this method has already been described above.

In this way a panel of radiation hybrids containing differing amounts of the DM region of chromosome 19 were selected for. The resultant hybrids were characterized for the presence or absence of a panel of the markers in the genetic and physical maps by Southern hybridization (Brook *et al.*, 1992). These hybrids divide the DM region into physically small intervals and so allow the rapid assignment of new markers to defined locations on 19q13.3.

The assignment of probes to such tightly defined physical intervals of chromosome 19q13.3 enabled restriction maps of the region to be constructed by use of pulsed-field gel electrophoresis (PFGE) (Myklebost and Rogne, 1988; Shaw *et al.*, 1989; Smeets *et al.*, 1990; Buxton and Johnson, unpublished data). These results coupled with the radiation hybrid data enabled the physical map of the region shown in Figure 4.4 to be constructed. This shows all of the tightly linked flanking markers and locates the key recombination events that define the 250 kb DM 'critical region'. These recombinants are in individuals in different families (Smeets *et al.*, 1991; Tsilfidis *et al.*, 1991a). As detailed in the next section, the isolation of yeast artificial chromosome clones (YACs) containing genomic inserts of several hundreds of kb, led to the generation of a detailed, long-range restriction map of the entire 'critical region'.

A number of probes on the physical map have been isolated because of their association with sites for rarely cutting restriction enzymes. This means that they have been selectively cloned by restricting (cutting) genomic DNA with a rare-cutter restriction enzyme, such as *Not*I, and another, more frequently cutting restriction enzyme, commonly *Eco*RI or *Bam*HI. Rare-cutter sites, like those for *Not*I and *Sac*II, tend to cluster in the genome at regions enriched for the dinucleotide CpG, which is the site of methylation in mammalian DNA (Bird, 1986). These clusters of CpGs frequently coincide with the presence of expressed sequences (genes) and are an extremely important clue to the location of coding sequences in the physical map. These rare-cutter end clone probes were therefore used to screen cDNA libraries for expressed sequences mapping to the DM 'critical region', as discussed later.

4.6 ISOLATION OF THE DM REGION AS A SERIES OF OVERLAPPING GENOMIC CLONES

Once the physical map of the region had been determined and the size of the 'critical region' defined, the next aim was to isolate the entire region as a series of overlapping genomic clones. This would enable the DM region to be screened for the presence of expressed sequences which could be systematically investigated as candidate genes. There were several key resources which were available to the various groups within the MDA/Piton Foundation working group for this phase of the project.

The major physical mapping group involved was at the Lawrence

Livermore Laboratories, California. The aim of this group was to physically map the whole of chromosome 19 by fingerprinting cosmid clones generated from flow-sorted material (Carrano *et al.*, 1989). They have constructed cosmid and bacteriophage libraries containing genomic inserts from chromosome 19 and, more recently, YAC clones as well.

The fingerprinting strategy started by analysing clones at random, using an automated DNA sequencer to generate data that could be stored in digitized form. A computer algorithm determines overlaps between clones and then derives a confidence limit for that overlap. This method is derived from that first used to generate a physical map of the *C. elegans* genome (Coulson *et al.*, 1986). In this way data from over 12 000 cosmid clones from chromosome 19 are now in the database and more than 90% of the chromosome has been placed into contigs (the term given to a group of overlapping clones identified in this way).

Additionally, cosmid clones containing probes from chromosome 19 were identified and these small 'seed' contigs of known positions were used to anchor the map. Also, as contigs were identified, clones from them were taken for fluorescent *in situ* hybridization (FISH) mapping, both in metaphase and interphase nuclei (Brandriff *et al.*, 1991). The contigs were rapidly assigned to cytogenetically determined regions using this method.

The Livermore cosmid resource was used by the Ottawa group to generate a 400 kb chromosome walk distal from the *ERCC-1* locus, which at the time they started was the closest proximal flanking marker for DM (Shutler *et al.*, 1990). The basic principles of this method are illustrated in Figure 4.5. From the walk clones they isolated a number of new polymorphic markers which were useful both for antenatal and presymptomatic diagnosis, and also for the investigation of linkage disequilibrium (see genetic mapping section and Figure 4.4) (Korneluk *et al.*, 1991; Shutler *et al.*, 1991a, 1991b, 1991c; Tsilfidis *et al.*, 1991b, 1991c).

Other groups began to isolate YACs containing markers from the DM region beginning with the markers known to be flanking *DM*, *CKMM*, *ERCC-1* and *D19S51*. The strategy was based on the fact that YACs can contain large (up to 1 Megabase) inserts of genomic DNA, and, because of the known physical and genetic proximity of the flanking markers (Figure 4.4) it was possible that these YAC clones might overlap. They would be a valuable resource for the isolation of new markers from the DM 'critical region', and from which to begin isolating candidate gene sequences.

As it turned out, the DM region was not going to succumb to these strategies so easily. The 400 kb cosmid walk from *ERCC-1* terminated at an 'unclonable region'. This is genetic parlance for a region of DNA that can not be isolated from a given library. The most likely explanation, based on later findings, is that the region that was missing from the Livermore library

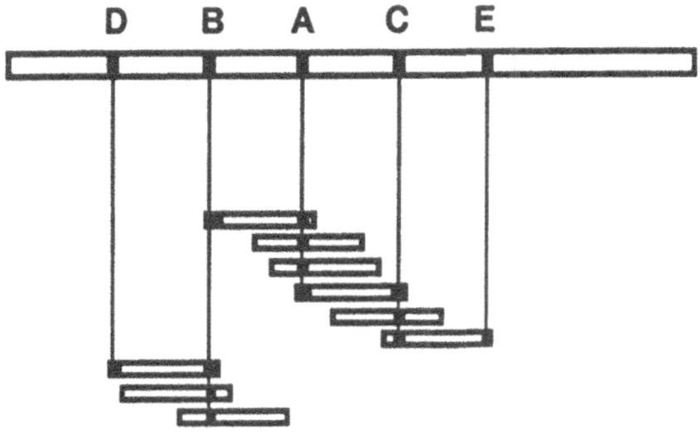

Figure 4.5 This cartoon represents the principles of bidirectional chromosome walking from a position corresponding to probe A to probes B–E. Probe A, obtained as a cDNA or anonymous genomic fragment is used to screen a library containing randomly generated genomic inserts in bacteriophage, cosmid or YAC vectors. The clones that hybridize will all have probe A in common and will overlap to differing degrees. By restriction mapping of the overlapping probe A-positive clones it is possible to identify the ends of the contiguous segment of DNA (contig) represented by these clones. These end fragments, probes B and C, are then used to screen the same library to identify other clones that overlap with the region defined by each of these probes (clones 5, 6 and 7 for probe B and clones 8 and 9 for probe C). In this way the contig is extended bidirectionally. By repeating the process in a stepwise manner the contig is extended to include probes D and E. The distance walked in each step is totally dependent upon the size of insert contained in the library clones.

contains multiple repeat sequences. Bacteria are very efficient at deleting repeat sequences by recombination mechanisms. To overcome this problem, recombination deficient host strains have to be used in the library construction.

The YAC cloning approach did not fare much better, but instead of this region proving to be unclonable, the clones isolated proved to be highly unstable and rapidly deleted probes for which they had earlier been positive in the initial screen. These unexpected findings did not mean that the YAC and cosmid efforts were in vain, for, as I have already described, the Ottawa group isolated a number of useful new genetic markers from their walk and at least one candidate gene. Eventually, by changing the yeast host strain in which the YACs were propagated it became possible to isolate stable clones and the physical map was closed (Aslanidis *et al.*, 1992).

The YACs proved useful for screening other libraries, both genomic and cDNA. A contig of YACs spanning some 700 kb was isolated around *D19S51* (Buxton *et al.*, 1992a). This contig contained other markers known to be tightly linked to DM and placed them accurately on the physical map, including *D19S62* and the breakpoint of one of the radiation hybrids (Brook *et al.*, 1992). Similarly a contig spanning some 750 kb around

ERCC-1 and including *CKMM*, *ERCC-2* and a number of the probes from Ottawa and Cardiff was generated (Perryman *et al.*, manuscript in preparation).

As it turned out, these YAC contigs did not overlap, as we had previously hoped they would, and so further YACs were looked for from the intervening region to close the gap. These were screened for by using *D19S63* because it is located between *ERCC-1* and *D19S51* in the physical map (see Figure 4.4). In this way a panel of YACs originating from several libraries was obtained and characterized by hybridizing them with probes from the DM 'critical region'. Analysis of this type enabled the key YACs to be identified with which to screen cDNA libraries and a chromosome 19 only bacteriophage library.

YACs obtained in this way from the St Louis library were used to generate further overlapping sets of clones, this time bacteriophage, by screening the chromosome 19 library from Livermore with the gel-purified YAC (Buxton *et al.*, 1993). In this way parts of the region from the gap in the middle were isolated. A new probe, X75 (*D19S112*), identified by the Nijmegen group (Hermens *et al.*, 1991), was also used to screen for YACs. Cosmid contigs of a YAC positive for this probe were used to extend the overlapping set of clones proximally towards the distal end of the walk from *ERCC-1*. The eventual map of cosmid clones identified from the Livermore library by the Nijmegen and Ottawa groups, when aligned with the physical map of the region still contained a small gap. This gap was bridged by YAC clones and was finally walked by the Livermore group to close the physical map of the region (Aslanidis *et al.*, 1992). (Chromosome walking is explained in Figure 4.5.) This now left the way open for the isolation and characterization of all of the coding sequences in this interval as candidates for the DM gene.

4.7 IDENTIFYING CANDIDATE GENES

The strategy for finding genes in genomic DNA fragments relies on several properties of coding sequences that distinguish them from the non-coding sequences that comprise the vast majority (> 90%) of the human genome. For a recent review of this area see Wicking and Williamson (1991).

The 6×10^9 bp of DNA that comprise the diploid human genome are made up of sequences that are referred to simply as either single-copy or repeated sequences (Jelinek and Schmid, 1982). These categories are roughly equally represented in the genome. The genes are almost entirely contained within the single-copy fraction whereas the repetitive fraction comprises low, middle and highly repetitive sequences which can occur as either clustered or interspersed repeats (Jelinek and Schmid, 1982; Moyzis *et al.*, 1989). Clustered repeats of short, highly repetitive sequences, are found mainly at telomeres and centromeres and are structural in function. Other clustered repeats occur as satellite or alphoid DNA sequences in chromosome-specific arrays (Waye and Willard, 1987). The interspersed repeats, found on all chromosomes, are divided into three main classes: the short interspersed repeats (SINES), the long interspersed repeats (LINES) and the simple sequence motifs (SSMs). The commonest human repeat, a member of the SINES category, is the Alu repeat. This repeat is 0.3 kb in length and accounts for about 6% of the human genomic DNA, or about 9×10^5 copies per diploid genome (Jelinek and Schmid, 1982).

In order to screen for expressed sequences within the single-copy fraction of the DNA, genomic clones are screened by hybridizing them with labelled total human genomic DNA. Fragments that contain repeats will hybridize strongly because of the high copy number of the repeat sequences within the labelled probe. Fragments that do not hybridize will contain single- or low-copy number sequences. Restriction fragments of this type within a genomic clone can readily be identified by performing a small number of single and double restriction digests followed by Southern blotting. The filter with the bound restriction fragments is then hybridized to a repetitive probe and non-hybridizing bands are gel-purified for further study.

When single-copy fragments are identified in this way they are further characterized by exploiting another key feature of genes, namely that they are conserved in evolution. This means that if a genomic fragment identified as single-copy contains part of a gene, it will hybridize to its homologous, conserved target sequence in genomic DNA of related species. This approach is referred to as zoo-blotting, and has been used successfully to identify many genes from genomic DNA clones (Estivill *et al.*, 1987). The method usually employs genomic DNA from several species, mammalian and otherwise. The resulting Southern blot is washed 'post-hybridization' to a relatively low-stringency (1 × SSC) to allow related sequences to be detected. In this way a genic sequence is identified by the detection of a specific hybridizing restriction fragment in one or more of the animal DNAs. A band of silver grains is seen on the autoradigraph which corresponds to the area of the filter to which the restriction fragment bearing the homologous sequence was bound.

Additionally there are CpG islands within the human genome which are detected as clusters of rare-cutter restriction sites. These islands are dispersed and commonly occur very close to coding sequences within the genome (Lindsay and Bird, 1986). They are therefore used as landmarks indicating the presence of genes within genomic clones. A restriction fragment containing a putative CpG island can be used as a probe to screen zoo blots to test for conservation of the sequences across species. The detection of a CpG island and evidence for its conservation is almost certain proof that the fragment is part of a gene.

Once a conserved fragment has been identified it is used to screen Northern blots of mRNA from different tissues in order to determine the site of expression and size of transcript for that gene. One problem with this approach is that the sensitivity of Northern blots is relatively low and only fairly abundant transcripts are detected. For this reason it is usual to screen cDNA libraries made from mRNA isolated from those tissues where the candidate genes is expected to be expressed. In the case of DM this is muscle, heart, lens and brain, as these are the tissues most frequently and severely affected in patients. In circumstances where there is no preferred tissue, the library of choice is fetal brain because it is estimated that > 50% of all genes are expressed in this tissue and a commercially available library can be used.

Candidate genes for DM, based on their map position in the DM critical region, have been isolated using these techniques (Aslanidis *et al.*, 1992; Buxton *et al.*, 1992b). One of these, *cDNA25* (Buxton *et al.*, 1992b) detects an altered fragment in restriction digests of genomic DNA from patients that is not seen in control samples (see Figure 4.6). The sequence of this gene from both normal and DM tissues will need to be determined in order to identify potential mutations that are responsible for conferring the disease phenotype. However, preliminary data indicate that there is a great deal of similarity between the mutation detected with *cDNA25* and that recently reported for the fragile-X mental retardation (*FraX*) gene (Verkerk *et al.*, 1991) and the X-linked spinal bulbar muscular atrophy gene (La Spada *et al.*, 1991). In both these diseases the mutation arises as the result of amplification of a trinucleotide repeat within the coding region of the gene. The *FraX* gene increases in size by between a few hundred and several thousand base pairs in patient DNA (Ying-Hui *et al.*, 1991). The mutations detected in DM patients appear to fall into a similar size range (Buxton *et al.*, 1992b), although it remains to be shown by genomic DNA sequencing whether the increase is due to the amplification of a repeat sequence.

4.8 FUTURE PROSPECTS

The hunt for the DM gene is now concentrated on a very small region of the genome that contains a few genes, one of which, *cDNA25*, appears to detect mutations in affected individuals. The next phase of the work is to determine the sequences and possible physiological roles of the predicted gene product of *cDNA25* in order to begin to unravel the mechanism of the mutation and how it causes the phenotypic changes characteristic of DM.

Eventually we will have to show that the mutations we detect alter the expression or the function of the protein encoded by *cDNA25* in tissues in a way which mimics the disease phenotype. It may be that the DM phenotype is due to alterations in the temporal or spatial expression of the gene

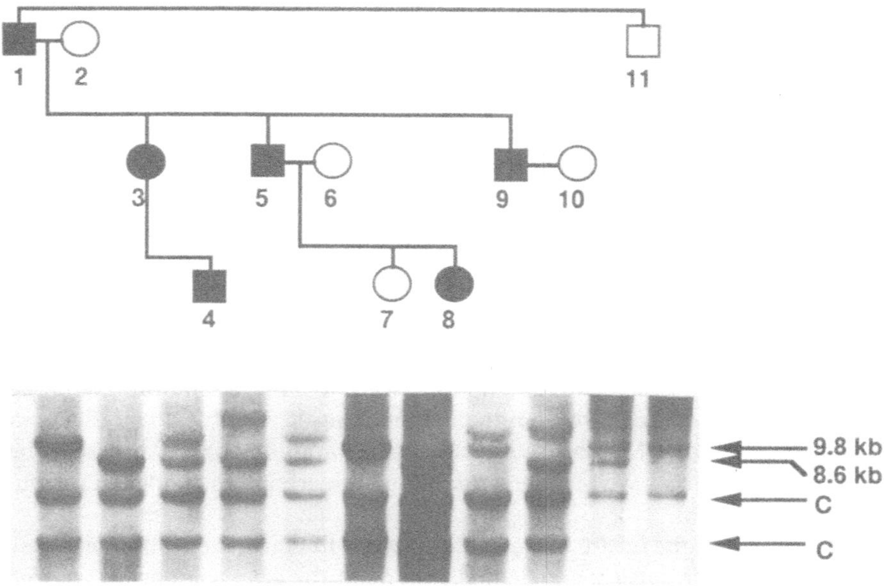

Figure 4.6 The detection of an unstable DNA fragment in myotonic dystrophy patients by cDNA25. This figure shows a genomic blot of *Eco*RI digested genomic DNA of DM family members hybridized with cDNA25 (Buxton *et al.*, 1992c). This probe detects a polymorphism with *Eco*RI in the normal population, with alleles of 9.8 kb and 8.6 kb arrowed, it also detects constant bands indicated with a C beside the arrow. Additional higher molecular weight DNA fragments specific to individuals with DM can be seen clearly in lanes 4, 8 and 9. The length of this fragment correlates very well with the severity of symptoms seen in the affected individuals. It particularly shows the molecular basis of the phenomenon of anticipation, with the fragment increasing in size in successive generations.

product, in which case embryological studies will be required. It is quite likely from the physiological and biochemical data that the gene product of *cDNA25* will be an ion-channel or a regulator of an ion channel. Therefore, detailed studies of cell membranes using ultrastructural and electrophysiological approaches will be required to unravel the normal function of the candidate gene and the effect of the mutation.

The next step will be to generate an animal model of DM, most probably by a transgenic approach (see Capecchi, 1989 for a review). The isolation of human disease genes raises the possibility of gene therapy (for review see Johnson, 1991b), however in the case of DM this is most unlikely. This is because biochemical studies indicate that the DM mutation affects all tissue types. It is difficult to imagine how, using current approaches, a method could be found by which all the different cell types within an individual can be treated by a gene therapy regimen. It is therefore more likely that any improvement in treatment for patients with this disorder will be via a pharmacological approach which is why an animal model of DM is so vital.

The animal model could be generated using one of several routes. A functional copy of the DM version of the gene could be injected into a pronucleus of a fertilized mouse ovum to create a transgenic animal. This was the approach first used to introduce globin mRNA into mouse ova (Brinster *et al.*, 1980). Another way would be to transfect embryonic stem (ES) cells with a construct containing the DM version of the gene in order to produce chimeric mice. ES cells are pluripotential stem cell lines derived by a modification of the procedure originally described by Evans and Kaufman (1981). In this way the dominant nature of the mutation could be introduced into the germ-line of mice, making it possible to produce an animal model of DM. This would allow the pathology of the disease to be studied in depth as well as the embryology underlying the congenital form of this disease.

In summary, the future for research into this human genetic disease is very bright. With the eagerly anticipated cloning of the gene responsible for this phenotype having now been achieved, it should be possible, given time, to unravel the mysteries of this disease that is so fascinating at both the clinical and molecular level.

ACKNOWLEDGEMENTS

This work was, and still is, supported by the Muscular Dystrophy Group of Great Britain through grant number RA3/257/1 and by the MDA/Piton Foundation through a grant under its Task Force in Genetics. Students in the group have been supported by grants from SERC and MRC (HGMP). I am extremely grateful to Jessica Buxton, June Davies, Clare Jones, Peggy Shelbourne and Tracey Van Tongeren for all of their efforts, both on the research side and for their critical appraisal of this manuscript.

REFERENCES

Ashizawa, T. and Epstein, H. (1991) Ethnic distribution of myotonic dystrophy gene. *Lancet*, **338**, 642–3.

Aslanidis, C., Jansen, G., Amemiya, C. *et al.* (1992) Cloning of the essential myotonic dystrophy region: mapping of the putative defect. *Nature*, **355**, 548–51.

Bartlett, R., Pericak-Vance, M., Yamaoka, L. *et al.* (1987) A new probe for the diagnosis of myotonic muscular dystrophy. *Science*, **235**, 1648–50.

Bird, A.P. (1986) CpG-rich islands and the function of DNA methylation. *Nature*, **321**, 209–13.

Botstein, D., White, R.L., Skolnick, M. and Davis, R.W. (1980) Construction of a genetic linkage map in man using restriction fragment length polymorphisms. *Am. J. Hum. Genet.*, **32**, 314–31.

Brandriff, B., Gordon, L. and Trask, B. (1991) A new system for high resolution DNA sequence mapping in interphase pronuclei. *Genomics*, **10**, 75–82.

Brinster, R.L., Chen, H.Y., Trumbauer, M.E. and Avarbock, M.R. (1980) Translation of globin mRNA by the mouse ovum. *Nature*, **283**, 499–501.

Brook, J.D., Harley, H.G., Rundle, S.A. *et al.* (1990a) RFLP for a DNA clone which maps to 19q13.2-qter (D19S63). *Nucl. Acids Res.*, **18**, 1085.

Brook, J.D., Walsh, K.V., Harley, H.G. *et al.* (1990b) A polymorphic DNA clone which maps to 19q13.2-qter (D19S62). *Nucl. Acids Res.*, **18**, 1086.

Brook, J.D., Knight, S.J.L., Roberts, S.H. *et al.* (1991) The physical map of chromosome arm 19q: some new assignments, confirmations and re-assessments. *Hum. Genet.*, **87**, 65–72.

Brook, J.D., Zemelman, B.V., Haddingham, K. *et al.* (1992) Radiation reduced hybrids for the myotonic dystrophy locus. *Genomics*, **13**, 243–50.

Brunner, H., Smeets, H., Lambermon, H.M.M. *et al.* (1989) A multipoint linkage map around the locus for myotonic dystrophy on chromosome 19. *Genomics*, **5**, 589–95.

Bundey, S. and Carter, C.O. (1972) Genetic heterogeneity for dystrophia myotonica. *J. Med. Genet.*, **9**, 311–15.

Butterfield, D.A., Chestnut, D.B., Appel, S.H. and Roses, A.D. (1976) Spin label study of erythrocyte membrane fluidity in myotonic and Duchenne muscular dystrophy and congenital myotonia. *Nature*, **263**, 159–61.

Buxton, J., Shelbourne, P., Davies, J. *et al.* (1992a) Characterisation of a YAC and cosmid contig around the markers D19S51 and D19S62 tightly linked to the myotonic dystrophy locus on chromosome 19. *Genomics*, **13**, 526–31.

Buxton, J., Davies, J., Shelbourne, P. *et al.* (1993) Isolation of bacteriophage genomic clones using purified YACs from 19q13.3: rapid construction of the corresponding contig using inter-Alu PCR. *Mol. Cell. Probes*, **7**, 78–80.

Buxton, J., Davies, J., Shelbourne, P. *et al.* (1992b) Detection of an unstable fragment of DNA specific to individuals with myotonic dystrophy. *Nature*, **355**, 547–8.

Capecchi, M. (1989) The new mouse genetics: altering the genome by gene targetting. *Trends Genet.*, **5**, 70–6.

Carrano, A.V., Lamerdin, J., Ashworth, L.K. *et al.* (1989) A high resolution, fluorescence-based, semiautomated method for DNA fingerprinting. *Genomics*, **4**, 129–36.

Cattenach, B.M. and Kirk, M. (1986) Differential activity of maternally and paternally derived chromosome regions in mice. *Nature*, **315**, 496–8.

Chen, K-M., Brody, J. and Kurland, L. (1968) Patterns of neurologic diseases on Guam. *Arch. Neurol.*, **19**, 573–8.

Coulson, A., Sulston, J., Brenner, S. and Karn, J. (1986) Towards a physical map of the genome of the nematode *Caenorhabditis elegans*. *Proc. Natl. Acad. Sci. USA*, **83**, 7821–5.

Cox, D., Pritchard, C.A., Uglum, E. *et al.* (1989) Segregation of the Huntington's disease region of human chromosome 4 in a somatic cell hybrid. *Genomics*, **4**, 397–407.

Doucette-Stamm, L.A., Riba, L., Handelin, B. *et al.* (1991) Generation and characterisation of Goss-Harris hybrids of chromosome 4. *Somat. Cell Mol. Genet*, **17**, 471–80.

Eiberg, H., Mohr, J., Nielsen, L.S. and Simonsen, N. (1981) Linkage relationships between the locus for C3 and 50 polymorphic systems: assignment of C3 to the DM-Se-Lu linkage group: confirmation of the C3-LES linkage. (Abstract) 6th World Congress of Genetics, Jerusalem.

Emery, A. (1991) Population frequencies of inherited neuromuscular diseases – A world survey. *Neuromus. Dis.*, **1**, 19–29.

Erickson, R.P. (1985) Chromosomal imprinting and the parent transmission specific variation in expressivity of Huntington disease. *Am. J. Hum. Genet.*, **37**, 827–9.

Estivill, X., Farrall, M., Scambler, P.J. *et al.* (1987) A candidate for the cystic fibrosis locus isolated by selection for methylation-free islands. *Nature*, **326**, 840–5.

Evans, M.J. and Kaufman, M.H. (1981) Establishment or culture of pluripotential cells from mouse embryos. *Nature*, **292**, 154–6.

Fearon, E.R., Cho, K.R., Nigro, J.M. *et al.* (1990) Identification of a chromosome 18q gene that is altered in colorectal cancers. *Science*, **47**, 49–56.

Fenton, J., Garner, S. and McComas, A. (1991) Abnormal M-wave responses during exercise in myotonic muscular dystrophy: A Na^+-K^+ pump defect? *Muscle Nerve*, **14**, 79–84.

Franke, C., Hatt, H., Iaizzo, P. and Lehmann-Horn, F. (1990) Characteristics of Na^+ channels and Cl^- conductance in resealed muscle fibre segments from patients with myotonic dystrophy. *J. Physiol.*, **425**, 391–405.

Glaser, T., Rose, E., Morse, H. *et al.* (1990) A panel of irradiation-reduced hybrids selectively retaining human chromosome 11p13: Their structure and use to purify the WAGR gene complex. *Genomics*, **6**, 48–64.

Goss, S.J. and Harris, H. (1975) A new method for mapping genes in human chromosomes. *Nature*, **255**, 680–4.

Griggs, R., Wood, D. and the MDA Working Group on myotonic dystrophy (1989) Criteria for establishing the validity of genetic recombination in myotonic dystrophy. *Neurology*, **39**, 420–1.

Gusella, J.F., Wexler, N.S., Conneally, P.M. *et al.* (1983) A polymorphic DNA marker genetically linked to Huntington's disease. *Nature*, **306**, 234–8.

Harley, H., Brook, J., Jackson, C. *et al.* (1988) Localisation of a human Na^+, K^+-ATPase alpha subunit gene to chromosome 19q12-13.2 and linkage to the myotonic dystrophy locus. *Genomics*, **3**, 380–4.

Harley, H.G., Brook, D.J., Floyd, J. *et al.* (1991) Detection of linkage disequilibrium between the myotonic dystrophy locus and a new polymorphic DNA marker. *Am. J. Hum. Genet.*, **49**, 68–75.

Harper, P. (1975) Congenital myotonic dystrophy in Britain II. Genetic basis. *Arch. Dis. Childh.*, **50**, 514–21.

Harper, P. (1989) *Myotonic dystrophy*, Saunders, Philadelphia.

Harper, P.S. and Dyken, P.R. (1972) Early onset dystrophia myotonica: evidence supporting a maternal environmental factor. *Lancet*, **2**, 53–5.

Hermens, R., Coerwinkel, M., Brunner, H. *et al.* (1991) MspI RFLP at 19q13.3 identified by the anonymous DNA sequence pX75B (D19S112). *Nucl. Acids Res.*, **19**, 1726.

Howeler, C.J. (1986) A clinical and genetic study in myotonic dystrophy. Thesis, University of Rotterdam.

Howeler, C.J., Bush, H.F.M., Geraedts, J.P.M. *et al.* (1989) Anticipation in myotonic dystrophy: fact or fiction. *Brain*, **12**, 779–97.

Hull, K. and Roses, A. (1976) Stoichiometry of sodium and potassium transport in erythrocytes from patients with myotonic muscular dystrophy. *J. Physiol. London*, **254**, 169–81.

Human Gene Mapping 10.5 (1990) Update to the 10th International workshop on human gene mapping. *Cytogenet. Cell Genet.*, **55**, 1–785.

Humphries, S.E., Jowett, N.I., Williams, L. *et al.* (1984) A DNA polymorphism adjacent to the human apolipoprotein CII gene. *Mol. Biol. Med.*, **1**, 463–71.

Jansen, G., de Jong, P., Voovs, M. *et al.* (1991) Genetic and physical definition of the myotonic dystrophy gene region. Abstract 405 Proc. 8th Intl. Cong. Hum. Genet., 6–11th Oct 1991. Washington. *Am. J. Hum. Genet. Suppl.*, **49:4**, 82.

Jelinek, W.R. and Schmid, C.W. (1982) Repetitive sequences in eukaryotic DNA and their expression. *Ann. Rev. Biochem.*, **51**, 813–44.

Johnson, K.J. (1991) Prospects for gene therapy. *Chem. and Ind.*, **18**, 664–7.

Johnson, K.J., Nimmo, E., Jones, P. *et al.* (1988a) Segregation of linked probes to myotonic dystrophy in a family demonstrating that 152 and APOC2 are on the same side of DM on 19q. *Hum. Genet.*, **80**, 379–81.

Johnson, K.J., Jones, P., Spurr, N. *et al.* (1988b) Linkage relationships of the protein kinase C gamma-gene which exclude it as a candidate for myotonic dystrophy. *Cytogenet. Cell Genet.*, **48**, 13–15.

Johnson, K.J., Shelbourne, P., Davies, J. *et al.* (1989) Recombination events that locate myotonic dystrophy distal to APOC2 on 19q. *Genomics*, **5**, 746–51.

Johnson, K.J., Shelbourne, P., Davies, J. *et al.* (1990) New polymorphic probes which define the region of chromosome 19 containing the myotonic dystrophy locus. *Am. J. Hum. Genet.*, **46**, 1073–81.

Kakehi, T., Kuzuyu, H., Kosaki, A. *et al.* (1990) Binding activity and autophosphorylation of the insulin receptor from patients with myotonic dystrophy. *J. Lab. Clin. Med.*, **115**, 688–95.

Kerem, B-S., Rommens, J.M., Buchanan, J.A. *et al.*, (1989) Identification of the cystic fibrosis gene: genetic analysis. *Science*, **245**, 1073–80.

Koch, M.C., Grimm, T., Harley, H. and Harper, P.S. (1991) Genetic risks for children of women with myotonic dystrophy. *Am. J. Hum. Genet.*, **48**, 1084–91.

Koenig, M., Hoffman, E.P., Bertelson, C.J. *et al.* (1987) Complete cloning of the Duchenne muscular dystrophy (DMD) cDNA and preliminary genomic organization of the DMD gene in normal and affected individuals. *Cell*, **50**, 509–17.

Korneluk, R.G., MacKenzie, A.E., Nakamura, Y. *et al.* (1989a) A reordering of human chromosome 19 long-arm markers and identification of markers flanking the myotonic dystrophy locus. *Genomics*, **5**, 596–604.

Korneluk, R.G., MacLeod, H., McKeithan, T. *et al.* (1989b) A chromosome 19 clone from a translocation breakpoint shows close linkage and linkage disequilibrium with myotonic dystrophy. *Genomics*, **4**, 146–51.

Korneluk, R.G., Tsilfidis, C., Shutler, G. *et al.* (1991) A three allele insertion polymorphism is identified by the human chromosome 19q13.3 probe pKBE0.8 (D19S119). *Nucl. Acids Res.*, **15**, 6769.

Kuwabara, T., Yuasa, T., Ohno, T. *et al.* (1991) Study on the erythrocytes from myotonic dystrophy with multinuclear NMR. *Muscle Nerve*, **14**, 57–63.

La Spada, A., Wilson, E., Lubahn, D. *et al.* (1991) Androgen receptor gene mutations in X-linked spinal and bulbar atrophy. *Nature*, **352**, 77–9.

Lavedan, C., Hofmann, H., Shelbourne, P. *et al.* (1991) Prenatal diagnosis of myotonic dystrophy using closely linked flanking markers. *J. Med. Genet.*, **28**, 89–91.

Le Beau, M.M., Ryan, D. and Pericak-Vance, M. (1989) Report of the committee on the genetic constitution of chromosomes 18 and 19. *Cytogenet. Cell Genet.*, **51**, 338–57.

Lindsay, S. and Bird, A. (1986) Use of restriction enzymes to detect potential gene sequences in mammalian DNA. *Nature*, **327**, 336–8.

Lotz, B. and van der Meyden, C. (1985) Myotonic dystrophy Part 1: A genealogical study in the northern Transvaal. *S. Afr. Med. J.*, **67**, 812–14.

MacKenzie, A.E., MacLeod, H.L., Hunter, A.G.W. and Korneluk, R.G. (1989) Linkage analysis of the apolipoprotein C2 gene and myotonic dystrophy on human chromosome 19 reveals linkage disequilibrium in a French-Canadian population. *Am. J. Hum. Genet.*, **44**, 140–7.

Mathieu, J., De Braekeleer, M. and Prevost, C. (1990) Genealogical reconstruction of myotonic dystrophy in the Saguenay-Lac-Saint-Jean area (Quebec, Canada). *Neurology*, **40**, 839–42.

Meredith, A., Huson, S.M., Lunt, P.W. *et al.* (1986) Application of closely linked polymorphism of restriction fragment length to counselling and prenatal testing in families with myotonic dystrophy. *Br. Med. J.*, **293**, 1353–6.

Merril, C.R. and Harrington, M.G. (1985) The search for mitochondrial inheritance of human disease. *Trends Genet.*, **1**, 140–4.

Mohr, J. (1954) *A Study of Linkage in Man*, Munksgaard, Copenhagen.

Moxley, R., Corbett, A., Minaker, K. and Rowe, J. (1984) Whole body insulin resistance in myotonic dystrophy. *Ann. Neurol.*, **15**, 157–62.

Moyzis, R.K., Torney, D.C., Meyne, J. *et al.* (1989) The distribution of interspersed repetitive DNA sequences in the human genome. *Genomics*, **4**, 273–89.

Myklebost, O. and Rogne, S. (1988) A physical map of the apolipoprotein gene cluster on human chromosome 19. *Hum. Genet.*, **78**, 244–7.

Nimmo, E. and Johnson, K. (1991) RFLPs: Uses and methods of detection, in *Methods in Gene Technology*, Volume 1, (eds J.W. Dale and P.G. Sanders), JAI Press, London pp. 267–83.

Nokelainen, P., Alanen-Kurki, L., Winquist, R. *et al.* (1991) Linkage disequilibrium detected between dystrophia myotonica and APOC2 locus in the Finnish population. *Hum. Genet.*, **85**, 541–5.

Oberle, I., Rousseau, F., Heitz, D. *et al.* (1991) Instability of a 550-base pair DNA segment and abnormal methylation in fragile X syndrome. *Science*, **252**, 1097–102.

Penrose, L.S. (1948) The problems of anticipation in pedigrees of dystrophia myotonica. *Ann. Eugen.*, **14**, 125–232.

Renaud, J-F., Desnuelle, C., Schmid-Antomarchi, H. *et al.* (1986) Expression of apamin receptor in muscles of patients with myotonic muscular dystropy. *Nature*, **319**, 678–80.

Riordan, J.R., Rommens, J.M., Kerem, B-S. *et al.* (1989) Identification of the cystic fibrosis gene: cloning and characterisation of complementary DNA. *Science*, **245**, 1066–73.

Rommens, J.M., Iannuzzi, M.C., Kerem, B-S. *et al.* (1989) Identification of the cystic fibrosis gene: chromosome walking and jumping. *Science*, **245**, 1059–65.

Rudel, R. and Lehmann-Horn, F. (1985) Membrane changes in cells from myotonia patients. *Physiol. Rev.*, **65**, 310–46.

Schonk, D., Coerwinkel-Driessen, M., van Dalen, I. *et al.* (1989) Definition of subchromosomal intervals around the myotonic dystrophy region at 19q. *Genomics*, **4**, 384–96.

Shaw, D.J., Meredith, A., Sarfarazi, M. *et al.* (1985) The apolipoprotein CII gene: subchromosomal localisation and linkage to the myotonic dystrophy locus. *Hum. Genet.*, **70**, 271–3.

Shaw, D., Meredith, A., Brook, J. *et al.* (1986) Linkage relationships of the insulin receptor gene with the complement component 3, LDL receptor, apolipoprotein C2 and myotonic dystrophy loci on chromosome 19. *Hum. Genet.*, **74**, 267–9.

Shaw, D.J., Harley, H.G., Brook, J.D. and McKeithan, T.W. (1989) Long-range restriction map of a region of human chromosome 19 containing the apolipoprotein genes, a CLL-associated translocation breakpoint, and two polymorphic MluI sites. *Hum. Genet.*, **83**, 71–4.

Shutler, G., LeBlond, S., Earle, J. *et al.* (1990) Rapid chromosome walking towards the myotonic dystrophy gene by an ALU based PCR technique. *Am. J. Hum. Genet.*, **47**, A262.

Shutler, G., LeBlond, S., Bailly, J. *et al.* (1991a) An insertion polymorphism identified by the probe pE0.8 (D19S115) at 19q13.3. *Nucl. Acids Res.*, **19**, 1159.

Shutler, G., MacKenzie, A.E., Brunner, H. *et al.* (1991b) Physical and genetic mapping of a novel chromosome 19 ERCC-1 marker showing close linkage with myotonic dystrophy. *Genomics*, **9**, 500–4.

Shutler, G., Tsilfidis, C., LeBlond, S. and Korneluk, R.G. (1991c) RFLP identified by the probe pKE0.6 (D19S117) at human chromosome 19q13.3. *Nucl. Acids Res.*, **19**, 1158.

Smeets, H., Bachinski, L.L., Coerwinkel, M. *et al.* (1990) A long-range restriction map of the human chromosome 19q13 region: Close physical linkage between CKMM and the ERCC1 and ERCC2 genes. *Am. J. Hum. Genet.*, **46**, 492–501.

Smeets, H.J., Hermens, R., Brunner, H.G. *et al.* (1991) Identification of variable simple sequence motifs (VSSMs) in 19q13.2-qter: Markers for the myotonic dystrophy locus. *Genomics*, **9**, 257–63.

Stallings, R.L., Olsen, E., Strauss, A.W. *et al.* (1988) Human creatine kinase genes on chromosomes 15 and 19, and proximity of the gene for the muscle form to the genes for apolipoprotein C2 and excision repair. *Am. J. Hum. Genet.*, **43**, 144–51.

Takemoto, Y., Miki, T., Nishikawa, K. *et al.* (1990) The locus of the Japanese myotonic dystrophy gene is also linked to D19S19 on the long arm of chromosome 19. *Genomics*, **6**, 195–6.

Thompson, L.H., Bachinski, L.L., Stallings, R.L. *et al.* (1989) Complementation of repair gene mutations on the hemizygous chromosome 9 in CHO: a third repair gene on human chromosome 19. *Genomics*, **5**, 670–9.

Tsilfidis, C., McKenzie, A.E., Shutler, G. *et al.* (1991a) A recombinant event which confirms the location of D19S51 distal to myotonic dystrophy. *Am. J. Hum. Genet*, **49**, 961–5.

Tsilfidis, C., Shutler, G., LeBlond, S. and Korneluk, R.G. (1991b) An SstI RFLP detected by the probe pKE2.1 (D19S116) localised to human chromosome 19q13.3. *Nucl. Acids Res.*, **19**, 1158.

Tsilfidis, C., Shutler, G., Mahadevan, M. and Korneluk, R.G. (1991c) A frequent HincII polymorphism identified by the human chromosome 19q13.3 probe pKEX0.8 (D19S118). *Nucl. Acids Res.*, **19**, 1157.

Vanier, T. (1960) Dystrophic myotonica in childhood. *Br. Med. J.*, **ii**, 1284–8.

Verkerk, A.J.M.H., Pieretti, M., Sutcliffe, J.S. *et al.* (1991) Identification of a gene (FMR-1) containing a CGG repeat coincident with a breakpoint cluster region exhibiting length variation in fragile X syndrome. *Cell*, **65**, 905–14.

Viskochil, D., Buchberg, A.M., Xu, G. *et al.* (1990) Deletions and a translocation interrupt a cloned gene at the neurofibromatosis type I locus. *Cell*, **62**, 187–92.

Wallace, M.R., Marchuk, D.A., Andersen, L.B. *et al.* (1990) Type I neurofibromatosis gene: Identification of a transcript disrupted in three NFI patients. *Science*, **249**, 181–6.

Walsh, F., Moore, S.E. and Dickson, J.G. (1988) Expression of membrane antigens in myotonic dystrophy. *J. Neurol. Neurosur. Psych.*, **51**, 136–8.

Waye, J.S. and Willard, H.F. (1987) Nucleotide sequence heterogeneity of alpha-satellite repetitive DNA: a survey of alphoid sequences from different chromosomes. *Nucl. Acids Res.*, **15**, 7549–69.

Whitehead, A., Solomon, E., Chambers, S. *et al.* (1982) Assignment of the structural gene for the third component of complement to chromosome 19. *Proc. Natl. Acad. Sci. USA*, **79**, 5021–5.

Wicking, C. and Williamson, R. (1991) From linked marker to gene. *Trends Genet.*, **7**, 288–93.

Williamson, R., Bowcock, A., Kidd, K. *et al.* (1990) Report of the DNA committee and catalogues of cloned and mapped genes and DNA polymorphisms, Human Gene Mapping 10.5: Update to the 10th international workshop on human gene mapping. *Cytogenet. Cell Genet.*, **55**, 457–778.

Ying-Hui, F., Kuhl, D., Pizzuti, A. *et al.* (1991) Variation of the CGG repeat at the fragile X site results in genetic instability: resolution of the Sherman paradox. *Cell*, **67**, 1–20.

Yu, S., Pritchard, M. Kremer, E. *et al.* (1991) Fragile X genotype characterised by an unstable genotype. *Science*, **252**, 1179–81.

5

Genetic mapping of facioscapulohumeral muscular dystrophy

CISCA WIJMENGA, MARTEN H. HOFKER, GEORGE W. PADBERG and RUNE R. FRANTS

5.1 INTRODUCTION

Facioscapulohumeral muscular dystrophy (FSHD) is a well described neuromuscular disorder with an autosomal dominant inheritance pattern. The molecular defect underlying FSHD is unknown and lack of biochemical and physiological markers precluded a reliable presymptomatic diagnosis. Therefore, chromosomal localization of the FSHD gene by linkage analysis seemed the most straightforward first step towards unravelling the FSHD gene defect. By application of the recently developed microsatellite markers, the FSHD gene was mapped to chromosome 4 within a period of six weeks, demonstrating the general utility of these efficient markers.

5.2 POSITIONAL CLONING

Defining the molecular defect of a particular genetic disorder requires the isolation and detailed characterization of the corresponding gene. Until recently a candidate gene could only be identified on the basis of biochemical characteristics of the gene product, the protein. The availability of amino acid sequence data allows the synthesis of oligonucleotide probes. These probes can be used for isolating the gene from libraries of cloned DNA. Also, purified protein can be used to raise specific antibodies, which can be employed to screen expression libraries. However, for many disease genes the (mutant) gene product has not been identified, delaying basic research and hampering the possibility of presymptomatic diagnosis.

The 'positional cloning' approach employs the chromosomal localization

Molecular and Cell Biology of Muscular Dystrophy
Edited by Terence Partridge
Published in 1993 by Chapman & Hall, London. ISBN 0 412 43440 7

Table 5.1 Autosomal single gene disorders mapped by linkage analysis. List of disorders for which linkage analysis was a first step towards cloning of the gene

Disease	*Chromosomal localization*
Acoustic neuroma, bilateral	22q
Adenomatous polyposis coli*	5q
Alzheimer disease (one form)*	21q
Aniridia*	2p, 11p
Ataxia-telangiectasia	11q
Batten disease	16p
Breast cancer (early-onset)	17q
Cataract, Marner type	16q
Charcot-Marie-Tooth disease*	1q, 17
Craniocynostosis	4p
Cystic fibrosis*	7q
Epidermolysis bullosa dystrophica	3p, 21
Facioscapulohumeral muscular dystrophy	4q
Friedreich ataxia	9q
Greig craniopolysyndactyly syndrome*	7p
Haemochromatosis	6p
von Hippel-Lindau syndrome	3p
Huntington disease	4p
Junvenile myoclonic epilepsy	6p
Langer-Giedion syndrome	8q
Malignant hyperthermia*	19q
Marfan syndrome*	15q
Multiple endocrine neoplasia, type I	11q
Multiple endocrine neoplasia, type II	10q
Myotonic dystrophy*	19q
Nail-patella syndrome	9q
Neurofibromatosis, type I*	17q
Paraganglioma (head and neck)	11q
Polycystic kidney disease	16p
Retinoblastoma*	13q
Spinal muscular atrophy, several types	5q
Spinocerebellar ataxia (one form)	6p
Torsion dystonia	9q
Tuberous sclerosis	9q, 11q
Usher syndrome	1q
Waardenburg syndrome*	2q
Wilms tumour*	11q
Wilson disease	13q
van der Woude lip-pit syndrome	1q

A few disease genes already have been cloned; they are indicated by an asterisk (*)

of a gene for its isolation. A very accurate knowledge of the position of the gene sometimes exists, when patients can be found where the disease is caused by a chromosomal translocation or a small deletion. For example, chromosomal abnormalities causing Duchenne muscular dystrophy (DMD) were instrumental in the cloning of the dystrophin gene (Chapter 1). More often, in the absence of cytogenetic clues, genes can be mapped by linkage analysis. The paucity of suitable genetic markers limited the success of this approach for a long time. This situation has changed dramatically following Kan and Dozy's (1978) discovery that restriction endonucleases can be used to reveal the genetic variation residing in arbitrary DNA sequences. Soon it was realized that an almost unlimited source of variation was available in the human genome (Jeffreys, 1979). It was proposed by Botstein and colleagues (1980) that these DNA polymorphisms could be used to create a complete linkage map of the human genome allowing almost every genetic trait to be mapped.

At present, the human gene map is well advanced and many gene disorders in man have already been mapped by linkage analysis (for a summary of mapped autosomal disorders see Table 5.1). The cloning of disease genes on the basis of their localization is beginning to be successful and has led to the identification of 13 genes, including many autosomal ones (Table 5.1). However, the task of starting with a chromosomal localization and arriving at the isolation of the gene itself remains a difficult one. Due to a worldwide interest, many technological advances have been made recently, including the preparation of genomic DNA libraries, *in situ* hybridization (Harper and Saunders, 1981), pulse field gel electrophoresis (Schwarz and Cantor, 1984), and cloning in yeast artificial chromosomes (Murray and Szostak, 1983; Burke *et al.*, 1987). These techniques were developed to facilitate the analysis of large genomic regions identified by linkage studies to harbour the gene of interest.

Another significant advance is the availability of enzymatic amplification of DNA by the polymerase chain reaction (PCR). Originally, the technique was developed for the rapid detection of point mutations in genes (Saiki *et al.*, 1985). Improvements in the PCR procedure (Saiki *et al.*, 1988) have led to many unforeseen applications. In particular, the development of a new kind of genetic markers, i.e. dinucleotide repeats of the $(CA)_n$ type (Weber, 1990a) was made possible by the PCR technology. These $(CA)_n$ markers are extremely useful for gene mapping projects due to their high polymorphism content and their technological advantages. This chapter will describe in detail how these and other markers have been applied to the successful localization of FSHD to chromosome 4qter, which is the first step towards isolating the FSHD gene itself.

5.2.1 Recombination

Exchange (recombination) of genetic material between two homologous chromosomes, known as crossing-over (Figure 5.1), is a fundamental feature of meiosis. This phenomenon can be used for the development of a genetic map of the human genome. The probability of a crossing-over event increases with the distance between two loci along the chromosome. The genetic distance is commonly expressed in centimorgans (cM). One cM corresponds to a crossing-over frequency of 1 in 100 meioses (Renwick, 1971) and corresponds to a physical distance of about 1 000 000 basepairs (bp). The total human genome measures approximately 3300 cM (Renwick, 1969).

However, the relationship between recombination fraction (genetic distance) and the distance in base pairs (physical distance) is not constant along the chromosomes. First, on average the recombination frequencies seem to be higher in females (Rao *et al.*, 1979), resulting in longer genetic maps for females (Donis-Keller *et al.*, 1987). Second, several subtelomeric chromosomal regions show increased recombination frequencies (Buetow *et al.*, 1991).

5.2.2 Linkage analysis

Linkage analysis involves the study of segregation of genetic markers (and disease loci in case of pathology) in families. When two loci are located far apart on the same chromosome or on separate chromosomes, they will recombine freely during meiosis. Such loci show 50% recombination and are therefore unlinked. When two loci tend to cosegregate, they are considered to be linked and are probably located on the same chromosome. In principle, linkage analysis can be used to map any disease gene of unknown location to a particular chromosomal region on the basis of cosegregation with a genetic marker of known chromosomal localization (Ott, 1985).

By following the segregation of a genetic marker within a family with a monogenic disorder, the probability of the disease and marker loci being

Figure 5.1 Crossing over between homologous chromosomes during meiosis. Initially, gametes resulting from these chromosomes would be AB or ab. After a recombination event, the gametes would be AB, Ab, aB or ab.

linked can be estimated. An example is shown in Figure 5.2 representing a part of a family segregating for an autosomal dominant disease gene. The affected parent carries two distinguishable alleles (A and B; = heterozygous) of a marker locus. Allele A appears to be co-segregating with the disease locus except in individual II-3 where a recombination event has occurred between the disease and the marker locus. Such a recombinant can only be identified when the parent is heterozygous for both the disease and the marker locus. As a consequence, many families will turn out to be uninformative because the inability to distinguish two different loci in the affected parent makes it impossible to follow the segregation through the pedigree. The frequency of recombination between the disease and the marker loci is a relative measure of the distance between them and is expressed by the symbol θ. The strength of evidence for linkage between the disease and the marker locus and an estimation for θ can be calculated by using statistical techniques (LOD score method, described below).

When the marker is contained within the disease gene, no recombinants will be observed. However, in most cases markers are arbitrary DNA segments mapping well outside the disease gene proper.

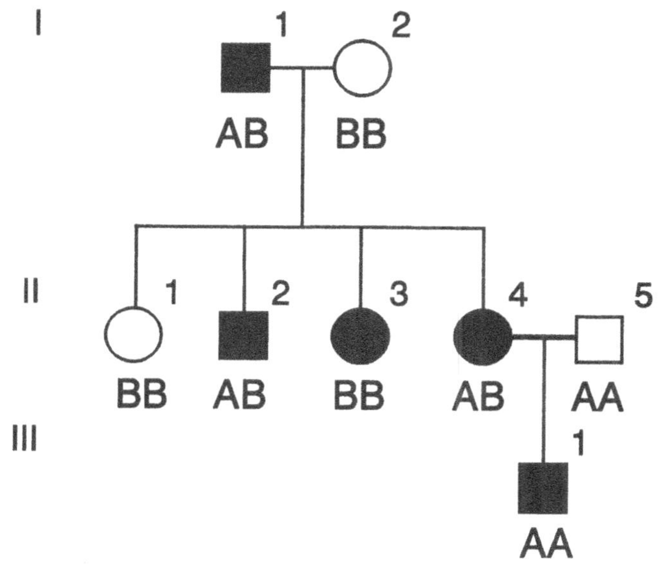

Figure 5.2 Part of a family segregating an autosomal dominant disease gene (indicated by filled symbols). The A allele of the grandfather (I-1) is inherited by the affected children and the grandchild, except in individual II-3 due to a recombination event between the disease and the marker loci.

5.2.3 DNA markers

A large set of highly polymorphic marker loci randomly distributed over the genome is the most important prerequisite for localizing new genes by linkage analysis. Until 1980 only protein markers were available and the human gene map contained about 400 loci (McKusick, 1991). However, the proportion of parents heterozygous for these markers was very low, and, as a consequence, almost no genetic data were available. Since the discovery of the DNA markers (Kan and Dozy, 1978) the number of assigned loci has increased rapidly. By 1990, almost 1900 genes and 4500 DNA segments (probes) had been mapped to specific chromosomal locations (McKusick 1991). About half of the DNA segments have been shown to be polymorphic and therefore they are useful as markers in linkage studies. Efficient mapping of disease genes will be possible when these DNA markers become available at intervals of 5–10 cM throughout each chromosome. For many chromosomes such a marker set has recently become available (Bowden *et al.*, 1989; Buetow *et al.*, 1990, 1991; Lathrop *et al.*, 1989; Petersen *et al.*, 1991).

The most commonly used DNA markers are the 'restriction fragment length polymorphisms' (RFLPs; Kan and Dozy, 1978). RFLPs are usually due to nucleotide changes that create or destroy a cleavage site for a specific restriction enzyme. The presence or absence of a restriction site causes a change in the length of a DNA fragment, which can be detected after digestion of genomic DNA with a particular restriction enzyme (Figure 5.3) and subsequent hybridization with a specific probe (Southern blotting; Southern, 1975).

The frequency of heterozygotes (= heterozygosity) for a marker determines its usefulness for linkage studies. This informativeness is directly related to the number and frequencies of different alleles present in the population. An alternative measure of informativeness is represented by the polymorphism information content (PIC) value (Botstein *et al.*, 1980). This PIC value reflects the proportion of informative parental matings that can be obtained with a particular marker. The majority of RFLPs consists of two-allele polymorphisms, giving a maximum heterozygosity of 50% and a maximum PIC value of 0.375, which indicates that, on average, about 63% of the potential pedigree information is being lost. Recently however, two new DNA marker systems have been discovered, which are even more polymorphic and may overcome this disadvantage of the two-allele RFLP systems. These markers are the minisatellite and microsatellite polymorphisms.

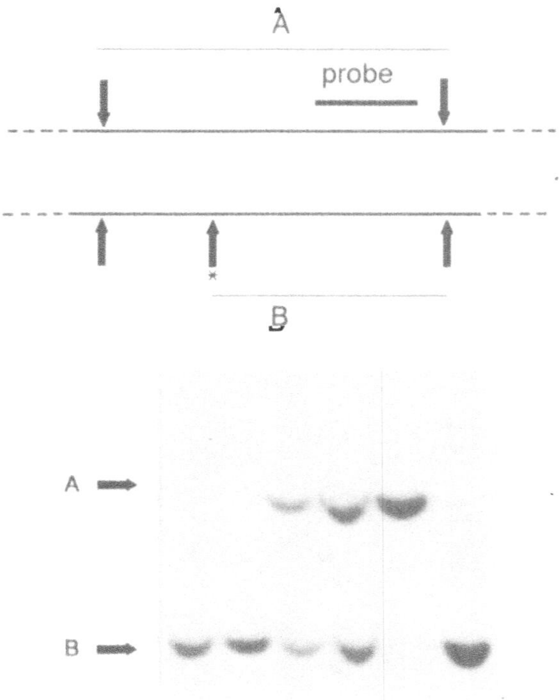

Figure 5.3 A restriction fragment length polymorphism (RFLP). One of the two chromosomes has an additional restriction site (arrow with an asterisk *) resulting in a short fragment B in comparison with fragment A. Southern blotting results from six different individuals are shown. The two different alleles (A and B) are indicated; only two out of these five individuals are heterozygous. For example, the far left lane shows an individual homozygous for the B allele (BB).

5.2.4 Minisatellite markers or variable number of tandem repeats (VNTR)

VNTR markers are characterized by restriction fragments containing relatively short tandemly repeated DNA segments. The unit length of each of these repeated DNA segments is in the order of 11 to 60 bp, and the number of copies of these segments in a tandem array ranges from 10 to more than 100 (Jeffreys *et al.*, 1985). The length of the restriction fragment that contains a VNTR is a function of the number of copies of the repeat unit. The VNTRs can easily be detected by Southern analysis (Figure 5.4). A marker system based on this type of polymorphism is highly informative in linkage studies (Nakamura *et al.*, 1987). Often, all individuals in a family will be heterozygous and all individual chromosomes can be followed throughout the entire family. The PIC values of these VNTRs can be as high

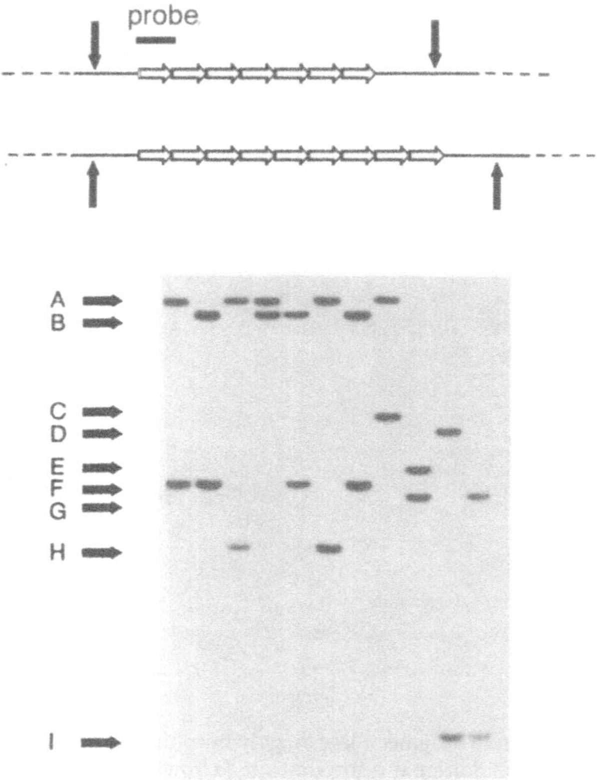

Figure 5.4 A 'variable number of tandem repeat' (VNTR) polymorphism. The tandemly repeated sequences are indicated by arrows. A VNTR polymorphism is caused by a variation in the number of repeat units. From 11 individuals, nine different alleles (A–I) of a VNTR fragment are seen on a Southern blot; all individuals are heterozygous. The far left lane shows an AF heterozygous individual.

as 0.9. Unfortunately, a disadvantage of the VNTRs is their apparent clustering to terminal regions of the chromosomes (Royle *et al.*, 1988). This clustering precludes reliable screening of a large fraction of the genome with VNTR markers.

5.2.5 Microsatellite markers or simple sequence polymorphisms

One of the most abundant human interspersed repetitive DNA families are the simple di-, tri- or tetranucleotide repeats. These DNA markers have collectively been named microsatellite polymorphisms (Litt and Luty, 1989) and have been proven to be an extremely abundant new source of highly polymorphic markers (Litt and Luty, 1989; Weber and May, 1989). Although all types of microsatellites are likely to be useful in linkage studies, blocks of $(CA)_n$-repeats are particularly abundant and almost all microsatellite polymorphisms known at this moment are based on these sequences.

Probably 50 000–100 000 copies of these blocks of $(CA)_n$-repeats exist in the human genome (Litt and Luty, 1989; Weber 1989). *In situ* hybridization of human metaphase chromosomes with a $(CA)_n$-oligonucleotide probe showed a uniform distribution over all chromosomes (Stallings *et al.*, 1990). Most of the $(CA)_n$-repeats are short in length: the number of repeats (n) may vary (approximately) from 5 to about 30. Sequences containing up to 10 CA-dinucleotides often turned out to have a low level of polymorphism, sequences with more than 16 CA-dinucleotides usually have high polymorphism frequencies (Weber 1990b). The average heterozygosity of $(CA)_n$-repeat markers commonly used for linkage analysis is 60–70% (PIC value about 0.6).

The stretches of dinucleotide repeats are relatively short and different alleles may differ in size by as little as one repeat unit (i.e. only 2 bp). Therefore these polymorphisms cannot easily be detected by Southern blotting analysis. The development of the polymerase chain reaction (PCR; Figure 5.5) has made the application of the $(CA)_n$-repeat markers for linkage analysis feasible. Amplification of particular blocks of $(CA)_n$-repeats by PCR will yield short fragments which can be resolved by electrophoresis on standard denaturing sequencing gels. By incorporating α^{32}P-dCTP during the PCR, the different alleles can be visualized directly by autoradiography of the gel (Figure 5.6). Up to four different loci can be amplified in a single PCR reaction ('multiplex' PCR; Figure 5.7). By using 'multiplex' PCR it becomes possible to screen a large part of the genome for linkage in a relatively short period of time.

Microsatellite markers are an important class of DNA markers because of their random distribution along the genome, the prospect of fast genotyping by PCR and the fact that only small amounts of DNA are required. Currently more than 500 $(CA)_n$-repeat markers are available and the development of these and other microsatellite markers such as tri- and tetranucleotide markers is in progress.

5.2.6 Two-point linkage analysis; LOD scores

The probability that a marker locus is located near a particular disease locus is determined by the use of statistical techniques, of which the LOD score method (Morton, 1955) is most commonly used. By this method, the probability that a given segregation pattern can be observed when two loci are linked with a given frequency of recombination is calculated. This

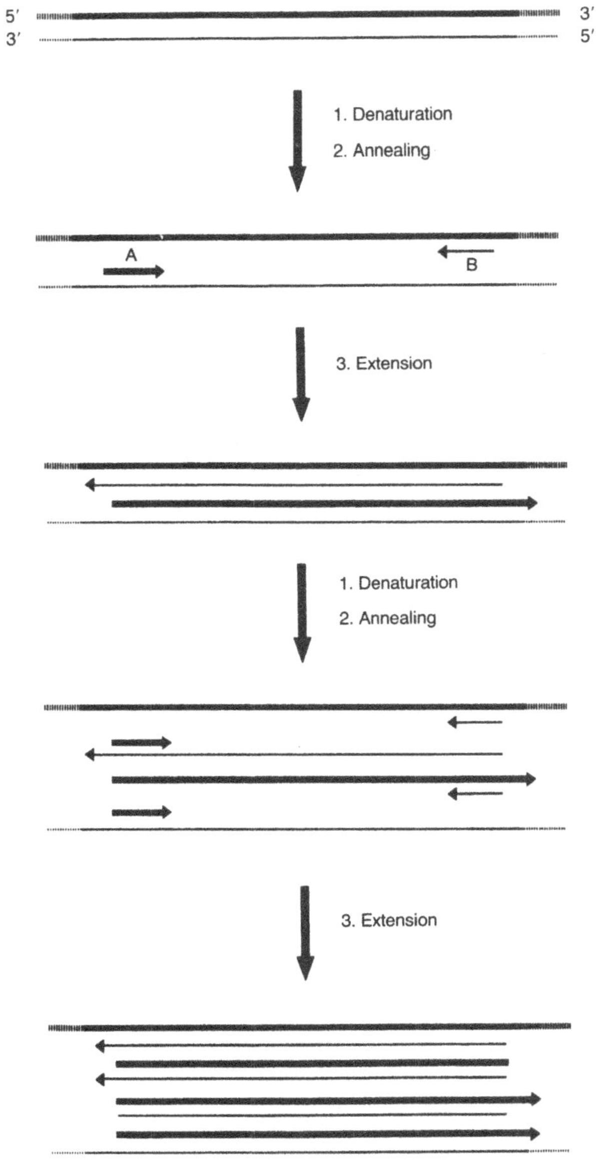

Figure 5.5 Principle of the polymerase chain reaction (PCR). It is necessary to synthesize two short oligonucleotide primers (A and B) complementary to known unique sequences flanking the region to be amplified. In every cycle of the PCR process, three steps are involved: 1. Denaturation of double-stranded DNA into single-stranded DNA; 2. Annealing of the primers to the single-stranded DNA; each primer will hybridize with its complementary sequence; 3. Synthesis of a new DNA-strand from the annealed primer by DNA polymerase, using the genomic DNA as template (= extension). The amount of material doubles each amplification cycle. After 30 cycles the DNA in the region between the two primers has been amplified about 1 billion-fold. By using the thermostable *Taq* polymerase the PCR process can be fully automated.

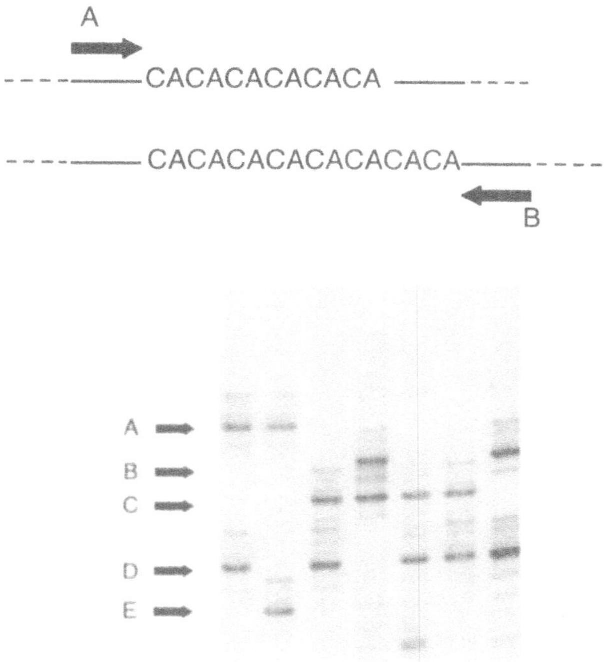

Figure 5.6 A microsatellite polymorphism based on a $(CA)_n$-repeat. The number of CA-dinucleotides varies between individuals in the population, but follows Mendelian inheritance. DNA from seven individuals has been amplified by PCR and resolved on a denaturing polyacrylamide gel. Five different alleles (A–E) can be distinguished; all individuals are heterozygous. For example, the individual in the far left lane is heterozygous AD. The different alleles are not seen as a single band but are represented by multiple bands for the following reasons: 1. Different mobilities of the two strands of the amplified fragment in the denaturing gel. 2. The *Taq* polymerase adds a single noncomplementary base to the 3′-end of a fraction of newly synthesized fragments. 3. *Taq* polymerase may skip (or add) repeats during the elongation step.

probability is compared with the probability observed in the absence of linkage. The LOD score (Z) is the ^{10}log of the likelihood ratio between these two probabilities. This ratio is computed for several recombination frequencies. A LOD score of +3 or higher (likelihood ratio of 1000 : 1) is taken as evidence for the presence of linkage, a LOD score of –2 (likelihood ratio 1 : 100 or lower) indicates that linkage is excluded, while a LOD score between –2 and +3 provides no conclusive evidence for or against linkage. LOD scores can easily be calculated by using computer programs such as LINKAGE (Lathrop *et al.*, 1984) and LIPED (Ott, 1974) which are suitable for a wide range of computers.

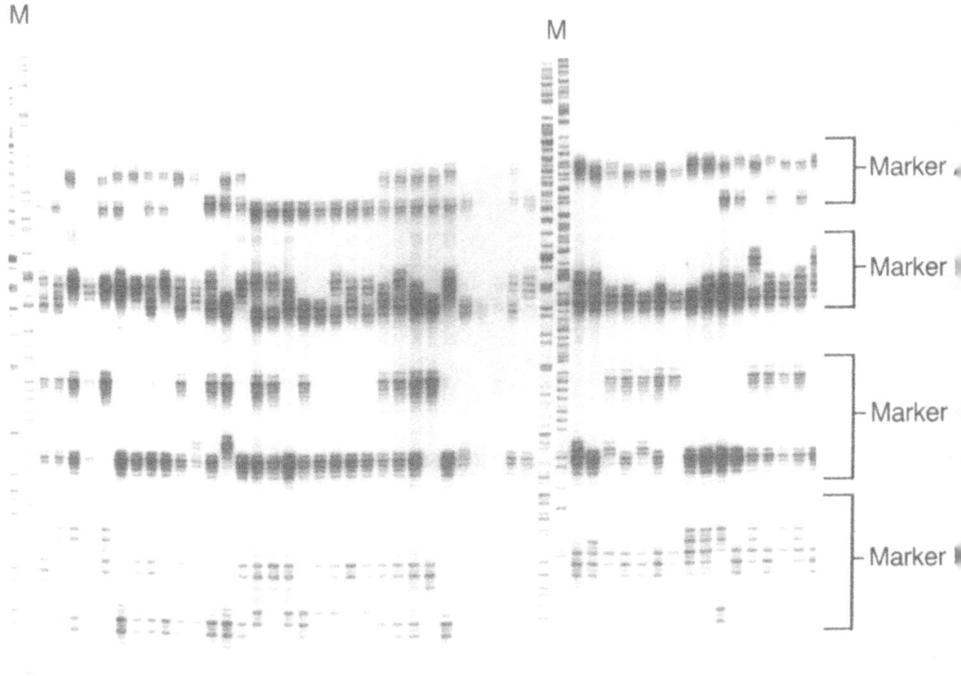

Figure 5.7 'Multiplex' PCR of DNA of 48 individuals with four different $(CA)_n$-repeat markers (A–D) and resolution by electrophoresis on a denaturing polyacrylamide gel. Per amplification only 30 ng of DNA per individual has been used, compared with 10 μg of DNA required for Southern blotting analysis. In this experiment 192 genotypes could be analysed. The length of the simultaneously amplified fragments has to be chosen in such a way that they will not co-migrate. DNA sequencing ladders were used for determination of allele sizes (indicated by M).

The ideal circumstances for detecting linkage are the following:

1. the disease shows complete penetrance, implying that all carriers of the mutant gene are affected;
2. the marker locus is heterozygous in the key patients, so the segregation of both chromosomes can be followed within the family (as in Figure 5.2);

3. for single gene disorders the families should be large and should contain at least three generations.

These conditions favour the detection of linkage within an individual family.

Most families used in linkage studies are not large enough to generate LOD scores of +3 or higher by themselves. Therefore, LOD scores from different families with the same genetic disorder have to be added to get significant evidence for or against linkage. However, in the case of genetic heterogeneity (mutations at different loci) a pooled data set may give rise to an incorrect LOD score, severely reducing the possibility of detecting linkage. Therefore, the use of large families for linkage analysis would be the strategy of choice.

If no linkage is found between a marker locus and a disease locus, a portion of the region surrounding the marker locus can be excluded as a likely site for the disease gene. The excluded area (in cM) is given by the recombination frequency at which the LOD score is –2 or less. All exclusion data together can be combined into a comparative map of the areas in the genome where the gene of interest is not located. These exclusion maps can help to focus on those areas in the genome which still have to be investigated. Exclusion maps can easily be generated by using the computer program EXCLUDE, which uses information against linkage to some loci to support linkage to other loci (Edwards, 1987). However, these exclusion maps are only reliable for those chromosomal regions where a marker to marker linkage map of the genome exists.

5.2.7 Multipoint linkage analysis

After obtaining the tentative map position of a disease gene, the evidence for linkage will be stronger when the disease gene can be linked to additional genetic markers residing in the same chromosomal region. Consequently sections of families uninformative for one marker can become informative for another marker located in the same area. To make optimal use of these data, three or more linked loci can be analysed simultaneously by multipoint linkage analysis. This method is more efficient than two-point linkage analysis, because multipoint linkage analysis allows the cumulative information from the different genetic markers.

Multipoint linkage analysis is extremely useful for constructing genetic maps; the order of the different closely linked genetic markers can be established as well as the location of the disease gene with respect to these markers. The evidence in favour of a particular gene order will be estimated by comparing the probability of a particular order with the probability of an alternative one. Repeating this procedure for all different gene orders will give the order with the largest likelihood.

The simultaneous analysis of more than two loci is, in general, a complex matter that can be resolved by using available computer programs (Lathrop *et al.*, 1985).

5.3 FACIOSCAPULOHUMERAL MUSCULAR DYSTROPHY: CLINICAL FEATURES

5.3.1 Symptoms and signs

FSHD is a neuromuscular disorder which was described in detail for the first time in 1884 by the two French physicians, Landouzy and Dejerine. The onset of the disease is characterized by weakness and atrophy of the facial and shoulder-girdle muscles (Figure 5.8). Where there is progression of the disease, the weakness and atrophy spreads to the foot extensor, upper arm and pelvic-girdle muscles subsequently (Padberg, 1982; Munsat, 1986). Often there is an asymmetrical involvement of muscles in the face as well as in the shoulder girdle and extremities (Padberg, 1982).

The intrafamilial and interfamilial expression of the disease is quite variable. The clinical course and rate of progression of the disease may also vary considerably and ranges from almost asymptomatic patients (30% of cases) (Padberg, 1982) to patients who will be wheelchair bound (about 10%) (Lunt 1989).

In addition to the muscle weakness, retinal vasculopathy (Fitzsimons *et al.*, 1987; Padberg *et al.*, 1990) and neurosensory hearing loss have been repeatedly observed in approximately 50% of the cases (Voit *et al.*, 1986; Brouwer *et al.*, 1991).

5.3.2 Laboratory tests

The diagnosis of FSHD requires a distinctive set of standard laboratory examinations. In the majority of cases the serum creatine kinase levels are mildly raised and rarely exceed five times the upper limit of normal. Electromyography usually shows action potentials of low amplitude and short duration while motor and sensory nerve conduction velocities are within the normal range. The muscle biopsy, when taken from an involved muscle, will show dystrophic changes often with ‘moth-eaten’ fibres and occasionally mononuclear infiltrates. Large group atrophy is not an aspect of the disease, but small angular fibres may be found. These fibres are due to muscle regeneration and express fetal myosin; they are not indicative of neurogenic changes.

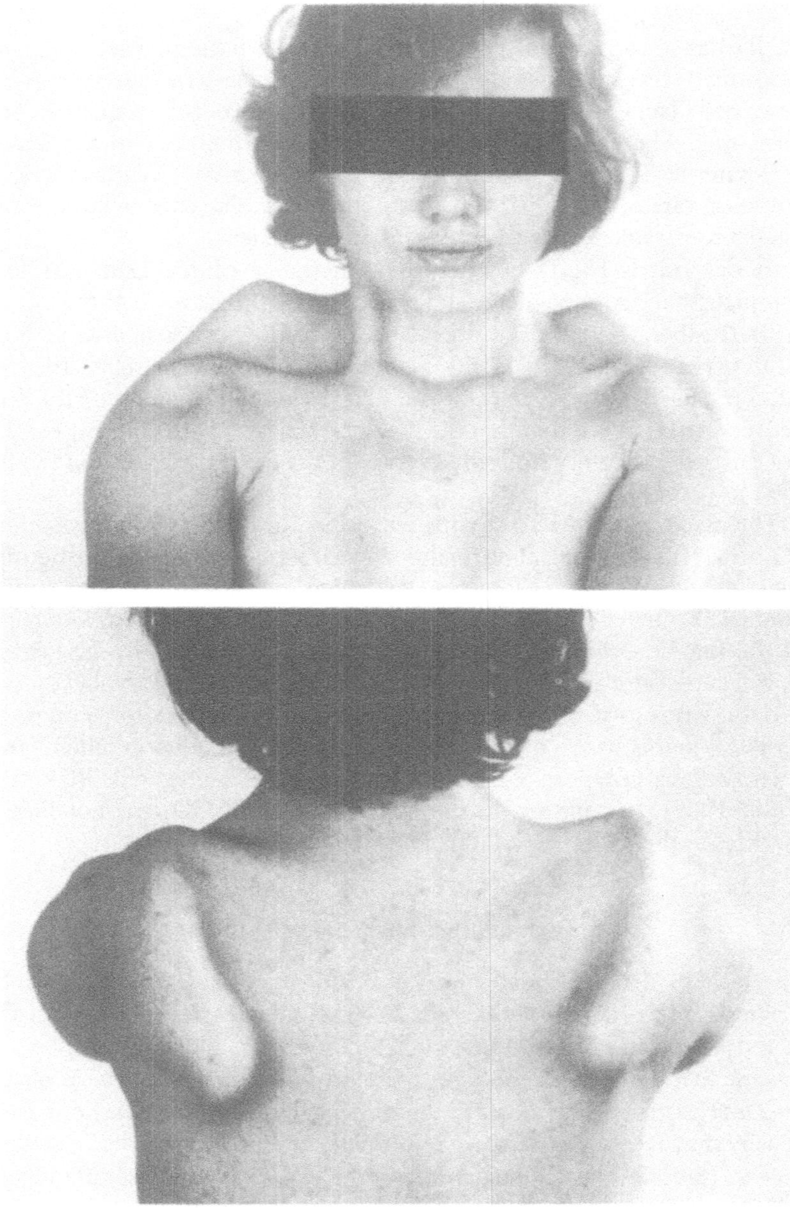

Figure 5.8 Facioscapulohumeral muscular dystrophy. Extreme winging of both scapulae due to loss of scapular fixation. The high raise of the scapulae and the partial atrophy of the deltoid muscles are characteristic of the disease. At this extreme stage asymmetry is hardly visible.

5.3.3 Inheritance

FSHD has an autosomal dominant inheritance pattern. Families with an autosomal recessive or X-linked inheritance pattern have never been described. Isolated or sporadic cases, however, occur frequently. These cases might be caused by a low penetrance of the gene in the ancestry, non-paternity, a new mutation, or environmental circumstances. The mutation rate of the FSHD gene is still unknown due to many uncertainties related to prevalence, penetrance and ascertainment.

In the Dutch FSHD families the penetrance of the gene was almost complete, but age dependent, and was estimated to be 95% at the age of 20 years (Padberg, 1982). The prevalence of FSHD was estimated to be 1 in 20 000 (Padberg, 1982) in the Netherlands. In these patients the age at onset was usually at the end of the first or in the beginning of the second decade. It was also found that the life expectancy of patients with FSHD did not differ significantly from the average of the general population (Padberg, 1982).

The association of FSHD with sensorineural hearing loss (Fitzsimons *et al.*, 1987) and retinal abnormalities (Voit *et al.*, 1986) could be due to genetic heterogeneity. However, extended studies in the Dutch FSHD families (Brouwer *et al.*, 1991; Padberg *et al.*, 1990) show that in almost all of the families, some of the FSHD patients are affected by sensorineural hearing loss and retinal abnormalities. The most likely explanation is a variable expression of a particular gene defect. Another example of possible genetic heterogeneity could be the finding of two families in which familial adenomatous polyposis coli (FAP) and FSHD cosegregated (Blake *et al.*, 1988, 1989). In contrast, in the Dutch families FSHD was not linked to markers in the FAP region (Wijmenga *et al.*, 1990a).

5.3.4 Other biochemical studies

The lack of detailed biochemical and morphological studies in FSHD is striking. Freeze fracture studies demonstrate a decrease of orthogonal arrays in muscle plasma membranes of FSHD patients. However, this is not a specific feature of FSHD because a similar phenomenon has been observed in DMD patients. This observation of decreased intramembranous particle density may indicate a decrease in protein content and metabolic activity in skeletal muscle plasma membranes of FSHD patients (Schotland *et al.*, 1981).

The retinal vasculopathy does not suggest an obvious relationship with the muscle disease, as the retinal capillaries showing telangiectasia and microaneurysms do not contain contractile elements (Fitzsimons *et al.*, 1987). In a similar manner the hearing loss has led only to speculations about a possible relationship with the muscle disease on the basis of actin filaments in the cochlea (Voit *et al.*, 1986).

All of these incomplete results and speculations underline the need to gain more insight in the molecular basis of FSHD.

5.4 FSHD: LINKAGE ANALYSIS AND 'POSITIONAL CLONING'

The gene responsible for FSHD could not be identified by a 'candidate gene' approach and therefore linkage analysis with random markers was the only alternative to find the gene. Because of its autosomal dominant inheritance pattern and almost complete penetrance, FSHD was a suitable target for gene mapping through linkage analysis.

In 1980 the linkage studies were started by collecting blood from ten Dutch multigenerational FSHD families showing an autosomal dominant inheritance pattern. The families contained 69 affected and 58 non-affected sibs and 25 spouses. All family members underwent a physical examination by the same neurologist. Due to the age-dependent penetrance, young asymptomatic persons cannot be excluded as carriers of the FSHD gene and therefore our study includes only affected individuals or individuals older than 20 years. At least one patient from each family had a muscle biopsy and electromyography to establish an accurate diagnosis. All sibs underwent audiometry to look for the hearing loss associated with this condition, and in every family at least the proband was examined by fluorescein angiography of retinal vessels. The findings of this examination strongly suggested that retinal vasculopathy and FSHD are associated conditions in these families.

Initially a total of 35 different blood group markers, enzyme isoforms and protein antigens were tested in the Dutch FSHD families. None of these markers turned out to be linked to FSHD. Weak positive LOD scores were found for the immunoglobin heavy chain gene cluster (IGH) on chromosome 14q32 (Padberg *et al.*, 1984). This possible linkage was supported by a maximum LOD score of 1.428 at a recombination fraction of 0.2 between IGH and FSHD. Subsequent study with the α_1-antitrypsin gene, located in the same chromosomal region, could not confirm the mapping of FSHD to chromosome 14q (Berriche *et al.*, 1988; Padberg *et al.*, 1988).

Once RFLPs became generally applicable for linkage analysis, these markers were used for more extensive studies in the Dutch FSHD families. In order to facilitate the search for the FSHD gene an international effort was initiated in 1988 by several groups (Lunt, 1989). Results were collected every three months and agreements were made about the markers to be tested. This international consortium produced a first exclusion map in 1988. More than 80% of the genome was excluded as a possible site for the FSHD gene by using 57 markers (Sarfarazi *et al.*, 1989). These markers mapped to all chromosomes, except the chromosomes 3, 10 and 15. The search for the FSHD gene proceeded rapidly and at the beginning of 1990 a total of 225 markers were tested and almost 95% of the genome was excluded by using the computer program EXCLUDE (Sarfarazi, personal communication).

However, this estimate should be viewed with caution. Overestimation of the actual excluded area is possible for several reasons.

a) The computer program EXCLUDE treats all imported loci as being unlinked.
b) By using EXCLUDE it was assumed that the disease gene did not map within an interval of 10 cM from a particular marker locus, even when the LOD scores at 10 cM were not significantly negative (i.e. below –2, Sarfarazi *et al.*, 1989).
c) EXCLUDE requires the location of the markers to be imported into the program. This approach is justified for some precisely mapped marker loci. However, the majority of the markers are not mapped accurately enough, leading to the exclusion of inappropriate regions.
d) Since no complete (telomere–telomere) linkage maps, which include telomere derived markers, are available, the actual length of most chromosomes cannot be estimated correctly. In addition, regions close to the telomeres show high recombination rates leading to an underestimation of the number of required probes.

Despite these drawbacks, the computer program EXCLUDE helps to focus on regions that are under-represented in the gene search, which proved to be successful in the case of Freidreich's ataxia (Chamberlain *et al.*, 1988) and von Recklinghausen neurofibromatosis (Sarfarazi *et al.*, 1987). However, since we decided to turn to a more convenient marker system (described below), we continued our search for the FSHD gene in a random fashion and disregarded the exclusion data.

5.4.1 Localization of the FSHD gene to chromosome 4 using the microsatellite marker Mfd22

To overcome a few of the problems described above, a new kind of highly polymorphic marker was applied in our effort to map the FSHD gene: microsatellite polymorphisms of the $(CA)_n$-type. These loci were not well mapped to chromosomal regions at that time, precluding the construction of an exclusion map. However, $(CA)_n$-repeats were attractive to use, because of their high information content and because they seem to be located in areas that were poorly covered by RFLPs and VNTRs at that time. Their high variability allowed the use of only a few Dutch kindreds, since in the case of linkage, even a single family would give a significant LOD score. In this case, even if FSHD were a genetically heterogeneous entity, the detection of linkage would be possible.

A large number of $(CA)_n$-repeat markers had just been isolated at that time (Weber and May, 1989). About 80 different $(CA)_n$-repeats were available, randomly distributed across all 22 autosomes. Within just six weeks, two key families were analysed with a total of 60 markers (i.e. more than 6000 genotypes). This amount of data could be produced by a single person, because the workload could be reduced significantly by amplifying four different $(CA)_n$-repeats simultaneously ('multiplex' PCR; Figure 5.7). The high proportion of informative FSHD carriers yielded a total 'excluded' area of approximately 1500 cM, compared with the total length of the human genome of 3300 cM. Thus approximately 47.5% had been screened for location of the gene. Fortunately, one of the microsatellite markers, Mfd22, showed linkage with the FSHD gene at a distance of 13 cM (Table 5.2) with odds in favour of linkage of more than 10^6 to 1 (Wijmenga *et al.*, 1990). The microsatellite marker Mfd22 (corresponding to the locus *D4S171*) was assigned to chromosome 4 by using a somatic cell hybrid panel (Weber and May, 1990). A more detailed regional localization required additional experiments.

Table 5.2 Two-point LOD scores for linkage between *D4S171* and FSHD

	Recombination fraction				
Family	0.10	0.15	0.20	0.25	0.30
1	1.21	1.78	1.95	1.87	1.62
2	0.21	0.29	0.32	0.32	0.29
3	0.47	0.43	0.38	0.32	0.27
5	0.58	0.49	0.40	0.32	0.24
7	0.94	0.82	0.69	0.56	0.42
9	0.42	0.32	0.28	0.15	0.10
12	0.15	0.12	0.09	0.06	0.04
14	1.16	1.02	0.87	0.72	0.55
19	0.24	0.39	0.46	0.47	0.44
20	1.86	1.62	1.36	1.09	0.81
21	0.11	0.15	0.15	0.14	0.12
Total	7.35	7.43	6.95	6.02	4.90

This table shows LOD scores at various recombination fractions for FSHD and the microsatellite marker Mfd22 (*D4S171*). The maximum total LOD score is 7.43 at a recombination fraction of 0.15. As some additional families (12 and 19) and individuals have been typed for Mfd22, the LOD scores for this marker are slightly higher than those published before (Wijmenga *et al.*, 1990b)

5.4.2 FSHD maps to the distal long arm of chromosome 4

In order to get additional probes in the vicinity of the FSHD gene, a series of chromosome 4 markers were investigated. A more closely linked marker could be identified: the VNTR marker pH30, corresponding to the locus *D4S139*, which had very recently been isolated (Milner *et al.*, 1989). The regional localization of this marker had not been established with certainty, although some evidence had been obtained assigning *D4S139* to the distal portion of the long arm of chromosome 4 (Milner *et al.*, 1989). In our laboratory *D4S139* was mapped by fluorescent *in situ* hybridization to the most distal part of the long arm of chromosome 4 (Figure 5.9; Wijmenga *et*

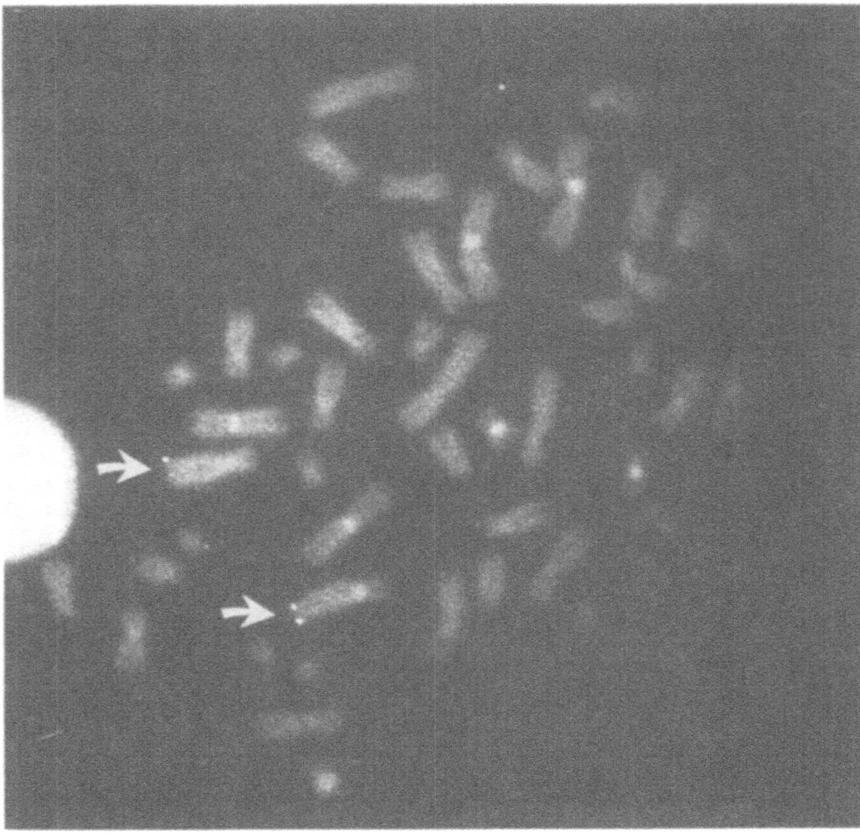

Figure 5.9 Fluorescent *in situ* hybridization of pH30 (locus *D4S139*). The marker maps to the terminal region, q35-qter, of the long arm of chromosome 4 (Wijmenga *et al.*, 1991).

al., 1991). Accordingly, the most likely position of the FSHD gene is 4q35-qter.

After localization of the FSHD gene to the subtelomeric region of 4q, it became clear that only a very few loci had been mapped to this region. The most distal polymorphic locus was the MNSs blood group (glycophorins A and B) on 4q28-q31 (Buetow *et al.*, 1991). The factor XI gene (locus F11) had been localized to 4q35 by *in situ* hybridization (Kato *et al.*, 1989); however no polymorphisms were described. The paucity of established markers at the subtelomeric region of chromosome 4q explains why the search for the gene took such a long time.

5.4.3 Multipoint linkage; the FSHD consortium

Several additional markers became available forming a linkage group consisting of five polymorphic loci (*D4S171, F11, D4S187, D4S163* and *D4S139*) and spanning about 20 cM (Mills *et al.*, 1992). This linkage group is currently being investigated by typing 55 large pedigrees made available through CEPH (Centre d'Etude Polymorphisme Humaine). Thorough genetic analysis of the 4q markers within these reference families is necessary to define an accurately established order of the loci as well as the orientation of the linkage group with respect to the telomere. Preliminary data suggest the following locus order: *Cen-D4S171-F11-D4S187-D4S163-D4S139-Tel* (Mills *et al.*, 1992; Weiffenbach *et al.*, 1992).

A combined effort of the FSHD consortium was initiated to analyse the homogeneity of the family material and to determine the exact location of FSHD gene within this linkage group. The combined data obtained from 65 families indicate that FSHD is a genetically homogenous disorder (Sarfarazi *et al.*, 1992) and strongly suggest a location of the FSHD gene approximately 5 cM distal to *D4S139* (Figure 5.10) (Sarfarazi *et al.*, 1992). A closely linked flanking marker will be required to confirm the latter result. In addition, new flanking markers are essential for reliable presymptomatic diagnosis of FSHD.

At this stage the mapping of FSHD is being carried out by multipoint linkage analysis and the position of FSHD depends on the detection of recombination events between FSHD and closely linked markers (Figure 5.11).

5.4.4 The future

It will still be a formidable, but not impossible, task to obtain the actual coding sequence of the FSHD gene and to discover the mutation responsible for FSHD. Currently, no obvious clues are available that will allow selection

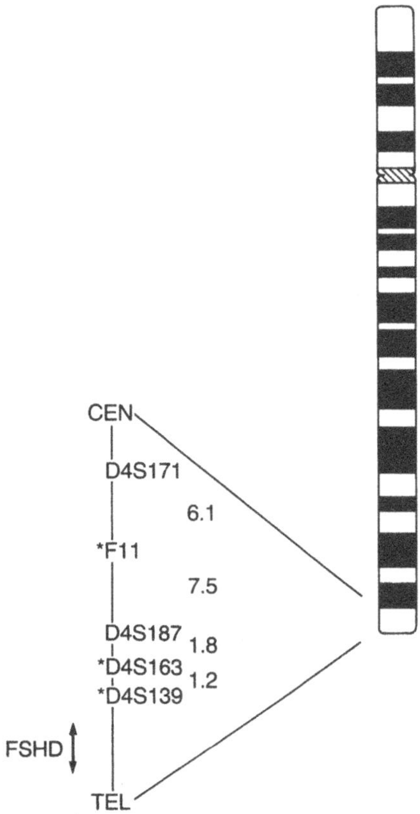

Figure 5.10 Map of the FSHD region on 4q35. On the basis of recombination data, the most likely position of the FSHD gene is approximately 5 cM distal to *D4S139*. The recombination frequency between consecutive loci (given in cM) is indicated in the figure. Loci assigned to 4q35 by *in situ* hybridization are indicated by an asterisk (*).

of a specific strategy to find the FSHD gene. Sometimes clues are provided by the nature of the gene defect which aid the identification of the disease gene.

When the gene defect is caused by large chromosomal rearrangements, the search for the gene can be straightforward. Milestones of this approach are the discovery of the DMD gene (Chapter 1), the retinoblastoma (Rb) gene (Friend *et al.*, 1986), and more recently neurofibromatosis type I (NFI) (Viskochil *et al.*, 1990) and the gene for familial adenomatous polyposis (FAP) (Kinzler *et al.*, 1991). So far, no gene rearrangements have been found for FSHD, although one interesting chromosomal translocation has been

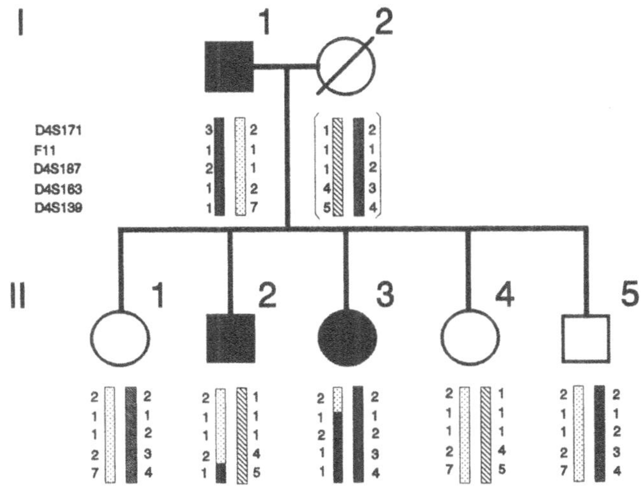

Figure 5.11 A more definite location of the FSHD gene with respect to the markers can be given by recombination events between the disease gene and one or more of the marker loci. This figure shows a part of an FSHD pedigree analysed with the five different markers of the linkage group. In this family the black chromosome of individual I-1 is segregating with FSHD. However, in the individuals II-2 and II-3 recombination events between FSHD and the marker loci narrow down the area where the FSHD gene will be located. Since the black chromosome will be the chromosome carrying the FSHD gene, FSHD will be distal to *D4S163*. These recombinants also can help in the assignment of order to the markers. The recombination in individual II-2 separates *D4S139* from the other markers. The recombination in individual II-3 separates *D4S171* from *D4S187, D4S163* and *D4S139*. However, it is impossible to find out if crossing-over occurred between *D4S171* and *F11* or between *F11* and *D4S187*. These two recombinants already provided a tentative ordering of the markers: *D4S171(F11)–(F11)D4S187/D4S163–D4S139*.

described. A female patient was identified carrying a translocation t(X;4)(p21;q35) (Bodrug *et al.*, 1990) and displaying an early onset muscular dystrophy. The X-chromosomal breakpoint involved the DMD-gene. The 4q breakpoint was situated in the subtelomeric region 4q35-qter. The patient showed some features of DMD, but in addition also some symptoms of FSHD. The simplest explanation for the phenotype would be that this patient displays a combination of both DMD and FSHD, whereby the translocation had disrupted the DMD gene on Xp21 as well as the FSHD gene on 4q35. Unfortunately, present linkage data map a probe derived from the junction region (pSBU10; locus *D4S187*) within the *D4S171*-*D4S139* interval (Figure 5.9). The combined consortium data strongly suggest that FSHD maps outside this interval (Sarfarazi *et al.*, 1992). However, the results from our laboratory cannot exclude FSHD mapping in the vicinity of *D4S187* (Wijmenga *et al.*, 1992). Although the role of this translocation t(X;4) remains uncertain, it will be valuable for mapping new probes in the 4q35-4qter region in order to saturate the subtelomeric region with markers.

In the absence of DNA rearrangements, a more difficult approach may be required, one that was instrumental in the discovery of the cystic fibrosis (CF) gene (Kerem *et al.*, 1989). This strategy makes use of linkage disequilibrium (see Vogel and Motulsky, 1986) between the disease and marker loci. In the case of linkage disequilibrium a particular allele of a marker and the disease mutation are associated. Such an allelic association can emerge when the disease has a neglectable new mutation frequency and recombination events between the marker and the disease locus are extremely rare. A stronger linkage disequilibrium usually corresponds to a shorter physical distance. Presently, the mutation rate for FSHD is unknown. Also, none of the markers seem sufficiently close to show linkage disequilibrium, because all markers already show recombination in the relatively small sample studied so far.

Because neither chromosomal rearrangements nor linkage disequilibrium can be used yet to isolate the FSHD gene, the isolation of additional markers is essential. These markers are required to improve the resolution of the genetic map and to create a long range restriction map (Van Ommen *et al.*, 1986). Recombinants will confine the chromosomal region that harbours the gene. The long range restriction map will define the amount of DNA that needs to be examined. It is possible that a sufficient density of markers may be available but none of the conditions described above are fulfilled. In this case the gene search may require exhaustive analysis of a large amount of DNA for the presence of coding sequences and their potential involvement in FSHD.

Current genetic analysis and screening of familial and sporadic FSHD cases for chromosomal rearrangements may give important hints as to how to overcome the last difficult step of the positional cloning pathway: isolation of the FSHD gene itself. When this aim has been reached, the molecular basis of FSHD will become attainable for research, allowing reliable presymptomatic diagnosis, evaluation of intriguing questions on genetic heterogeneity, estimation of mutation frequencies and a better general insight in the aetiology of FSHD.

ACKNOWLEDGEMENTS

We would like to thank Dr L.A. Sandkuijl for his invaluable support. The European Neuromuscular Center and the Vereniging Spierziekten Nederland are gratefully acknowledged for stimulating the family studies. C.W. is supported by the Prinses Beatrix Fonds (88-2837). M.H.H. receives a fellowship of the Royal Netherlands Academy of Arts and Sciences.

REFERENCES

Berriche, S., Guettair, N., Intrator, S. *et al.* (1988) Le locus de susceptibilité à la dystrophie musculaire facio-scapulo-humérale n'est past lié à celui des chaînes lourdes d'immunoglobulines. *Bio-sciences*, 7, 90–2.

Blake, D., Gilliam, T.C., Warburton, D. and Rowland, L.P. (1988) Possible clue for chromosomal assignment of the gene for facioscapulohumeral muscular dystrophy: a family with polyposis. *Ann. Neurol.*, **1**, 178.

Blake, D., Brown, R., Gilliam, T.C. *et al.* (1989) The second family with facioscapulohumeral muscular dystrophy and familial polyposis coli. *Neurol.*, **39** (S1), 404.

Bodrug, S.E., Roberson, J.R., Weiss, L. *et al.* (1990) Prenatal identification of a girl with a t(X;4) (p21;35) translocation: molecular characterisation, parental origin, and association with muscular dystrophy. *J. Med. Genet.*, **27**, 426–32.

Botstein, D., White, R., Skolnick, M. and Davis, R. (1980) Construction of a genetic linkage map in man using restriction fragment length polymorphisms. *Am. J. Hum. Genet.*, **32**, 314–31.

Bowden, N.W., Gravius, T.C., Green, P. *et al.* (1989) A genetic linkage map of 32 loci on human chromosome 10. *Genomics*, **5**, 718–26.

Brouwer, O.F., Ruys, C.J.M., Brand, R. *et al.* (1991) Hearing loss in facioscapulohumeral muscular dystrophy. *J. Neurol.*, **56S**, 237.

Buetow, K.H., Nishimura, D., Green, P. *et al.* (1990) A detailed multipoint gene map of chromosome 1q. *Genomics*, **8**, 13–21.

Buetow, K.H., Shiang, R., Yang, P. *et al.* (1991) A detailed multipoint map of human chromosome 4 provides evidence for linkage heterogeneity and position specific recombination rates. *Am. J. Hum. Genet.*, **48**, 911–25.

Burke, D.T., Carle, D.T. and Olson, M.V. (1987) Cloning of large segments of exogenous DNA into yeast by means of artificial chromosome vectors. *Science*, **236**, 806–12.

Chamberlain, S., Shaw, J., Rowland, A. *et al.* (1988) Mapping of mutation causing Friedreich's ataxia to human chromosome 9. *Nature*, **334**, 248–50.

Donis-Keller, H., Green, P., Helms, C. *et al.* (1987) A genetic linkage map of the human genome. *Cell*, **51**, 319–37.

Edwards, J.H. (1987) Exclusion mapping. *J. Med. Genet.*, **24**, 539–43.

Fitzsimons, R.B., Gurwin, E.B. and Bird, A.C. (1987) Retinal vascular abnormalities in facioscapulohumeral muscular dystrophy. *Brain*, **110**, 631–48.

Friend, S.H., Bernards, R., Rogelj, S. *et al.* (1986) A human DNA segment with properties of the gene that predisposes to retinoblastoma and osteosarcoma. *Nature*, **323**, 643–6.

Harper, M.A. and Saunders, G.F. (1981) Localization of single copy DNA sequences on G-banded human chromosomes by *in situ* hybridization. *Chromosoma*, **83**, 431–9.

Jeffreys, A.J. (1979) DNA sequence variants in the G_γ-, A_γ-, δ- and β-globin genes of man. *Cell*, **18**, 1–10.

Jeffreys, A.J., Wilson, V. and Thein, S.L. (1985) Hypervariable 'minisatellite' regions in human DNA. *Nature*, **314**, 67–73.

Kato, A., Asakai, R., Davie, E.W. and Aoki, N. (1989) Factor XI gene (F11) is located on the distal end of the long arm of human chromosome 4. *Cytogenet. Cell Genet.*, **52**, 77–8.

Kan, Y.W. and Dozy, A.M. (1978) Polymorphisms of DNA sequence adjacent to human β-globin structural gene: relation to sickle cell mutation. *Proc. Natl. Acad. Sci. USA*, **75**, 5631–5.

Kerem, B.-S., Rommens, J.M., Buchanan, J.A. *et al.* (1989) Identification of the cystic fibrosis gene: genetic analysis. *Science*, **245**, 1073–80.

Kinzler, K.W., Nilbert, M.C., Su, L.-K. *et al.* (1991) Identification of FAP locus genes from chromosome 5q21. *Science*, **253**, 661–5.

Landouzy, L. and Dejerine, J. (1984) De la myopathy atrophique progressive (myopathy héréditaire débutant dans l'efance, par la face sans altération du système nerveux). *C. R. Seances Acad. Sci.* (Paris), **98**, 53–5.

Lathrop, G.M., Lalouel, J.M., Julier, C. and Ott, J. (1984) Strategies for multilocus linkage analysis in humans. *Proc. Natl. Acad. Sci. USA*, **81**, 3443–6.

Lathrop, G.M., Lalouel, J.M., Julier, C. and Ott, J. (1985) Multilocus linkage analysis in humans: detection of linkage and estimation of recombination. *Am. J. Hum. Genet.*, **37**, 482–98.

Lathrop, G.M., O'Connell, P., Leppert, M. *et al.* (1989) Twenty-five loci form a continuous linkage map of markers for human chromosome 7. *Genomics*, **5**, 866–73.

Litt, M. and Luty, J.A. (1989) A hypervariable microsatellite revealed by in vitro amplification of a dinucleotide repeat within the cardiac muscle actin gene. *Am. J. Hum. Genet.*, **44**, 397–401.

Lunt, P.W. (1989) A workshop on facioscapulohumeral (Landouzy-Dejerine) disease. *J. Med. Genet.*, **26**, 535–7.

McKusick, V.A. (1991) Current trends in mapping human genes. *FASEB J*, **5**, 12–20.

Milner, E.C.B., Lotshaw, C.L., Willems, van Dijk K. *et al.* (1989) Isolation and mapping of a polymorphic DNA sequence pH30 on chromosome 4 [HGM provisional no. D4S139]. *Nucleic Acids Res.*, **17**, 4002.

Mills, K.A., Mathews, K.D., Xu, Y. *et al.* (1992) Genetic and physical mapping on chromosome 4 narrows the localization of the gene for FSHD. *Am. J. Hum. Genet*, **57**, 432–9.

Morton, N.E. (1955) Sequential tests for the detection of linkage. *Am. J. Hum. Genet.*, **7**, 277–318.

Munsat, T.L. (1986) Facioscapulohumeral dystrophy and the scapuloperoneal syndrome, in *Myology*, McGraw Hill, New York.

Murray, A.W. and Szostak, J.W. (1983) Construction of artificial chromosomes in yeast. *Nature*, **305**, 189–93.

Nakamura, Y., Leppert, M., O'Connell, P. *et al.* (1987) Variable number of tandem repeat (VNTR) markers for human gene mapping. *Science*, **235**, 1616–22.

Ott, J. (1974) Estimation of the recombination fraction in human pedigrees: Efficient computation of the likelihood for human linkage studies. *Am. J. Hum. Genet.*, **26**, 588–97.

Ott, J. (1985) *Analysis of Human Genetic Linkage*, Johns Hopkins Univ. Press, Baltimore, MD.

Padberg, G.W. (1982) Facioscapulohumeral disease. Thesis, Leiden University.

Padberg, G., Eriksson, A.W., Volkers, W.S. *et al.* (1984) Linkage studies in autosomal dominant facioscapulohumeral muscular dystrophy. *J. Neurol. Sci.*, **65**, 261–8.

Padberg, G.W., Klasen, E.C., Volkers, W.S. *et al.* (1988) Linkage studies in facioscapulohumeral muscular dystrophy. *Muscle Nerve*, **11**, 833–5.

Padberg, G.W., Brouwer, O.F., de Keizer, R.J.W. and Wijmenga, C. (1990) Retinal vascular disease and perception deafness in facioscapulohumeral muscular dystrophy. *J. Neurol Sci.*, **98S**, 196–7.

Petersen, M.B., Slaugenhaupt, S.A., Lewis, J.G. *et al.* (1991) A genetic linkage map of 27 markers on human chromosome 21. *Genomics*, **9**, 407–19.

Rao, D.C., Keats, B.J.B., Lalouel, J.-M. *et al.* (1979) A maximum likelihood map of chromosome 1. *Am. J. Hum. Genet.*, **31**, 680–96.

Renwick, J.R. (1969) Genetic linkage in man, in *Computer Applications in Genetics*, Univ. Hawaii Press, Honolulu.

Renwick, J.H. (1971) The mapping of human chromosomes. *Ann. Rev. Genet.*, **5**, 81–120.

Royle, N.J., Clarkson, R.E., Wong, Z. and Jeffreys, A.J. (1988) Clustering of hypervariable minisatellites in the proterminal regions of human autosomes. *Genomics* **3**, 352–60.

Saiki, R.K., Scharf, S., Faloona, F. *et al.* (1985) Enzymatic amplification of β-globin genomic sequences and restriction site analysis for diagnosis of sickle cell anaemia. *Science*, **230**, 1350–4.

Saiki, R.K., Gelfand, D.H., Stoffel, S. *et al.* (1988) Primer-directed enzymatic amplification of DNA with a thermostable DNA polymerase. *Science*, **239**, 487–91.

Sarfarazi, M., Huson, S.M. and Edwards, J.H. (1987) An exclusion map for Von Recklinghausen neurofibromatosis. *J. Med. Genet.*, **24**, 515–20.

Sarfarazi, M., Upadhyaya, M., Padberg, G. *et al.* (1989) An exclusion map for facioscapulohumeral (Landouzy-Dejerine) disease. *J. Med. Genet.*, **26**, 481–4.

Sarfarazi, M., Wijmenga, C., Upadhyaya, M. *et al.* (1992) Regional mapping of facioscapulohumeral disease on 4q35. Combined analysis of an international consortium. *Am. J. Hum. Genet*, **57**, 396–403.

Schotland, D.L., Bonilla, E. and Wakayama, Y. (1981) Freeze fracture studies of muscle plasma membrane in human muscular dystrophy. *Acta Neuropathol.* (Berl) **54**, 189–97.

Schwarz, D.C. and Cantor, C.R. (1984) Separation of yeast chromosome-sized DNAs by pulsed field gradient gel electrophoresis. *Cell*, **37**, 67–75.

Southern, E. (1975) Detection of specific sequences among DNA fragments separated by gel electrophoresis. *J. Mol. Biol.*, **98**, 503–17.

Stallings, R.L., Torney, D.C., Hildebrand, C.E. *et al.* (1990) Physical mapping of human chromosomes by repetitive sequence fingerprinting. *Proc. Natl. Acad. Sci. USA*, **87**, 6218–22.

Van Ommen, G.J.B., Verkerk, J.M.H., Hofker, M.H. *et al.* (1986) A physical map of

4 million bp around the Duchenne muscular dystrophy gene on the human X-chromosome. *Cell*, **47**, 499–504.

Viskochil, D., Buchberg, A.M., Xu, G. *et al.* (1990) Deletions and a translocation interrupt a cloned gene at a neurofibromatosis type I locus. *Cell*, **62**, 187–92.

Vogel, F. and Motulsky, A.G. (1986) *Human Genetics*, Springer-Verlag, Berlin, Heidelberg, New York.

Voit, T., Lamprecht, A., Lenard, H.G. and Goebel, H.H. (1986) Hearing loss in facioscapulohumeral dystrophy. *Eur. J. Pediatr.*, **145**, 280–5.

Weber, J.L. and May, P.E. (1989) Abundant class of human DNA polymorphisms which can be typed using the polymerase chain reaction. *Am. J. Hum. Genet.*, **44**, 388–96.

Weber, J.L. (1989) Length polymorphisms in $(dC\text{-}dA)_{n.}$ $(dG\text{-}dT)_n$ sequences detected using the polymerase chain reaction, in *Current Communications in Molecular Biology*, Cold Spring Harbor Laboratory Press, Cold Spring Harbor, NY.

Weber, J. (1990a) Microsatellites and genetic mapping, in *Genetic and Physical Mapping*, Cold Spring Harbor Laboratory Press, Cold Spring Harbor, NY.

Weber, J.L. (1990b) Informativeness of human $(dC\text{-}dA)_{n.}$ $(dG\text{-}dT)_n$ polymorphisms. *Genomics*, **7**, 524–30.

Weber, J.L. and May, P.E. (1990) Dinucleotide repeat polymorphism at the D4S171 locus. *Nucleic Acids Res.*, **18**, 2202.

Weiffenbach, B., Bagley, R., Falls, K., *et al.* (1992) Linkage analysis of five 4q markers in 24 facioscapulohumeral muscular dystrophy (FSHD) families. *Am. J. Hum. Genet*, **57**, 416–23.

Wijmenga, C., Frants, R.R., Brouwer, O.F. *et al.* (1990a) Facioscapulohumeral muscular dystrophy gene not linked to markers for familial polyposis coli on the long arm of chromosome 5. *J. Neurol. Sci.*, **95**, 225–9.

Wijmenga, C., Frants, R.R., Brouwer, O.F. *et al.* (1990b) Location of facioscapulohumeral muscular dystrophy gene on chromosome 4. *Lancet*, **336**, 651–3.

Wijmenga, C., Padberg, G.W., Moerer, P. *et al.* (1991) Mapping of facioscapulohumeral muscular dystrophy gene to chromosome 4q35-qter by multipoint linkage analysis and in situ hybridization. *Genomics*, **9**, 570–5.

Wijmenga, C., Sandkuijl, L.A., Moerer, P. *et al.* (1992) Genetic linkage map of Facioscapulohumeral muscular dystrophy and five polymorphic loci on chromosome 4q35-qter. *Am. J. Hum. Genet.*, **57**, 411–5.

6

Dystrophin-associated glycoproteins: their possible roles in the pathogenesis of Duchenne muscular dystrophy

JAMES M. ERVASTI and KEVIN P. CAMPBELL

6.1 INTRODUCTION

The application of 'positional cloning' to the analysis of inherited disorders has led to the identification of an impressive and rapidly growing list of gene defects which are responsible for some of the most debilitating and scientifically perplexing human diseases (Collins, 1992). For example, Duchenne muscular dystrophy (DMD) is caused by a defective gene found on the X chromosome which was identified by two variations of positional cloning (Monaco *et al.*, 1985; Ray *et al.*, 1985). The DMD gene encodes for a large protein named dystrophin which has a predicted primary structure that is similar to a number of cytoskeletal proteins (Koenig *et al.*, 1988). Knowledge of dystrophin's predicted primary sequence has enabled the production of specific antibodies, which in turn have been very useful in confirming its size (Hoffman *et al.*, 1987), determining its tissue distribution (Hoffman *et al.*, 1988b) and cellular location (Zubrzycka-Gaarn *et al.*, 1988; Arahata *et al.*, 1988; Bonilla *et al.*, 1988; Watkins *et al.*, 1988). Dystrophin has been localized to the cytoplasmic face of the sarcolemma in skeletal muscle (Zubrzycka-Gaarn *et al.*, 1988; Arahata *et al.*, 1988; Bonilla *et al.*, 1988; Watkins *et al.*, 1988), including the neuromuscular junction (Ohlendieck *et al.*, 1991a; Yeadon *et al.*, 1991; Byers *et al.*, 1991). Dystrophin is completely absent in skeletal muscle of DMD patients, *mdx* mice and *GRMD* dogs (Hoffman *et al.*, 1987; Zubrzycka-Gaarn *et al.*, 1988; Watkins *et al.*, 1988; Hoffman *et al.*, 1988a; Cooper *et al.*, 1988).

Thus, positional cloning techniques have enabled scientists to identify

Molecular and Cell Biology of Muscular Dystrophy
Edited by Terence Partridge
Published in 1993 by Chapman & Hall, London. ISBN 0 412 43440 7

abnormal genes, pinpoint the mutations that occur and make powerful predictions concerning the structure of the gene products. However, identification of the defect of a genetic disorder may not explain the function of the normal gene product or the complete molecular mechanism by which the disorder progresses. In the case of DMD, neither the exact function of dystrophin, nor the precise mechanism of fibre necrosis which occurs in its absence in dystrophic muscle, has been determined.

The function of a protein may be better understood through its interactions with other proteins. For example, previous studies of genetic diseases involving cytoskeletal proteins (Alloisio *et al.*, 1985; Mueller and Morrison, 1981) have demonstrated that the absence of one component of the cytoskeleton is sometimes accompanied by the loss of an associated component of the membrane cytoskeleton. Thus, in order to understand the function of dystrophin and the molecular pathogenesis of DMD, it is imperative to identify the proteins which are associated with dystrophin and to characterize the status of these proteins in muscle where dystrophin is absent.

In this chapter, we review recent advancement in the understanding of the structure of dystrophin, its relative subcellular abundance, the identification and characterization of six proteins that form a complex with dystrophin, and the fate of these dystrophin-associated proteins in dystrophic tissues. Of particular interest is the marked reduction of dystrophin-associated proteins in muscle from *mdx* mice and DMD patients. These results indicate that the first step in the molecular pathogenesis of Duchenne muscular dystrophy is the loss of the dystrophin-associated glycoproteins, which leads to the loss of linkage between the sarcolemmal cytoskeleton and the extracellular matrix, ultimately rendering muscle fibres more susceptible to necrosis.

6.2 MEMBRANE PROPERTIES OF DYSTROPHIN

Based on its deduced primary structure, dystrophin was originally predicted to consist of four distinct regions, dominated by a large rod-shaped domain with a length of 125 nm (Koenig *et al.*, 1988). Immunogold labelling studies of skeletal muscle using site-specific antibodies have reported sarcolemmal periodicities ranging from 100 to 140 nm (Watkins *et al.*, 1988; Cullen *et al.*, 1990, 1991) while the length of rotary-shadowed images of dystrophin have varied from 100 to 180 nm (Murayama *et al.*, 1990; Pons *et al.*, 1990). The rod-shaped domain is flanked on its N-terminus by 240 amino acids with high homology to the actin binding domain of α-actinin, spectrin and *Dictyostelium* actin binding protein 120 (Koenig *et al.*, 1988; Karinch *et al.*, 1990; Bresnick *et al.*, 1990). From this sequence homology, it has been hypothesized that the N-terminal 240 amino acids comprise a filamentous actin binding site. In support of this hypothesis, Hemmings *et al.* (1992)

recently demonstrated that a fusion protein corresponding to the first 233 amino acids was able to bind filamentous actin *in vitro*. Immediately C-terminal to the rod-shaped domain of dystrophin is a cysteine-rich region with significant homology to a domain of *Dictyostelium* α-actinin that contains two potential Ca^{2+}-binding sites. However, there is currently no evidence that this putative Ca^{2+}-binding domain is functional in skeletal muscle dystrophin. The last C-terminal 420 amino acids comprise the fourth distinct domain of dystrophin and exhibit no homology with any known sequence. The lack of significant homology with proteins of known function has led to speculation that these last 420 amino acids may be involved in dystrophin's interaction with the sarcolemmal membrane.

Immunogold labelling studies (Cullen *et al.*, 1991) with an antibody against the extreme C-terminus of dystrophin suggest that the C-terminal domain is closely apposed or inserted into the plasma membrane of skeletal muscle, thus providing the strongest evidence to date that the C-terminal domain is the membrane-binding region of dystrophin. However, three patients clinically diagnosed as DMD have recently been found to express a truncated form of dystrophin lacking the cysteine-rich and C-terminal domains (Hoffman *et al.*, 1991; Recan *et al.*, 1992; Helliwell *et al.*, 1992). Indirect immunofluorescence analysis of biopsies from these patients have demonstrated sarcolemmal localization of the truncated dystrophin, indicating that its interaction with the membrane cytoskeleton is important in determining dystrophin's location in the muscle cell.

Dystrophin's predicted primary structure suggests that it shares many features with abundant structural proteins of the membrane cytoskeleton such as α-actinin and spectrin (Koenig *et al.*, 1988). However, dystrophin is an extremely minor component of skeletal muscle, representing only about 0.002% of the total muscle protein (Hoffman *et al.*, 1987). For this reason, dystrophin has not been considered to play a major structural role in the membrane cytoskeleton of skeletal muscle. On the other hand, sarcolemmal proteins constitute only a minute fraction of total muscle protein. Since it has long been apparent that the initial degenerative processes leading to DMD are associated with the surface membrane of skeletal muscle (Mokri and Engel, 1975), it was important to be able to study the structure of isolated sarcolemma in order to identify the normal protein composition of the sarcolemmal membrane from skeletal muscle and to determine the relative abundance of dystrophin to other sarcolemmal proteins.

For the biochemical characterization of surface membrane components, sarcolemmal vesicles have to be prepared in a sufficient yield and with a high degree of purity. A variety of procedures have been employed to isolate skeletal muscle fractions enriched in sarcolemma, most involving density gradient centrifugation (Barchi *et al.*, 1979; Moczydlowski and Latorre, 1983; Seiler and Fleischer, 1982, 1988). However, previous attempts to obtain pure sarcolemma have been hindered by the lack of specific and well defined markers for the sarcolemma and the low abundance of sarcolemma in comparison to other subcellular membranes in skeletal muscle. Wheat germ agglutinin is a homodimeric lectin which crosslinks terminal residues of *N*-acetyl-D-glucosamine and/or sialic acid glycoproteins (Bhavanandan and Katlic, 1979). Thus, lectin agglutination is expected to aggregate specifically sealed right-side-out sarcolemmal vesicles because the carbohydrate chains of membrane glycoproteins are extracellular (Charuk *et al.*, 1989). In fact, Charuk *et al.* (1989) employed a wheat germ agglutination procedure following density gradient centrifugation for the subfractionation of cardiac sarcolemma.

Our approach was to first identify specific markers for the sarcolemma using immunofluorescence localization of a library of monoclonal antibodies isolated from mice immunized with a crude preparation of rabbit skeletal muscle membranes. We then set out to isolate highly purified sarcolemmal vesicles from rabbit skeletal muscle, using sucrose density step gradient centrifugation and wheat germ agglutination, and to characterize the isolated sarcolemmal vesicles by immunoblot analysis using the subcellular membrane-specific monoclonal antibodies as probes (Ohlendieck *et al.*, 1991b). SDS-polyacrylamide gel analysis of the purified sarcolemma preparation revealed a protein band of approximately 400 kDa which was detectable with Coomassie Blue. This 400 kDa protein comigrated with dystrophin detected on immunoblots.

To establish that the 400 kDa protein band in isolated sarcolemma was exclusively dystrophin, the lectin agglutination procedure was used to isolate sarcolemma from control and *mdx* mouse muscle, which is known from immunological studies to be missing dystrophin (Bonilla *et al.*, 1988). The overall SDS-PAGE profile (Figure 6.1) of control and *mdx* sarcolemma was very similar (Ohlendieck and Campbell, 1991b). The major difference between the control and *mdx* sarcolemma is the absence in the latter of the 400 kDa protein band which is stained in immunoblotting by antiserum against the C-terminal decapeptide of dystrophin in normal mouse muscle sarcolemma (Figure 6.1). Restricted immunofluorescence labelling of the cell periphery in normal mouse muscle cryosections, in comparison to no staining of *mdx* muscle cells, established the specificity of the polyclonal rabbit antiserum against the C-terminal decapeptide of dystrophin (Figure 6.1). A faint Coomassie Blue stained protein band in *mdx* mouse sarcolemma with a slightly higher molecular weight than dystrophin was identified as dystrophin-related protein that is encoded by a different gene (Love *et al.*, 1989; Khurana *et al.*, 1990; Ohlendieck *et al.*, 1991a). Thus, the analysis of control and *mdx* sarcolemma demonstrated that the 400 kDa Coomassie Blue stained protein band in isolated sarcolemma was exclusively dystrophin. Furthermore, immunoadsorption experiments with

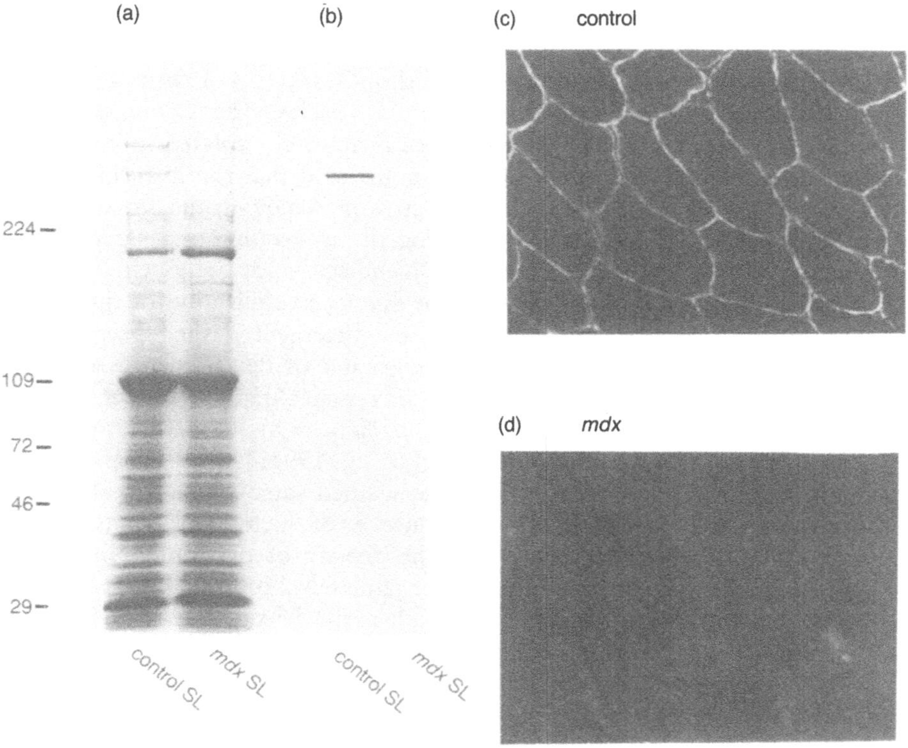

Figure 6.1 Dystrophin in control and *mdx* sarcolemma membranes. Shown are a Coomassie Blue stained gel (a) of isolated sarcolemma (SL) and an immunoblot (b) of an identical gel stained with polyclonal antiserum against the C-terminal decapeptide of dystrophin. Molecular weight standards ($\times$ 10^{-3}) are indicated on the left. Transverse cryosections of normal (c) and *mdx* (d) mouse skeletal muscle were labelled by indirect immunofluorescence with polyclonal antiserum against the C-terminal decapeptide of dystrophin. After Ohlendieck and Campbell (1991a).

anti-dystrophin immunoaffinity beads quantitatively removed the 400 kDa band from digitonin-solubilized sarcolemma, thus confirming the 400 kDa band as dystrophin (Ohlendieck *et al.*, 1991b).

To estimate the amount of dystrophin in the skeletal muscle plasma membrane, purified sarcolemma were analysed by densitometric scanning of Coomassie Blue stained gels. To account for possible variations in Coomassie Blue staining intensity relative to the amount of total protein separated by SDS-PAGE, gels with different amounts of sarcolemmal protein were analysed by densitometric scanning and the values averaged. Since thin section electron microscopy indicated entrapment of small sarcoplasmic reticulum vesicles within larger sarcolemma vesicles, lectin-agglutinated sarcolemma vesicles were additionally treated with low concentrations of the non-ionic detergent Triton X-100 (0.1%) with the aim of removing trapped sarcoplasmic reticulum vesicles. The Coomassie Blue stained protein pattern of crude surface membrane, isolated sarcolemma, and detergent-washed sarcolemma demonstrated that the low concentrations of Triton X-100 were very effective in removing the sarcoplasmic reticulum Ca^{2+}-ATPase while enriching the dystrophin and sarcolemma marker Na^{+}/K^{+} ATPase content (Ohlendieck and Campbell, 1991b). Removal of the Ca^{2+}-ATPase was not due to solubilization of the membranes since electron microscopy of the detergent-washed preparation demonstrated sealed vesicles. Peak integration of densitometric scans of isolated rabbit skeletal muscle sarcolemma revealed that the protein band of apparent 400 kDa accounted for 2.1 ± 0.7% (n = 8) of the total protein of this membrane preparation (Ohlendieck *et al.*, 1991b). Peak integration of the densitometric scan of the detergent-washed sarcolemma revealed that dystrophin accounted for 4.8 ± 0.8% (n = 6) of the total protein (Ohlendieck and Campbell, 1991b). Thus, the density of dystrophin in highly purified sarcolemma membranes is approximately 2400-fold higher than its density in whole muscle and is comparable to the density of spectrin in brain membranes (Bennett *et al.*, 1982).

One criterion for determining whether a protein is a component of the cytoskeleton is its relative insolubility in Triton X-100 (Salas *et al.*, 1988; Carraway and Carothers-Carraway, 1989). Extraction of plasma membranes with high concentrations (0.5%) of the non-ionic detergent Triton X-100 leaves the cytoskeleton as an insoluble residue while solubilizing the membrane proteins not associated with the cytoskeleton (Salas *et al.*, 1988; Carraway and Carothers-Carraway, 1989). Dystrophin was exclusively found in the Triton-insoluble pellet, comprising 5.1 ± 1.0% (n = 6) of the total cytoskeleton protein while Na^{+}/K^{+}-ATPase was found in the supernatant (Ohlendieck and Campbell, 1991b). In contrast to the treatment with Triton X-100, treatment of membranes with strong alkaline solutions is known to remove tightly associated cytoskeletal components (Korsgren and Cohen, 1986) from membranes while leaving the integral membrane proteins with the bilayer (Steck and Yu, 1973; Carraway and Carothers-Carraway, 1989). Dystrophin is completely extracted by alkaline treatment (Chang *et al.*, 1989; Ohlendieck and Campbell, 1991b; Ervasti and Campbell, 1991). Thus, established biochemical methods for the identification of cytoskeletal proteins demonstrate that dystrophin is an integral component of the cytoskeleton of the sarcolemma in skeletal muscle.

Although dystrophin is a minor protein when compared to the total muscle protein (Hoffman *et al.*, 1987), it appears to be a major component of the subsarcolemmal cytoskeletal network in skeletal muscle. These find-

ings throw new light on the possible function and relative abundance of dystrophin in the membrane skeleton of skeletal muscle cells, comparable to that of the major membrane skeleton component spectrin in other cell types (Bennett *et al.*, 1982).

6.3 DYSTROPHIN–GLYCOPROTEIN COMPLEX

The ability of alkaline treatment to extract dystrophin from membrane preparations (Chang *et al.*, 1989; Ohlendieck and Campbell, 1991b) indicated that dystrophin was tightly associated with the plasma membrane through strong protein–bilayer or protein–protein interactions. In order to understand the function of dystrophin and its role in the molecular pathogenesis of DMD, it was imperative to identify the proteins which are associated with dystrophin. Our approach to identifying dystrophin-associated proteins was to solubilize and purify dystrophin from rabbit skeletal membranes using medium stringency conditions which retain specific, high affinity protein–protein interactions while minimizing non-specific aggregation. The detergent digitonin, in combination with 0.5 M NaCl, had previously been shown to optimally solubilize intact skeletal muscle dihydropyridine receptor, a hetereotetrameric, integral membrane glycoprotein complex (Leung *et al.*, 1987). In contrast to dystrophin's insolubility in Triton X-100 concentrations as high as 1% (Chang *et al.*, 1989; Ohlendieck and Campbell, 1991b), dystrophin was quantitatively solubilized from rabbit skeletal muscle membranes using 1% digitonin and 0.5 M NaCl (Campbell and Kahl, 1989). Interestingly, it was discovered that dystrophin could be purified 17 000-fold from digitonin-solubilized skeletal muscle membranes using immobilized wheat germ agglutinin. The interaction of dystrophin with wheat germ agglutinin was disrupted by agents that dissociate cytoskeletal proteins from membranes, indicating that dystrophin itself was not a glycoprotein but, rather, was tightly linked to one or several integral membrane glycoproteins (Campbell and Kahl, 1989).

The dystrophin–glycoprotein complex was initially purified from digitonin-solubilized rabbit skeletal muscle membranes using wheat germ agglutinin–Sepharose, ion exchange chromatography and sucrose density gradient centrifugation (Ervasti *et al.*, 1990). It was later found (Ervasti *et al.*, 1991) that substituting succinylated wheat germ agglutinin–agarose for wheat germ agglutinin–Sepharose resulted in dystrophin–glycoprotein complex preparations of the same purity and yield as previously reported (Ervasti *et al.*, 1990) while obviating the sucrose gradient step. The size of the dystrophin complex was estimated to be ~18S by comparing its sedimentation through a sucrose density gradient to that of the standards beta-galactosidase (16S), thyroglobulin (19S) and the 20S dihydropyridine receptor (Ervasti *et al.*, 1990). Densitometric scanning of the peak dystro-

phin containing gradient fractions revealed several proteins which cosedimented with dystrophin: a broad, diffusely staining component with an apparent molecular weight of 156 kDa, a triplet of proteins centred at 59 kDa, a 50 kDa protein, a protein doublet at 43 kDa, a 35 kDa protein and a 25 kDa protein (Figure 6.2a). Immunoaffinity beads against dystrophin or the 50 kDa protein each selectively immunoprecipitated dystrophin, the 156 kDa, 59 kDa, 50 kDa, 43 kDa, 35 kDa and 25 kDa proteins (Ervasti *et al.*, 1990; Ervasti and Campbell, 1991). Since the 50 kDa dystrophin-associated glycoprotein-antibody matrix immunoprecipitated more of the 156 kDa dystrophin-associated glycoprotein than the dystrophin-antibody matrix, these data further suggested that the 156 kDa dystrophin-associated glycoprotein was directly linked to the 50 kDa glycoprotein rather than to dystrophin (Ervasti and Campbell, 1991). As

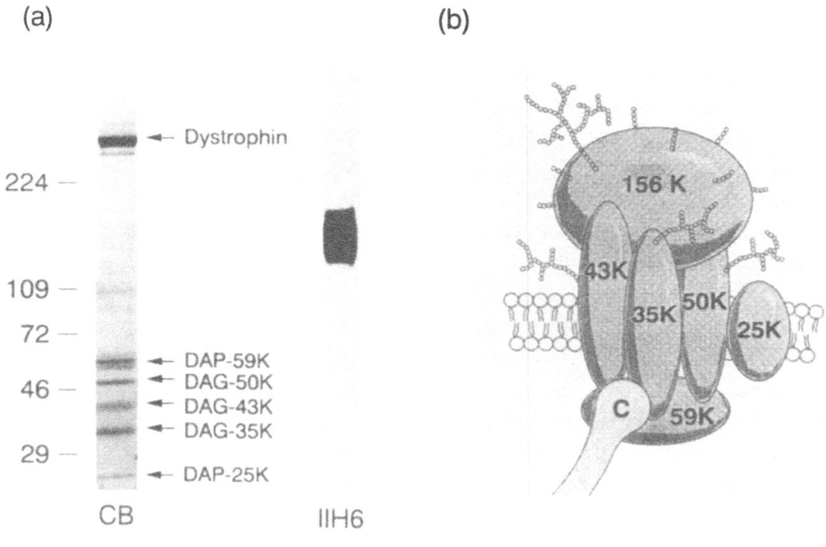

Figure 6.2 Dystrophin-glycoprotein complex: SDS PAGE analysis and proposed structural model (a) Shown are Coomassie Blue stained SDS polyacrylamide gel (CB) and corresponding nitrocellulose transfer stained with monoclonal antibody IIH6 against the 156 kDa dystrophin-associated glycoprotein of purified dystrophin glycoprotein complex. Dystrophin, the 59 kDa dystrophin associated protein (DAP-59K) and the 156 kDa, 50 kDa, 43 kDa and 35 kDa dystrophin associated glycoproteins (DAG-156K, 50K, 43K, 35K) are indicated on the right. Molecular weight standards ($\times\ 10^{-3}$) are indicated on the left. (b) Proposed structural model of the dystrophin–glycoprotein complex. The C denotes the cysteine-rich and C-terminal domains of dystrophin. After Ervasti and Campbell (1991).

expected from the alkaline extraction results (Chang *et al.*, 1989; Ohlendieck and Campbell, 1991b), the components of the alkaline-treated complex no longer cosedimented on sucrose gradients but each sedimented as much smaller entities (Ervasti *et al.*, 1991). These data indicate that the dystrophin–glycoprotein complex can be dissociated by alkaline treatment. The alkaline-treated dystrophin co-sedimented with the 11S standard catalase (Ervasti *et al.*, 1991) which is in good agreement with the sedimentation of tetrameric spectrin (Bennett *et al.*, 1982). This result indicates that alkaline-treated dystrophin sediments as a dimer. The 59 kDa dystrophin-associated protein sedimented near the top of the gradient and was completely separated from dystrophin (Ervasti *et al.*, 1991). The 156 kDa dystrophin-associated glycoprotein and the upper band of the 43 kDa doublet co-sedimented with a peak in fraction 5. However, the 50 kDa, the lower band of the 43 kDa doublet, 35 kDa and 25 kDa dystrophin-associated proteins co-sedimented even after alkaline treatment, possibly as a complex intermediate in size between the 43 kDa/156 kDa dystrophin-associated glycoprotein peak and dimeric dystrophin (Ervasti *et al.*, 1991).

To determine whether the 50 kDa, 43 kDa, 35 kDa and 25 kDa dystrophin-associated proteins remain complexed after alkaline dissociation, alkaline-treated dystrophin–glycoprotein complex was immunoprecipitated by anti-dystrophin or anti-50 kDa dystrophin-associated glycoprotein immunoaffinity beads and the void analysed by SDS-polyacrylamide gel electrophoresis and immunoblotting (Ervasti and Campbell, 1991). The dystrophin-antibody matrix immunoprecipitated dystrophin from the alkaline-treated dystrophin–glycoprotein complex but the 59 kDa, 50 kDa, 43 kDa and 35 kDa dystrophin-associated proteins remained largely in the void indicating that the interaction between dystrophin and the complex was disrupted by alkaline treatment. The 50 kDa dystrophin-associated glycoprotein-antibody matrix was not effective in immunoprecipitating dystrophin, the 156 kDa or 59 kDa dystrophin-associated proteins from the alkaline-treated complex. However, the 50 kDa, 43 kDa and 35 kDa dystrophin-associated glycoproteins were still immunoprecipitated from the alkaline-treated complex by the anti-50 kDa dystrophin-associated glycoprotein-antibody matrix. Neither of the immunoaffinity matrices precipitated the 25 kDa dystrophin-associated protein from alkaline-dissociated complex. Thus, these data demonstrate that the 50 kDa, 43 kDa and 35 kDa dystrophin-associated proteins alone form an alkali-stable complex.

Densitometric analysis of Coomassie Blue stained SDS-polyacrylamide gels containing the electrophoretically separated components of six different preparations of dystrophin–glycoprotein complex demonstrated that the 59 kDa, 50 kDa, 43 kDa, 35 kDa and 25 kDa dystrophin-associated proteins exhibited average stoichiometric ratios of 1.6 ± 0.22, 0.82 ± 0.11, 0.95 ± 0.14, 1.8 ± 0.19 and 0.36 ± 0.12 relative to dystrophin (Ervasti and Campbell, 1991). However, the stoichiometry of the 156 kDa dystrophin-associated glycoprotein relative to dystrophin could not be determined in this manner because it stains very poorly with Coomassie Blue (Ervasti *et al.*, 1990). Therefore, the dystrophin-associated glycoprotein-specific antibody staining intensity was quantitated from autoradiograms of immunoblots containing pure rabbit sarcolemma and dystrophin–glycoprotein complex after incubation with [^{125}I]-Protein A and was compared to the Coomassie Blue staining intensity of dystrophin. The 156 kDa, 59 kDa, 50 kDa, 43 kDa and 35 kDa dystrophin-associated proteins each possess unique antigenic determinants, as none of the antibodies specific for a particular component of the complex cross-reacts with any other component of the complex (Ervasti *et al.*, 1990; Ohlendieck *et al.*, 1991b; Campbell *et al.*, 1991; Ervasti and Campbell, 1991). Densitometric analysis of Coomassie Blue stained gels demonstrated that dystrophin was enriched 2.5-fold in dystrophin–glycoprotein complex versus sarcolemma. The ratios of autoradiographic densitometric intensities of dystrophin–glycoprotein complex versus sarcolemma for polyclonal antibodies against each of the dystrophin-associated glycoproteins varied between 2.2 and 3.0 (Ervasti and Campbell, 1991). These results suggest that all components of the dystrophin–glycoprotein complex quantitatively co-enrich and that the 156 kDa dystrophin-associated glycoprotein is stoichiometric with dystrophin.

The cellular localization of the dystrophin-associated proteins was determined by indirect immunofluorescence labelling of transverse cryostat sections of rabbit skeletal muscle. The 50 kDa glycoprotein was previously identified as a very convenient sarcolemma marker (Jorgensen *et al.*, 1990). In addition, antibodies specific for the 156 kDa, 59 kDa, 50 kDa, 43 kDa and 35 kDa dystrophin-associated proteins also exhibited immunofluorescent staining of the sarcolemmal membrane, demonstrating the unique association of these proteins with the muscle fibre plasma membrane or the intracellular cytoskeleton subjacent to the surface membrane (Ervasti *et al.*, 1990; Ohlendieck *et al.*, 1991b; Ervasti and Campbell, 1991). Immunoblot analysis of subcellular fractions from rabbit skeletal muscle confirmed that components of the dystrophin–glycoprotein complex are highly enriched in sarcolemma vesicles (Ohlendieck *et al.*, 1991b).

Fast-twitch skeletal muscle fibres of muscle from DMD patients are affected earlier than slow-twitch fibres (Webster *et al.*, 1988). Immunofluorescence localization studies are affected earlier (Schafer and Stockdale, 1987) identified sarcolemma-associated antigens with different distribution in fast and slow skeletal muscle fibres. Variability of staining intensity among fibres were also found for a sarcolemmal Na^+/K^+-ATPase in chicken muscle (Fambrough and Bayne, 1983). Therefore, the fibre type distribution of the dystrophin–glycoprotein complex was examined (Ohlendieck *et al.*, 1991b). Dystrophin and the 50 kDa component of the dystrophin–glycoprotein complex were equally distributed between fast- and slow-twitch fibres. It remains to be determined why the fibre type plays a role in the early steps of abnormal muscle protein degradation and fibre necrosis in dystrophic muscle.

The 156 kDa, 50 kDa, 43 kDa and 35 kDa dystrophin-associated proteins were found to contain Asn-linked oligosaccharides, as determined by specific lectin staining (Ervasti *et al.*, 1990; Ervasti and Campbell, 1991) and enzymatic deglycosylation (Ervasti and Campbell, 1991). In addition, the 156 kDa dystrophin-associated glycoprotein contained $\alpha(2,3)$-linked sialic acid residues and Ser/Thr-linked oligosaccharides (Ervasti and Campbell, 1991). Dystrophin, the 59 kDa and the 25 kDa dystrophin-associated proteins do not appear to be glycosylated (Ervasti *et al.*, 1990; Ervasti and Campbell, 1991).

Consistent with predictions that it is a cytoskeletal protein (Koenig *et al.*, 1988), dystrophin can be extracted from membranes in the absence of detergents by simple alkaline treatment (Chang *et al.*, 1989; Ohlendieck and Campbell, 1991b). The 59 kDa dystrophin-associated protein was also extracted by alkaline treatment while the 156 kDa, 50 kDa, 43 kDa and 35 kDa glycoproteins were retained in the membrane pellet after alkaline treatment (Ervasti and Campbell, 1991). Surprisingly, the 156 kDa dystrophin-associated glycoprotein, which was not extracted from membranes incubated at pH 11, was almost completely extracted from surface membranes incubated at pH 12 while the 50 kDa, 43 kDa and 35 kDa dystrophin-associated glycoproteins remained in the membrane pellet even after incubation of surface membranes at pH 12 (Ervasti and Campbell, 1991). That dystrophin, the 156 kDa dystrophin-associated glycoprotein and the 59 kDa dystrophin-associated protein can be extracted from skeletal muscle membranes by alkaline treatment in the absence of detergents demonstrates that these proteins are not integral membrane proteins. These data also suggest that the 50 kDa, 43 kDa and 35 kDa dystrophin-associated glycoproteins are integral membrane proteins. Since the 156 kDa dystrophin-associated glycoprotein remains membrane-bound under conditions which extract dystrophin, these data further suggest that the 156 kDa dystrophin-associated glycoprotein is linked to dystrophin by way of the 50 kDa, 43 kDa and/or 35 kDa components of the complex. The 50 kDa, 43 kDa and 35 kDa dystrophin-associated glycoproteins and the 25 kDa dystrophin-associated protein were further confirmed as integral membrane proteins by covalent labelling with a hydrophobic probe (Ervasti and Campbell, 1991).

We recently proposed a model of the dystrophin–glycoprotein complex

(Ervasti and Campbell, 1991) to aid in visualizing what is presently known about its structure (Figure 6.2b). Dystrophin was modelled as a bent, antiparallel dimer with the cysteine-rich and C-terminal domains linked to the transmembrane components of the complex and the amino terminus binding to the filamentous actin cytoskeleton. Dystrophin was depicted as such to conform to the results obtained from sequence analysis (Koenig *et al.*, 1988), protease mapping (Koenig and Kunkel, 1990), rotary shadowed images of purified dystrophin–glycoprotein complex (Murayama *et al.*, 1990), size estimates of purified dystrophin (Ervasti *et al.*, 1991), ultrastructural localization (Cullen *et al.*, 1991) and recent filamentous actin co-sedimentation results (Hemmings *et al.*, 1992).

We postulated that the 156 kDa dystrophin-associated glycoprotein is located on the extracellular side of the sarcolemma on the basis of the presence of Ser/Thr-linked oligosaccharides and its resistance to proteolysis (Ervasti and Campbell, 1991). By analogy with cell surface molecules containing densely Ser/Thr-linked glycosylated regions, such as NCAM (Walsh *et al.*, 1989; Moore *et al.*, 1987) and the LDL receptor (Cummings *et al.*, 1983), we conclude that the 156 kDa dystrophin-associated glycoprotein is an extracellular protein.

The extraction of the 156 kDa dystrophin-associated glycoprotein from membranes incubated at pH 12, but not pH 11 (Ervasti and Campbell, 1991) suggests that its association with the sarcolemma is distinct from that of dystrophin and the 59 kDa dystrophin-associated protein. It is interesting that proteoglycans were originally (Carney, 1986) extracted from connective tissues by incubation in 2% NaOH (i.e. > pH 12). This feature of the 156 kDa dystrophin-associated glycoprotein coupled with its failure to focus as a sharp band after enzymatic deglycosylation suggests that the 156 kDa dystrophin-associated glycoprotein may also contain glycosaminoglycan chains.

The placement of the 59 kDa dystrophin-associated protein in the cytoplasm in direct contact with dystrophin was based on its cross-linking to dystrophin (Yoshida and Ozawa, 1990), solubilization from skeletal muscle membranes by alkaline treatment and the absence of labelling by hydrophobic probe (Ervasti and Campbell, 1991). Placement of the 59 kDa dystrophin-associated protein in contact with the 50 kDa, 43 kDa and 35 kDa dystrophin-associated glycoproteins is solely by analogy with the 58 kDa of MAT-Cl ascite tumour cell microvilli, which is thought to stabilize the association of microfilaments with a glycoprotein complex located in the microvillar membrane (Carraway and Carothers-Carraway, 1989). Alternatively, the 59 kDa dystrophin-associated protein could be located near the predicted actin-binding domain of dystrophin (Koenig *et al.*, 1988) where it might stabilize dystrophin binding to actin filaments in a manner analogous to protein 4.1 promoting spectrin–actin association

(Bennett, 1990) or zyxin promoting α-actinin/actin association (Crawford and Beckerle, 1991).

That the 50 kDa, 43 kDa and 35 kDa dystrophin-associated glycoproteins form an integral membrane complex (Ervasti and Campbell, 1991) indicates that they are the components of the complex which span the sarcolemmal membrane and link dystrophin to the 156 kDa dystrophin-associated glycoprotein. The large amount of hydrophobic probe incorporation into the 25 kDa dystrophin-associated protein places this component of the complex in the sarcolemmal membrane as well.

The structural organization of the dystrophin–glycoprotein complex (Figure 6.2b) is strikingly similar to that of the cadherins (Takeichi, 1991) or integrins (Ruoslahti and Pierschbacher, 1987). The data accumulated thus far imply that the function of dystrophin is to link, by way of a transmembrane glycoprotein complex, the actin cytoskeleton of a muscle cell to the extracellular matrix of skeletal muscle. That dystrophin comprises 2% of sarcolemmal protein (Ohlendieck *et al.*, 1991b) and 5% of the sarcolemmal cytoskeleton (Ohlendieck and Campbell, 1991b) supports the role for the dystrophin–glycoprotein complex in maintaining skeletal muscle architecture.

6.4 STRUCTURE AND FUNCTION OF DYSTROGLYCAN (43/156 DYSTROPHIN-ASSOCIATED GLYCOPROTEIN)

The complete amino acid sequence of the 43 kDa and 156 kDa dystrophin-associated glycoproteins have been deduced from isolated cDNAs (Ibraghimov-Beskrovnaya *et al.*, 1992). A 0.6 kb cDNA clone, the protein product of which was recognized by two polyclonal antibodies against the 43 kDa dystrophin-associated glycoprotein, was isolated from a rabbit skeletal muscle cDNA expression library. This cDNA clone hybridized with a 5.8 kb transcript in mRNA preparations from a variety of rabbit tissues. Overlapping clones covering the entire coding region of the mRNA were isolated by rescreening cDNA libraries and the full length sequence determined. Sequence analysis of cDNAs revealed an open reading frame of 2685 bases encoding a precursor polypeptide of 895 amino acids, with a predicted molecular weight of 97 kDa, which exhibited no significant sequence similarity with any known proteins. The predicted amino acid sequence of the 43 kDa and 156 kDa dystrophin-associated glycoproteins revealed structural characteristics (Figure 6.3) that were in good agreement with the native proteins' biochemical properties (Ervasti and Campbell, 1991). The N-terminal portion of the precursor polypeptide encodes the 56 kDa core protein of the 156 kDa dystrophin-associated glycoprotein because antibodies specific for a fusion protein corresponding to this region of the message also recognize the native 156 kDa dystrophin-associated

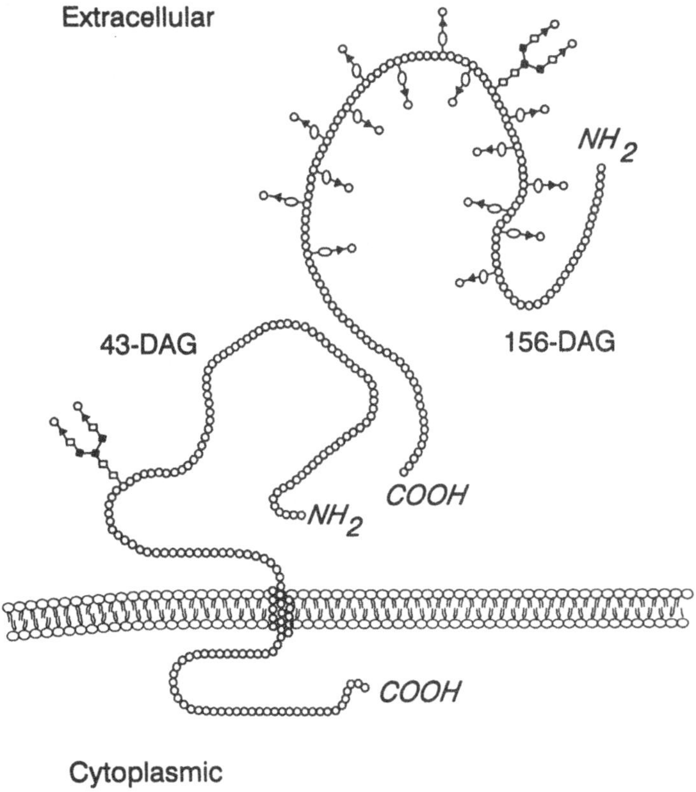

Figure 6.3 Model of dystroglycan.

glycoprotein. The 56 kDa core protein contains a single consensus sequence for Asn-linked glycosylation with many potential attachment sites for O-linked carbohydrates. Carbohydrate moieties appear to constitute up to two-thirds of the molecular mass of the 156 kDa dystrophin-associated glycoprotein which suggests that it may be a proteoglycan. The exact modifications involved in the processing of the N-terminal portion of the precursor polypeptide to the 156 kDa dystrophin-associated glycoprotein are not known but the 56 kDa core protein contains several Ser-Gly repeats for the possible addition of glycosaminoglycan chains (Bourdon *et al.*, 1987). The C-terminal portion of the precursor polypeptide is processed into the mature 43 kDa dystrophin-associated glycoprotein with four potential N-glycosylation sites, a single transmembrane domain and 120 amino acid cytoplasmic tail. The C-terminal half of the message was determined to encode for the 43 kDa dystrophin-associated glycoprotein because an 11 amino acid sequence determined directly from protein exactly matched the predicted sequence from cDNA.

Northern and Western blot analyses have demonstrated that the 43 kDa and 156 kDa dystrophin-associated glycoproteins are expressed in both muscle and non-muscle tissues. A prominent 5.8 kb transcript was detected in mRNA from rabbit skeletal muscle, cardiac muscle and lung (Ibraghimov-Beskrovnaya *et al.*, 1992). Thus, the 5.8 kb transcript for the 43/156 kDa dystrophin-associated glycoproteins is present in various muscle and non-muscle tissues, most likely originating from the one gene. Identification of the 43/156 kDa dystrophin-associated glycoproteins in muscle and non-muscle tissues was performed using immunoblots of membranes from different tissues and affinity-purified antibodies. The 43 kDa dystrophin-associated glycoprotein was detected in isolated membranes from skeletal muscle, brain, cardiac muscle and lung. The 156 kDa dystrophin-associated glycoprotein was detected in skeletal and cardiac muscle membranes, but was slightly lower in molecular weight in cardiac membranes. In brain and lung membranes, the molecular weight of the '156 kDa' dystrophin-associated glycoprotein reactive protein was ~120 kDa. The variability in molecular weight for the '156 kDa' reactive protein may be due to differential glycosylation of the core protein. The broad tissue distribution of the 43/156 kDa dystrophin-associated glycoprotein precursor argues for a significant role of both glycoproteins in membrane organization within different tissues and might indicate the existence of the entire glycoprotein complex in non-muscle tissues. Absence of significant amounts of dystrophin in the examined non-muscle tissues suggests that in non-muscle tissues 43/156 kDa dystrophin-associated glycoprotein is involved in formation of a different type of complex where dystrophin may be replaced by another cytoskeleton component.

The organization (Ervasti and Campbell, 1991), available primary sequence (Ibraghimov-Beskrovnaya *et al.*, 1992) and abundance in purified sarcolemma (Ohlendieck *et al.*, 1991b; Ohlendieck and Campbell, 1991b) of the dystrophin–glycoprotein complex suggest that the complex plays a structural role and functions to link the cytoskeleton with the extracellular matrix. To evaluate this hypothesis, we designed experiments which tested for an interaction between the 156 kDa dystrophin-associated glycoprotein and several well characterized proteins of the extracellular matrix. Rabbit skeletal muscle surface membranes and pure dystrophin–glycoprotein complex were electrophoretically separated, transferred to nitrocellulose membranes and overlaid with ^{125}I-labelled extracellular matrix proteins (Ibraghimov-Beskrovnaya *et al.*, 1992). A single laminin-binding band, corresponding to the 156 kDa dystrophin-associated glycoprotein, was detected in surface membranes and pure dystrophin–glycoprotein complex. Binding of ^{125}I-labelled laminin to the 156 kDa dystrophin-associated glycoprotein was completely inhibited by inclusion of an excess of unlabelled laminin to the incubation medium. ^{125}I-labelled fibronectin did not label any component of the dystrophin–glycoprotein complex, nor did an excess of unlabelled fibronectin have any effect on the binding of ^{125}I-labelled laminin to the 156 kDa dystrophin-associated glycoprotein. The interaction of the 156 kDa dystrophin-associated glycoprotein with laminin was also shown by co-immunoprecipitation using anti-laminin antibodies. These results suggest that the 156 kDa dystrophin-associated glycoprotein specifically binds laminin and may mediate interaction of the dystrophin–glycoprotein complex with the extracellular matrix.

A number of laminin binding proteins have previously been identified in skeletal muscle (Lesot *et al.*, 1983; Clegg *et al.*, 1988; Mecham, 1991) but the 156 kDa dystrophin-associated glycoprotein does not appear to be related to any of these. In addition, the sequence of the 43/156 kDa dystrophin-associated glycoprotein indicates that it is not related to integrins or cadherin. It is also interesting that the conditions used to identify the 156 kDa dystrophin-associated glycoprotein as a laminin binding protein are similar to those which have been used for the identification of cranin as a laminin binding protein (Smalheiser and Schwartz, 1987). The apparent molecular weight of cranin is also very similar to the protein we have identified in brain membranes with the '156 kDa'-specific antibody.

We have proposed the name 'dystroglycan' (Ibraghimov-Beskrovnaya *et al.*, 1992) because of the 43/156 kDa dystrophin-associated glycoproteins' identification via dystrophin and its extensive glycosylation.

6.5 MOLECULAR PATHOGENESIS OF DUCHENNE MUSCULAR DYSTROPHY

Early histopathological events in DMD are characterized by persistent skeletal muscle necrosis. A central question of current muscular dystrophy research is how the absence of dystrophin causes muscle cell necrosis. In comparison, the *mdx* mouse is completely missing dystrophin (Bonilla *et al.*, 1988; Hoffman *et al.*, 1987) and also exhibits necrosis of skeletal muscle fibres. The absence of dystrophin accompanied by skeletal muscle necrosis makes the *mdx* mouse a good model system in which to study how muscle fibre necrosis is caused by the absence of dystrophin. To learn more about the early events in the molecular pathogenesis of muscular dystrophy, we investigated the relative abundance of all of the components of the dystrophin–glycoprotein complex in skeletal muscle membranes from *mdx* mice. Initially, we only investigated whether the 156 kDa dystrophin-associated glycoprotein was affected by the absence of dystrophin, because the monoclonal antibody to it was the only non-dystrophin probe to the complex which cross-reacted with mouse and human (Ervasti *et al.*, 1990).

Immunoblots of skeletal muscle membranes were prepared from control and *mdx* mice and stained with the various antibodies. Staining with polyclonal antiserum against the C-terminal decapeptide of dystrophin revealed that this protein was completely absent from *mdx* mouse membranes. In addition, comparison of normal and *mdx* mouse by immunostaining with the monoclonal antibody against the 156 kDa glycoprotein revealed that this too was absent or greatly reduced in *mdx* mouse membranes. The absence of the 156 kDa glycoprotein was also confirmed using SDS muscle extracts, instead of isolated membranes, from control and *mdx* mice. Estimation of the amount of 156 kDa glycoprotein remaining in the *mdx* muscle membranes using ^{125}I-labelled secondary antibodies and total membrane preparations from four different control and four different *mdx* mice revealed an average reduction of 85% in *mdx* muscle (Ervasti *et al.*, 1990).

Total muscle extracts were also prepared from biopsy samples of normal controls and patients with Duchenne muscular dystrophy. The dystrophic samples exhibited no staining with antibodies against dystrophin by indirect immunofluorescence microscopy and immunoblotting. In contrast to the normal muscle extract the three DMD samples showed greatly reduced staining for the 156 kDa glycoprotein (Ervasti *et al.*, 1990). Identical immunoblots stained with monoclonal antibodies against the sarcoplasmic reticulum Ca^{2+}-dependent ATPase revealed no difference in the staining intensity between normal and dystrophic muscle samples. As in the case of *mdx* mouse muscle, the amount of 156 kDa glycoprotein was estimated to be reduced by approximately 90% in DMD samples. The drastic reduction of the 156 kDa dystrophin-associated glycoprotein (the component of the complex most distal to dystrophin) in muscle from *mdx* mice and DMD patients is evidence that alteration in dystrophin expression profoundly affects components external to the muscle cell.

In evaluating studies of other diseases, it is apparent that loss of a protein due to genetic defect often results in the loss of associated proteins. For example, spectrin deficiency in hereditary elliptocytosis is also associated with a reduced abundance in protein-4.1 and minor sialoglycoproteins (Alloisio *et al.*, 1985). Skeletal muscle phosphorylase kinase deficiency, which is caused by a single gene defect on the X chromosome, is characterized by the combined loss of all four subunits of this enzyme (Cohen *et al.*, 1976). However, a generalized loss of components in a protein complex is not observed in the genetic disease muscular dysgenesis. This disorder results in a complete absence of skeletal muscle contraction due to the failure of depolarization of the transverse tubular membrane to trigger calcium release from sarcoplasmic reticulum. Interestingly, only the α_1-subunit of the dihydropyridine receptor is absent in dysgenic mice while the alpha$_2$-subunit of the receptor is present (Knudson *et al.*, 1989). In view of these findings, it was important to investigate the status of all the dystrophin-associated proteins in *mdx* skeletal muscle. The relative abundance of each of the components of the dystrophin–glycoprotein complex in skeletal muscle was determined from normal and *mdx* mice using antibodies specific for each of the dystrophin-associated glycoproteins (Ohlendieck and Campbell, 1991a). Immunoblot analysis using total muscle membranes from control and *mdx* mice found that all of the dystrophin-associated proteins were greatly reduced in *mdx* mouse skeletal muscle. The specificity of the loss of the dystrophin-associated glycoproteins was demonstrated by the finding that the major glycoprotein composition of skeletal muscle membranes from normal and *mdx* mice was identical. Densitometric scanning of ^{125}I-Protein A-labelled immunoblots revealed an average 84% reduction for the 156 kDa, 59 kDa, 50 kDa, 43 kDa and 35 kDa dystrophin-associated glycoproteins in *mdx* muscle membranes when compared to control membranes (Ohlendieck and Campbell, 1991a). The comparative densitometric scanning was performed with individually isolated membranes from five 10-week-old control mice and five 10-week-old *mdx* mice. A similarly reduced expression of dystrophin-associated proteins was also observed in membranes isolated from 1, 2, 5, 20 and 30-week-old *mdx* mice as compared to age-matched control mice. Immunofluorescence microscopy confirmed that the density of dystrophin-associated proteins is greatly reduced in skeletal muscle cryosections from *mdx* mice. These findings strongly suggest that the deficiency of dystrophin-associated proteins in *mdx* mouse muscle is a primary event following the absence of dystrophin and that a reduction in dystrophin-associated proteins may initiate muscle cell necrosis.

The murine mutant *dystrophia muscularis dy/dy* which has an autosomal-recessive mode of inheritance is another animal model for muscular dystrophy which exhibits progressive and severe degeneration of skeletal muscle fibres (Bray and Banker, 1970). Coomassie Blue staining revealed no apparent differences between membranes isolated from control and *dy/dy* mouse skeletal muscle and the density of dystrophin and dystrophin-related protein is also comparable between both membrane preparations (Ohlendieck and Campbell, 1991a). Most importantly, antibodies to the different dystrophin-associated proteins showed approximately equal amounts of these proteins in skeletal muscle membranes from control and *dy/dy* mice (Ohlendieck and Campbell, 1991a). These findings demonstrate that dystrophin-associated proteins are not affected by secondary events in necrotic muscle and suggest that the reduced density of dystrophin-associated proteins in skeletal muscle membranes from *mdx* mice is most likely due to the absence of dystrophin from the membrane cytoskeleton of *mdx* muscle.

One could envision three different mechanisms to account for the loss of dystrophin-associated proteins from the cell surface of dystrophin-deficient muscle. First, point mutations, deletions or duplications in the DMD gene which result in the absence or abnormal structure of dystrophin could affect the transcription, processing or stability of dystrophin-associated protein mRNAs. Second, an absence or abnormality in dystrophin could cause a decrease in translation and/or assembly of the components of the dystrophin–glycoprotein complex. Third, loss of dystrophin-associated proteins could be due to an increase in degradative pathways.

To begin to address which of these three possible mechanisms may account for the loss of dystrophin-associated proteins in dystrophin-less tissues, Northern blots of skeletal muscle mRNA from control and *mdx* mice of different ages were probed using radiolabelled cDNA corresponding to the 43/156 kDa dystrophin-associated glycoprotein (Ibraghimov-Beskrovnaya *et al.*, 1992). Northern blot analysis revealed no reduction of 43/156 kDa dystrophin-associated glycoprotein mRNA in *mdx* mice vs. control mice. Thus, the absence of dystrophin causes no change in the mRNA for the 43/156 kDa dystrophin-associated glycoprotein but leads to dramatic reductions in the amount of the 43 kDa and 156 kDa dystrophin-associated glycoproteins in skeletal muscle. Analysis of mRNA from control and DMD skeletal muscle also showed no difference in 43/156 kDa dystrophin-associated glycoprotein mRNA expression. Thus, the 43/156 kDa dystrophin-associated glycoprotein encoding gene is transcribed and specific mRNA is still present at the normal level in dystrophic muscle, but the amount of 43 kDa and 156 kDa dystrophin-associated glycoproteins is greatly reduced in dystrophic muscle.

Since the 43/156 kDa dystrophin-associated glycoproteins are expressed in non-muscle tissues, we also examined expression of the 43 kDa dystrophin-associated glycoprotein in non-muscle tissues of control and *mdx* mice. Immunoblot analysis of brain and kidney membranes from control and *mdx* mice, stained with polyclonal anti-43 kDa dystrophin-associated glycoprotein antibodies, revealed no reduction in the amount of 43 kDa dystrophin-associated glycoprotein in these *mdx* tissues. Thus, the dramatic reduction of the 43 kDa dystrophin-associated glycoprotein that is found in *mdx* mice appears to be restricted to skeletal muscle and is not found in non-muscle tissues.

The loss of dystrophin-associated proteins from the muscle cell surface could principally occur in two different ways: (1) Translation of dystrophin-associated proteins may be downregulated or dystrophin-associated proteins could be synthesized in normal amounts but may not be properly assembled into an oligomeric complex due to the lack of dystrophin; (2) dystrophin-associated proteins may be synthesized and assembled correctly but due to the deficiency in dystrophin the membrane complex will lack the proper interaction with the actin cytoskeleton, resulting in greater mobility

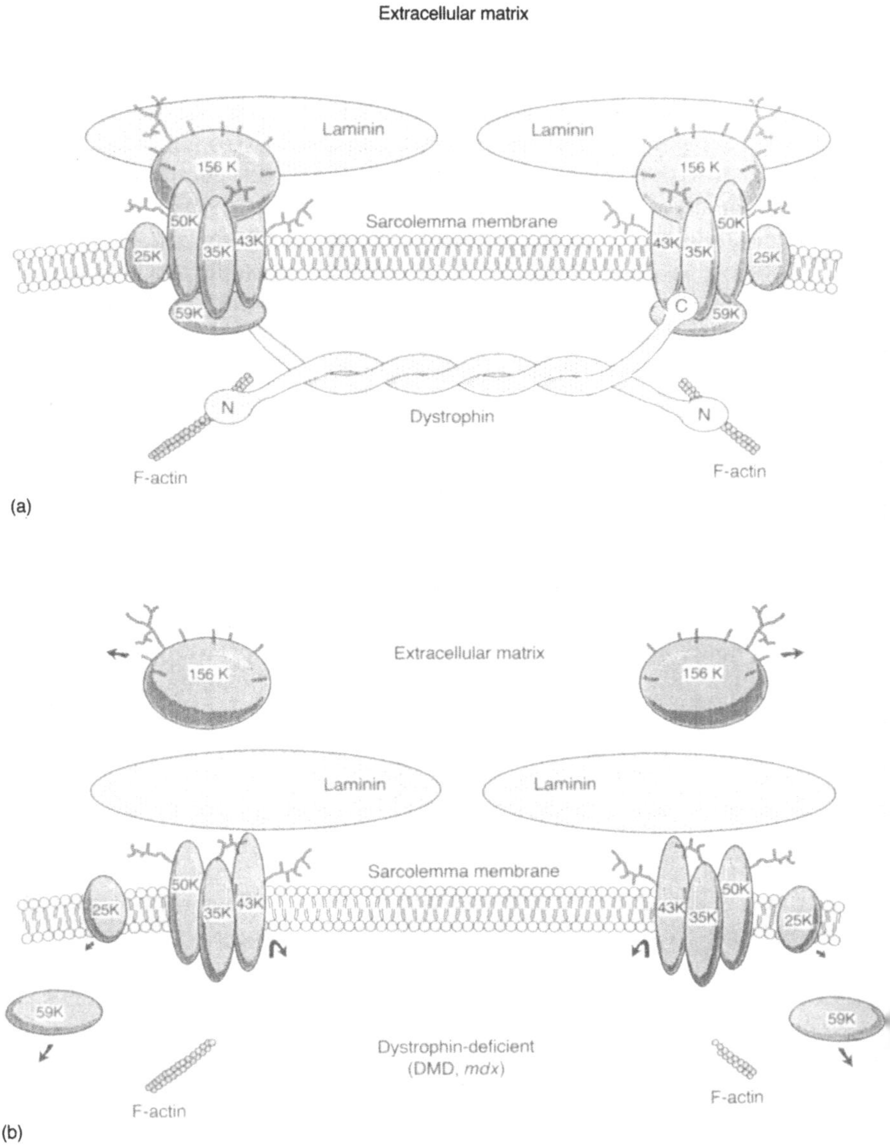

Figure 6.4 Model of dystrophin-associated glycoproteins in the absence of dystrophin. Shown is the proposed model of the dystrophin–glycoprotein complex in normal skeletal muscle (a) and in dystrophin-deficient skeletal muscle (b). In the absence of dystrophin, dystrophin-associated proteins are present in greatly reduced amounts as a result of either down-regulation of synthesis or increased protein degradation. After Ervasti and Campbell (1991).

of the membrane complex which may render the protein components of the complex more vulnerable to degradation (Figure 6.4).

Our findings suggest that the function of dystrophin is to link the subsarcolemma membrane cytoskeleton through a transmembrane complex to an extracellular glycoprotein which binds laminin. Since the absence of dystrophin leads to the loss of all the dystrophin-associated proteins (Ervasti *et al.*, 1990; Ohlendieck and Campbell, 1991a) our results suggest that dystrophin-deficient muscle fibres may lack the normal interaction between the sarcolemma and the extracellular matrix. Disruption of the various components involved in the structural link between subsarcolemmal cytoskeleton and extracellular matrix may severely weaken the flexibility of the sarcolemma membrane during skeletal muscle contraction (Figure 6.5).

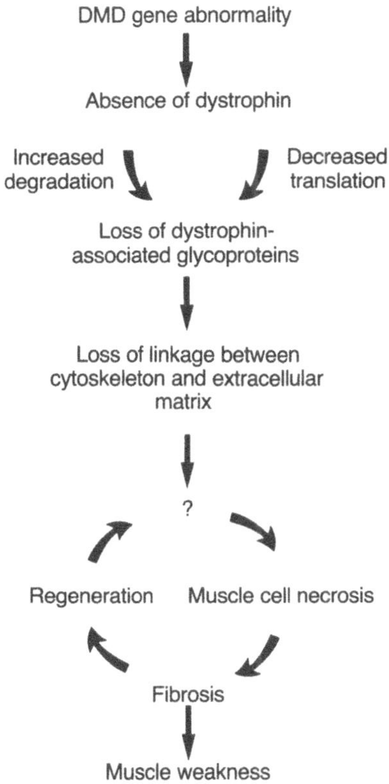

Figure 6.5 Pathway for molecular pathogenesis of Duchenne muscular dystrophy. Proposed sequence of events leading to the molecular pathogenesis of Duchenne muscular dystrophy. The sequence of events leading from loss of linkage between the cytoskeleton and the extracellular matrix to muscle cell necrosis (?) are presently unresolved. Please refer to the text for discussion of the current hypotheses regarding these unresolved steps.

This hypothesis is supported by the histopathological finding that DMD muscle fibres exhibit an early separation between muscle cell surface and basal lamina (Bonilla and Moggio, 1986).

Comprehensive analysis of DMD skeletal muscle shows that muscle cell necrosis is preceded by a breakdown of the plasma membrane (Carpenter and Karpati, 1979; Engel and Banker, 1986; Mokri and Engel, 1975). In addition, skeletal muscle fibres from *mdx* mice exhibited in contraction experiments an enhanced vulnerability which may render the sarcolemma more susceptible to suffer focal breaks (Weller *et al.*, 1990). Recent findings demonstrated that skeletal muscle fibres from *mdx* mouse are more fragile and have a decreased osmotic stability (Menke and Jockusch, 1991). It is not known if the osmotic fragility is based only on the absence of dystrophin in the sarcolemma membrane cytoskeleton or if it is also linked to the reduced density of dystrophin-associated proteins or other secondarily affected proteins in *mdx* skeletal muscle. Alternatively, the disruption of this linkage between the sarcolemma and the extracellular matrix may be responsible for the alteration of specific Ca^{2+} regulatory mechanisms (Franco and Lansman, 1990) which may lead to excessive influx of Ca^{2+} ions in dystrophic muscle (Turner *et al.*, 1991). Since the extracellular matrix of the adult tissue is a scaffold that is required to allow repair following injury, the absence of the interaction between the sarcolemma and the extracellular matrix may also render dystrophic muscle fibres more prone to injury and less able to repair such injury.

The normal production of the mRNA for the 43/156 kDa dystrophin-associated glycoprotein in dystrophic muscle is important for potential DMD therapies. Prior to these findings, it was unclear how the absence of dystrophin led to the loss of the dystrophin-associated glycoproteins. Our work now indicates that dystrophin-associated glycoproteins are produced in dystrophic muscle but that the dystrophin-associated glycoproteins may not be properly assembled and/or integrated into the sarcolemma or may be degraded. These results suggest that restoring dystrophin by myoblast transfer (Partridge *et al.*, 1989; Gussoni *et al.*, 1992) or gene therapy (Lee *et al.*, 1991; Ascadi *et al.*, 1991) may stabilize and restore normal dystrophin-associated glycoprotein levels in DMD muscle.

Finally, it would be interesting to examine the status of the dystrophin–glycoprotein complex in neuromuscular diseases which have yet to be characterized at the molecular level. For example, perhaps a deficiency or abnormality in a dystrophin-associated glycoprotein could explain the DMD-like symptoms observed in suspected autosomal recessive patients (Arahata *et al.*, 1989; Francke *et al.*, 1989; Vainzof *et al.*, 1990; Ben Jelloun-Dellagi *et al.*, 1990) that express apparently normal dystrophin.

SUMMARY

Dystrophin constitutes approximately 5% of the cytoskeletal protein of skeletal muscle sarcolemma, suggesting that dystrophin could play a major structural role in skeletal muscle. We have presented evidence for the existence of a large oligomeric complex containing dystrophin, a 59 kDa triplet, a 25 kDa protein and four sarcolemmal glycoproteins with apparent M_r of 156 kDa, 50 kDa, 43 kDa and 35 kDa. All components of the dystrophin–glycoprotein complex were localized to the skeletal muscle sarcolemma. Dystrophin, the 156 kDa and 59 kDa dystrophin-associated protein were found to be peripheral membrane proteins while the 50 kDa, 43 kDa, 35 kDa and 25 kDa dystrophin-associated proteins were confirmed as integral membrane proteins. The primary sequences of the 43 kDa and 156 kDa dystrophin-associated glycoproteins have been established by recombinant DNA techniques. Both the 43 and 156 kDa dystrophin-associated glycoproteins are encoded by a single 5.8 kb mRNA which is expressed in a variety of tissues in addition to skeletal muscle. The 156 kDa dystrophin-associated glycoprotein binds laminin, a well characterized component of the extracellular matrix. Finally, the dystrophin–glycoprotein complex is specifically and greatly reduced in Duchenne-afflicted and *mdx* mouse skeletal muscle, suggesting that the loss of dystrophin-associated proteins is due to the absence of dystrophin and not due to secondary effects of muscle fibre degradation. Taken together, these data support the hypothesis that the absence of dystrophin leads to a loss of the linkage between the subsarcolemmal cytoskeleton and extracellular matrix and that this may initiate muscle fibre necrosis.

ACKNOWLEDGEMENTS

Kevin P. Campbell is an Investigator of the Howard Hughes Medical Institute.

REFERENCES

Alloisio, N., Morle, L., Bachir, D. *et al.* (1985) Red cell membrane sialoglycoprotein B in homozygous and heterozygous 4.1(–) hereditary elliptocytosis. *Biochim. Biophys. Acta*, **816**, 57.

Arahata, K., Ishiura, S., Ishiguro, T. *et al.* (1988) Immunostaining of skeletal and cardiac muscle surface membrane with antibody against Duchenne muscular dystrophy peptide. *Nature*, 333, 861–6.

Arahata, K., Ishihara, T., Kamakura, K. *et al.* (1989) Mosaic expression of dystrophin in symptomatic carriers of Duchenne's muscular dystrophy. *N. Engl. J. Med.*, 320, 138–42.

Ascadi, G., Dickson, G., Love, D.R. *et al.* (1991) Human dystrophin expression in mdx mice after intramuscular injection of DNA constructs. *Nature*, **352**, 815–8.

Barchi, R.L. Weigele, J.B., Chalakian, D.M. and Murphy, L.E. (1979) Muscle surface membranes: preparative methods affect apparent chemical properties and neurotoxin binding. *Biochim. Biophys. Acta*, **550**, 59–76.

Ben Jelloun-Dellagi, S., Chaffey, P., Hentati, F. *et al.* (1990) Presence of normal dystrophin in Tunisian severe childhood autosomal recessive muscular dystrophy. *Neurology*, **40**, 40.

Bennett, V. (1990) Spectrin-based membrane skeleton: a multipotential adaptor between plasma membrane and cytoplasm. *Physiol. Rev.*, **70**, 1029–65.

Bennett, V., Davis, J. and Fowler, W.E. (1982) Brain spectrin, a membrane-associated protein related in structure and function to erythrocyte spectrin. *Nature*, **299**, 126–30.

Bhavanandan, V.P. and Katlic, A.W. (1979) The interaction of wheat germ agglutinin with sialoglycoproteins. *J. Biol. Chem.*, **254**, 4000–8.

Bonilla, E. and Moggio, E. (1986) *Neurology*, **36** (Suppl. 1) 171.

Bonilla, E., Samitt, C.E., Miranda, A.F. *et al.* (1988) Duchenne muscular dystrophy: deficiency of dystrophin at the muscle cell surface. *Cell*, **54**, 447–52.

Bourdon, M.A., Krusius, T., Campbell, S. *et al.* (1987) Identification and synthesis of a recognition signal for the attachment of glycosaminoglycans to proteins. *Proc. Natl. Acad. Sci. USA*, **84**, 3194–8.

Bray, G.M. and Banker, B.D. (1970) An ultrastructural study of degeneration and necrosis of muscle in the dystrophic mouse. *Acta Neuropathol.*, **15**, 34–44.

Bresnick, A.R., Warren, V. and Condeelis, J. (1990) Identification of a short sequence essential for actin binding by Dictyostelium ABP-120. *J. Biol. Chem.*, **265**, 9236–40.

Byers, T.J., Kunkel, L.M. and Watkins, S.C. (1991) The subcellular distribution of dystrophin in mouse skeletal, cardiac and smooth muscle. *J. Cell Biol.*, **115**, 411–421.

Campbell, K.P. and Kahl, S.D. (1989) Association of dystrophin and an integral membrane glycoprotein. *Nature*, **338**, 259–62.

Campbell, K.P., Ervasti, J.M., Ohlendieck, K. and Kahl, S.D. (1991) The dystrophin-glycoprotein complex: identification and biochemical characterization. In: *Frontiers in Muscle Research*, (eds E. Ozawa, T. Masaki and Y. Nabeshima), Elsevier, Amsterdam, pp. 321–40.

Carney, S.L. (1986) Proteoglycans. In: *Carbohydrate Analysis: A Practical Approach*, (eds M.F. Chaplin and J.F. Kennedy), IRL Press, Oxford, pp. 97–141.

Carpenter, S. and Karpati, G. (1979) Duchenne muscular dystrophy: plasma membrane loss initiates muscle cell necrosis unless it is repaired. *Brain*, **102**, 147–61.

Carraway, K.L. and Carothers-Carraway, C.A. (1989) Membrane-cytoskeleton interactions in animal cells. *Biochim. Biophys. Acta*, **988**, 147–71.

Chang, H.W., Bock, E. and Bonilla, E. (1989) Dystrophin in electric organ of *Torpedo californica* homologous to that in human muscle. *J. Biol. Chem.*, **264**, 20831–4.

Charuk, J.H.M., Howlett, S. and Michalak, M. (1989) Subfractionation of cardiac sarcolemma with wheat-germ agglutinin. *Biochem. J.*, **264**, 885–92.

Clegg, D.O., Helder, J.C., Hann, B.C. *et al.* (1988) Amino acid sequence and distribution of mRNA encoding a major muscle laminin binding protein: An extracellular matrix-associated protein with an unusual COOH-terminal poly-aspartate domain. *J. Cell Biol.*, **107**, 699–705.

Cohen, P.T.W., Burchell, A. and Cohen, P. (1976) The molecular basis of skeletal muscle phosphorylase kinase deficiency. *Eur. J. Biochem.*, **66**, 347–56.

Collins, F.S. (1992) Positional cloning: Let's not call it reverse anymore. *Nature Genetics*, **1**, 3–6.

Cooper, B.J., Winand, N.J., Stedman, H. *et al.* (1988) The homologue of the Duchenne locus is defective in X-linked muscular dystrophy of dogs. *Nature*, **334**, 154–6.

Crawford, A.W. and Beckerle, M.C. (1991) Purification and characterization of zyxin, an 82 000-dalton component of adherens junctions. *J. Biol. Chem.*, **266**, 5847–53.

Cullen, M.J., Walsh, J., Nicholson, L.V.B. and Harris, J.B. (1990) Ultrastructural localization of dystrophin in human muscle by using gold immunolabelling. *Proc. R. Soc. Lond.*, **240**, 197–210.

Cullen, M.J., Nicholson, L.V.B., Harris, J.B. *et al.* (1991) Immunogold labelling of dystrophin in human muscle using an antibody to the last 17 amino acids of the C-terminus. *Neuromuscular Disorders*, **1**, 113–19.

Cummings, R.D., Kornfeld, S., Schneider, W.J. *et al.* (1983) Biosynthesis of the N- and O-linked oligosacharides of the low density lipoprotein receptor. *J. Biol. Chem.*, **258**, 15261–73.

Engel, A.G. and Banker, B.Q. (1986) *Myology: Basic and Clinical*, McGraw-Hill Inc., New York, pp. 1–2159.

Ervasti, J.M. and Campbell, K.P. (1991) Membrane organization of the dystrophin-glycoprotein complex. *Cell*, **66**, 1121–31.

Ervasti, J.M., Ohlendieck, K., Kahl, S.D., *et al.* (1990) Deficiency of a glycoprotein component of the dystrophin complex in dystrophic muscle. *Nature*, **345**, 315–9.

Ervasti, J.M., Kahl, S.D. and Campbell, K.P. (1991) Purification of dystrophin from skeletal muscle. *J. Biol. Chem.*, **266**, 9161–5.

Fambrough, D.M. and Bayne, E.K. (1983) Multiple Forms of (Na+K)-ATPase in the chicken. *J. Biol. Chem.*, **358**, 3926–35.

Francke, U., Darras, B.T., Hersh, J.H. *et al.* (1989) Brother/sister pairs affected with early-onset, progressive muscular dystrophy: molecular studies reveal etiologic hetereogeneity. *Am. J. Hum. Genet.*, **45**, 63–72.

Franco, A. and Lansman, J.B. (1990) Calcium entry through stretch-inactivated ion channels in mdx myotubes. *Nature*, **344**, 670–3.

Gussoni, E., Pavlath, G.K., Lanctot, A.M. *et al.* (1992) Normal dystrophin transcripts detected in Duchenne muscular dystrophy patients after myoblast transplantation. *Nature*, **356**, 435–8.

Helliwell, T.R., Ellis, J.M., Mountford, R.C. *et al.* (1992) A truncated dystrophin lacking the C-terminal domain is localized at the muscle membrane. *Am. J. Hum. Genet.*, 58, 508–14.

Hemmings, L., Kuhlman, P.A. and Critchley, D.R. (1992) Analysis of the actin-binding domain of α-actinin by mutagenesis and demonstration that dystrophin contains a functionally homologous domain. *J. Cell Biol.*, **116**, 1369–80.

Hoffman, E.P., Brown, R.H. and Kunkel, L.M. (1987) Dystrophin: the protein product of the Duchenne muscular dystrophy locus. *Cell*, **51**, 919–28.

Hoffman, E.P., Fischbeck, K.H., Brown, R.H. *et al.* (1988a) Characterization of dystrophin in muscle-biopsy specimens from patients with Duchenne's or Becker's muscular dystrophy. *N. Engl. J. Med.*, **318**, 1363–8.

Hoffman, E.P., Hudecki, M.S., Rosenberg, P.A. *et al.* (1988b) Cell and fiber-type distribution of dystrophin. *Neuron*, **1**, 411–20.

Hoffman, E.P., Garcia, C.A., Chamberlain, J.S. *et al.* (1991) Is the carboxyl-terminus of dystrophin required for membrane association? A novel, severe case of Duchenne muscular dystrophy. *Ann. Neurol.*, **30**, 605–10.

Ibraghimov-Beskrovnaya, O., Ervasti, J.M., Leveille, C.J. *et al.* (1992) Primary structure of dystrophin-associated glycoproteins linking dystrophin to the extracellular matrix. *Nature*, **355**, 696–702.

Jorgensen, A.O., Arnold, W., Shen, A.C.-Y. *et al.* (1990) Identification of novel proteins unique to either transverse tubules (TS28) or the sarcolemma (SL50) in rabbit skeletal muscle. *J. Cell Biol.*, **110**, 1173–85.

Karinch, A.M., Zimmer, W.E. and Goodman, S.R. (1990) The identification and sequence of the actin-binding domain of human red blood cell beta-spectrin. *J. Biol. Chem.*, **265**, 11833–40.

Khurana, T.S., Hoffman, E.P. and Kunkel, L.M. (1990) Identification of a chromosome 6-encoded dystrophin-related protein. *J. Biol. Chem.*, **265**, 16717–20.

Knudson, C.M., Chaudhari, N., Sharp, A.H. *et al.* (1989) Specific absence of the a1 subunit of the dihydropyridine receptor in mice with muscular dysgenesis. *J. Biol. Chem.*, **264**, 1345–8.

Koenig, M. and Kunkel, L.M. (1990) Detailed analysis of the repeat domain of dystrophin reveals four potential hinge segments that may confer flexibility. *J. Biol. Chem.*, **265**, 4560–6.

Koenig, M., Monaco, A.P. and Kunkel, L.M. (1988) The complete sequence of dystrophin predicts a rod-shaped cytoskeletal protein. *Cell*, **53**, 219–28.

Korsgren, C. and Cohen, C.M. (1986) Purification and properties of human erythrocyte band 4.2. *J. Biol. Chem.*, **261**, 5536–43.

Lee, C.C., Pearlman, J.A., Chamberlain, J.S. and Caskey, C.T. (1991) Expression of recombinant dystrophin and its localization to the cell membrane. *Nature*, **349**, 334–6.

Lesot, H., Kuhl, U. and von der Mark, K. (1983) Isolation of a laminin-binding protein from muscle cell membranes. *EMBO J.*, **2**, 861–5.

Leung, A.T., Imagawa, T. and Campbell, K.P. (1987) Structural characterization of the dihydropyridine receptor of the voltage-dependent Ca^{2+} channel from rabbit skeletal muscle: evidence for two distinct high molecular weight subunits. *J. Biol. Chem.*, **2623**, 7943–6.

Love, D.R., Hill, D.F., Dickson, G. *et al.* (1989) An autosomal transcript in skeletal muscle with homology to dystrophin. *Nature*, **339**, 55–8.

Mecham, R.P. (1991) Receptors for laminin on mammalian cells. *FASEB J.*, **5**, 2538–46.

Menke, A. and Jockusch, H. (1991) Decreased osmotic stability of dystrophin-less muscle cells from the mdx mouse. *Nature*, **349**, 69–71.

Moczydlowski, E.G. and Latorre, R. (1983) Saxitoxin and ouabain binding activity of isolated skeletal muscle membrane as indicators of surface origin and purity. *Biochim. Biophys. Acta*, **732**, 412–20.

Mokri, B. and Engel, A.G. (1975) Duchenne dystrophy: Electron microscopic findings pointing to a basic or early abnormality in the plasma membrane of the muscle fiber. *Neurology*, **25**, 1111–20.

Monaco, A.P., Bertelson, C.J., Middlesworth, W. *et al.* (1985) Detection of deletions spanning the Duchenne muscular dystrophy locus using a tightly linked DNA segment. *Nature*, **316**, 842–5.

Moore, S.E., Thompson, J., Kirkness, V. *et al.* (1987) Skeletal muscle neural cell adhesion molecule (N-CAM): Changes in protein and mRNA species during myogenesis of muscle cell lines. *J. Cell Biol.*, **105**, 1377–86.

Mueller, T.J. and Morrison, M. (1981) Glycoconnectin (PAS 2), a membrane attachment site for the human erythrocyte cytoskeleton. In: *Erythrocyte Membrane 2: Recent Clinical and Experimental Advances*, Alan R. Liss, New York, pp. 95–112.

Murayama, T., Osamu, S., Kimura, S. *et al.* (1990) Molecular shape of dystrophin purified from rabbit skeletal muscle myofibrils. *Proc. Japan. Acad.*, **66**, 96–9.

Ohlendieck, K. and Campbell, K.P. (1991a) Dystrophin-associated proteins are greatly reduced in skeletal muscle from mdx mice. *J. Cell Biol.*, **115**, 1685–94.

Ohlendieck, K. and Campbell, K.P. (1991b) Dystrophin constitutes 5% of membrane cytoskeleton in skeletal muscle. *FEBS Letters*, **283**, 230–4.

Ohlendieck, K., Ervasti, J.M., Matsumura, K. *et al.* (1991a) Dystrophin-related protein is localized to neuromuscular junctions of adult skeletal muscle. *Neuron*, **7**, 499–508.

Ohlendieck, K., Ervasti, J.M., Snook, J.B. and Campbell, K.P. (1991b) Dystrophin-glycoprotein complex is highly enriched in isolated skeletal muscle sarcolemma. *J. Cell Biol.*, **112**, 135–48.

Partridge, T.A., Morgan, J.E., Coulton, G.R. *et al.* (1989) Conversion of mdx myofibres from dystrophin-negative to -positive by injection of normal myoblasts. *Nature*, **337**, 176–9.

Pons, F., Augier, N., Heilig, R. *et al.* (1990) Isolated dystrophin molecules as seen by electron microscopy. *Proc. Natl. Acad. Sci. USA*, **87**, 7851–5.

Ray, P.N., Belfall, B., Duff, C. (1985) Cloning of the breakpoint of an X;21 translocation associated with Duchenne muscular dystrophy. *Nature*, **318**, 672–5.

Recan, D., Chafey, P., Leturcq, F. *et al.* (1992) Are cysteine-rich and COOH-terminal domains of dystrophin critical for sarcolemmal localization? *J. Clin. Invest.*, **89**, 712–16.

Ruoslahti, E. and Pierschbacher, M.D. (1987) New perspectives in cell adhesion: RGD and integrins. *Science*, **238**, 491–7.

Salas, P.J.I., Vega-Salas, D.E., Hochman, J. *et al.* (1988) Selective anchoring in the specific plasma membrane domain: a role in epithelial cell polarity. *J. Cell Biol.*, 107, 2363–76.

Schafer, D.A. and Stockdale, F.E. (1987) Identification of sarcolemma-associated antigens with differential distribution on fast and slow skeletal muscle fibres. *J. Cell Biol.*, **104**, 967–79.

Seiler, S. and Fleischer, S. (1982) Isolation of plasma membrane vesicles from rabbit skeletal muscle and their use in ion transport studies. *J. Biol. Chem.*, **257**, 13862–71.

Seiler, S. and Fleischer, S. (1988) Isolation and characterization of sarcolemma vesicles from rabbit fast skeletal muscle. *Meth. Enzymol.*, **157**, 26–36.

Smalheiser, N.R. and Schwartz, N.B. (1987) Cranin: A laminin-binding protein of cell membranes. *Proc. Natl. Acad. Sci. USA*, **84**, 6457–61.

Steck, T.L. and Yu, J. (1973) Selective solubilization of proteins from red blood cell membranes by protein perturbants. *J. Supramol. Struct.*, **1**, 220–32.

Takeichi, M. (1991) Cadherin cell adhesion receptors as a morphogenetic regulator. *Science*, **251**, 1451–5.

Turner, P.R., Fong, P., Denetclaw, W.F. and Steinhardt, R.A. (1991) Increased calcium influx in dystrophic muscle. *J. Cell Biol.*, **115**, 1701–12.

Vainzof, M., Pavanello, R.C.M., Filho, I.P. *et al.* (1990) Dystrophin immunostaining in muscles from patients with different types of muscular dystrophy: a Brazilian study. *J. Neurol. Sci.*, **98**, 221–33.

Walsh, F.S., Parekh, R.B., Moore, S.E. *et al.* (1989) Tissue specific O-linked glycosylation of the neural cell adhesion molecule (NCAM). *Development*, **105**, 803–11.

Watkins, S.C., Hoffman, E.P., Slayter, H.S. and Kunkel, L.M. (1988) Immunoelectron microscopic localization of dystrophin in myofibres. *Nature*, **333**, 863–6.

Webster, C., Silberstein, L., Hays, A.P. and Blau, H.M. (1988) Fast muscle fibres are preferentially affected in Duchenne muscular dystrophy. *Cell*, **52**, 503–13.

Weller, B., Karpati, G. and Carpenter, S. (1990) Dystrophin-deficient mdx muscle fibers are preferentially vulnerable to necrosis induced by experimental lengthening contractions. *J. Neurol. Sci.*, **100**, 9–13.

Yeadon, J.E., Lin, H., Dyer, S.M. and Burden, S.J. (1991) Dystrophin is a component of the subsynaptic membrane. *J. Cell Biol.*, **115**, 1069–76.

Yoshida, M. and Ozawa, E. (1990) Glycoprotein complex anchoring dystrophin to sarcolemma. *J. Biochem.*, **108**, 748–52.

Zubrzycka-Gaarn, E.E., Bulman, D.E., Karpati, G. *et al.* (1988) The Duchenne muscular dystrophy gene product is localized in sarcolemma of human skeletal muscle. *Nature*, **333**, 466–9.

7

PCR analysis of muscular dystrophy in *mdx* mice

JEFFREY S. CHAMBERLAIN, STEPHANIE F. PHELPS, GREGORY A. COX, ANDREA J. MAICHELE and ALEX D. GREENWOOD

7.1 INTRODUCTION

Animal models for human genetic diseases provide a biological system in which many of the basic features of a disease can be studied and in which approaches for therapeutic intervention can be tested. In many cases it is difficult to obtain human disease tissue samples in sufficient quantity or at particular stages of development to use in identifying the underlying pathological basis of the disease or to study the expression and mutation of the disease gene. The availability of an appropriate animal model of a human genetic disease enables exploration of many of these issues in the laboratory and can provide a rich source of information that can be applied to understanding or treating the corresponding human disease. Duchenne and Becker muscular dystrophies (DMD/BMD) are among the most common human genetic diseases. These allelic, X-linked recessive disorders are prevalent due to a high frequency of new mutation, with one-third of all cases showing no prior family history (Moser, 1984). Several mouse models, referred to as *mdx* mice, are available for the study of DMD/BMD (Bulfield *et al.*, 1984; Chapman *et al.*, 1989). The *mdx* mutants have proved valuable in identifying the gene product of the DMD locus (dystrophin), have enabled an analysis of the expression and isoform diversity of the dystrophin gene, have provided a model system for testing mutation detection strategies, and are likely to prove useful as a model system for testing the feasibility of gene replacement therapy for DMD/BMD.

The dystrophin gene presents a number of challenges that have slowed analysis of the gene and the study of DMD. The human gene is approxi-

Molecular and Cell Biology of Muscular Dystrophy
Edited by Terence Partridge
Published in 1993 by Chapman & Hall, London. ISBN 0 412 43440 7

mately 2.4 Mb in size, and appears to be of similar large size in all vertebrate species examined (Lemaire *et al.*, 1988; Hoffman *et al.*, 1987b; Chamberlain *et al.*, 1989; den Dunnen *et al.*, 1989). The major dystrophin mRNA is 14 kb, the protein product is 427 kDa, and both are expressed at extremely low levels in muscle and nerve tissues [approximately 0.02% of the total mRNA or protein (Hoffman *et al.*, 1987b)]. A number of dystrophin isoforms have been described in muscle and nerve cells (Feener *et al.*, 1989; Bies *et al.*, 1992), and recently a non-muscle isoform of dystrophin has been identified in liver and testis (Bar *et al.*, 1990). These additional isoforms are expressed at lower levels than the major dystrophin isoform found in muscle, and can be difficult to detect with conventional recombinant DNA and immunological approaches. Transcripts of the dystrophin gene have been reported to be present at lower than normal levels in tissues of animals or humans with a mutant dystrophin gene, further reducing the accessibility of the transcript (Chelly *et al.*, 1990; Chamberlain *et al.*, 1988; Scott *et al.*, 1988). We have approached the study of DMD using normal and mutant mice as a model system for understanding the expression and mutation of dystrophin. Much of this work would not have been possible without the availability of polymerase chain reaction techniques for gene and mRNA amplification. This article describes the application of PCR methods to study the normal and mutant dystrophin genes and their pattern of expression in mice. These approaches are applicable to the study of other animal models as well as the corresponding human genetic diseases.

7.2 MOUSE DYSTROPHIN cDNA CLONES

The starting point for our studies of *mdx* mice was the isolation of cDNA clones for the murine dystrophin gene. A variety of approaches was used to isolate the 14 kb mRNA in overlapping fragments, determine their sequence, and construct a full-length version of the cDNA (Lee *et al.*, 1991). A combination of oligonucleotide probes for murine dystrophin gene exons and human cDNA probes were used to isolate the majority of the mouse cDNA (Figure 7.1). Most of the murine dystrophin cDNA sequence was deduced from analysis of these clones by direct sequencing of selected subclones using vector and insert-specific primers. Short gaps between some clones as well as regions of ambiguous cDNA sequence resulting from cloning artifacts were isolated and directly sequenced by PCR amplification of reverse transcribed RNA (Gibbs *et al.*, 1989).

Sequence analysis has revealed the dystrophin gene to be highly conserved between humans and mice. The mouse and human cDNAs share 91.7% amino acid identity, with a slightly lower degree of conservation at the DNA level (Koenig *et al.*, 1988; Chamberlain *et al.*, 1991b). This sequence identity is particularly well conserved at both the 5′ and 3′ ends of the

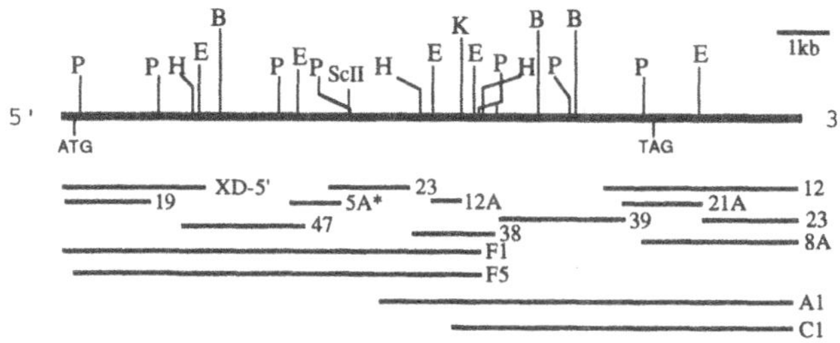

Figure 7.1 Restriction map of isolated dystrophin cDNA clones. Shown at the top is a restriction map of the murine dystrophin cDNA. Also indicated are the locations of the muscle translation start site and the adult translation stop site. B, *Bam*HI; E, *Eco*RI; H, *Hind*III; P, *Pst*I; K, *Kpn*I; ScII, *Sac*II. Below the map is indicated the approximate location of various cDNA clones that have been isolated. The clone designated with an asterisk was isolated by PCR. Clones C1, C5, F1 and F5 were used to prepare full-length cDNAs, and have been described (Lee *et al.*, 1991).

mRNA. Only the central part of the mRNA sequence between the region encoded by exons 42–51 displays a sequence identity of less than 90%, and this area shares 79% identity at the amino acid level (Chamberlain *et al.*, 1991b). The dystrophin gene displays a nearly identical pattern of expression when compared between mice and humans. Figure 7.2 shows the results of Northern analysis using a cDNA subclone from the 3′ portion of the mouse cDNA. The primary transcript detected is a 14 kb mRNA in skeletal muscle, heart and brain. Two smaller mRNAs of approximately 4.8 and 6.6 kb are also detected in a variety of tissues, particularly liver and kidney (Figure 7.2). These results illustrate a portion of the isoform diversity displayed by dystrophin transcripts. While Northern blots can reveal the tissue distribution of the dystrophin mRNAs, this method provides little information on the extensive pattern of alternative splicing displayed by the gene. Northern analysis also has not been able to provide much information about the types of mutations that occur in *mdx* mice. These types of analyses are more readily performed via PCR.

PCR analysis of dystrophin transcripts was not possible until the cDNA sequence had been determined. The resulting experiments have provided extensive information about the expression and mutation of the gene and will be described below. The cDNA sequence also proved to be important for the construction of a full-length cDNA clone for the mouse gene. Lee *et al.* used sequence data from the 3′ end and the middle of the cDNA to construct primers for reverse transcription of two large portions of the mRNA, which were subsequently ligated together to create four full-length cDNA clones [Figure 7.1; (Lee *et al.*, 1991)]. Expression of these cDNAs in lymphoblasts and COS cells revealed that only one of the four clones

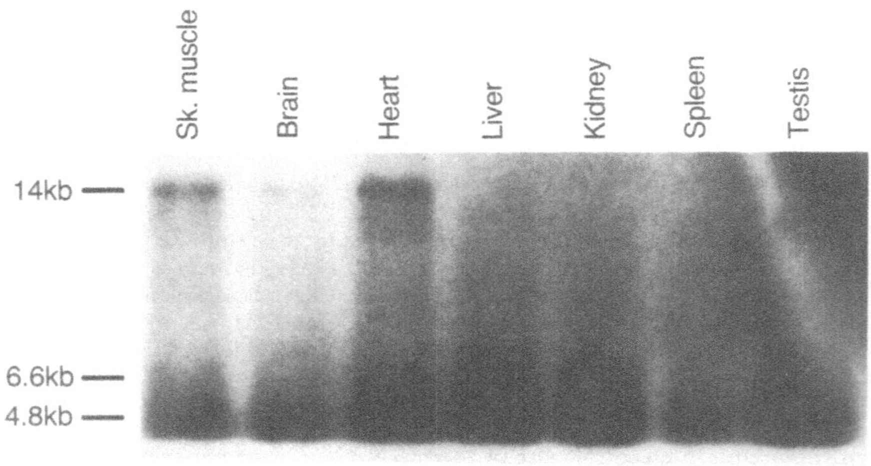

Figure 7.2 Northern analysis of mouse dystrophin mRNA. 20 μg of total RNA from the indicated tissues was separated on a 0.8% agarose/formaldehyde gel, transferred to a nylon membrane, and hybridized with a probe that spans bases 10 278 to 10 960 of the murine dystrophin cDNA. A major band of approximately 14 kb is observed in skeletal muscle, brain and heart RNAs. Smaller bands of 4.8 and 6.6 kb are observed in liver and kidney, while the 4.8 kb band is the predominant form in testis and can be observed at various levels in several of the other tissues.

encoded a full size dystrophin protein. As will be described below, PCR analysis of the four clones revealed three to have reverse transcriptase errors, rendering the clones non-functional. The fourth clone produces full-size dystrophin in transfected cells, and is currently the subject of experiments designed to identify structurally important segments of the protein and to explore the feasibility of gene therapy for DMD.

7.3 *mdx* MICE

One of the advantages to studying the dystrophin gene in mice is the availability of at least four murine models for DMD known as *mdx* mice (Bulfield *et al.*, 1984; Chapman *et al.*, 1989). Each of the four known *mdx* mutants are individually isolated strains that show non-complementarity of the mutations and apparently arise from independent lesions within the dystrophin gene. Unlike human DMD patients, the *mdx* mice display a mild muscle pathology that, after about six weeks of age, is not overtly progressive (Anderson *et al.*, 1988; Bridges, 1986). Characteristics of the *mdx* mutants include elevated serum creatine kinase levels, variation in fibre size in skeletal muscle, mild fibrosis and cardiomyopathy, centrally located nuclei in skeletal muscle, and small lesions of necrotic muscle tissue (Anderson *et al.*, 1988; Bridges, 1986; Chapman *et al.*, 1989; Hoffman, 1992). The original *mdx* isolate was a spontaneously arising mutant (Bulfield *et al.*, 1984), while the newer isolates (strains 467, 551 and 2019) were generated by N-ethylnitrosourea mutagenesis (Chapman *et al.*, 1989). Although the symptoms of muscular dystrophy in these mice are relatively mild, there is a variety of assays for the dystrophic phenotype and mdx mice are a versatile system with which to explore potential therapies for DMD/BMD.

7.3.1 Dystrophin expression in *mdx* mice

As a prerequisite to the development of therapeutic approaches for DMD/BMD it is important to understand both the normal pattern of dystrophin gene expression and the effect that mutations in the gene have upon the expression of various dystrophin isoforms in separate tissues. As previously mentioned, Northern analysis provides only a gross evaluation of the relative level of expression of dystrophin mRNA in separate tissues. Analysis of the *mdx* strains has revealed that dystrophin mRNA appears normal in size, but is reduced by approximately 80% below the normal levels in muscle and brain (Chamberlain *et al.*, 1988). As the mRNA is present at extremely low levels in normal mouse tissues, little information can be gained by RNA gel blot analysis of the *mdx* mutants.

Western analysis of dystrophin expression provides a more sensitive assay for expression of the gene (Hoffman *et al.*, 1988). These types of experiments have demonstrated that the *mdx* mutants produce only trace amounts of dystrophin in muscle tissue, and what dystrophin is present apparently arises from reversion of the mutation (Hoffman *et al.*, 1990). Southern analysis using murine cDNA clones does not detect any genomic rearrangements in DNA from the *mdx* mouse strains (Chamberlain *et al.*, 1989). The presence of apparently full-sized dystrophin mRNA in the *mdx* mice combined with the absence of dystrophin suggests that each of the *mdx* isolates contains a point mutation in the gene. PCR analysis of dystrophin mRNAs has identified a single base mutation in the original *mdx* mutant (Bulfield *et al.*, 1984) and supports the existence of such lesions in the newer isolates (see below).

7.3.2 Scanning for dystrophin DNA sequence variations in *mdx* mice

Identification of point mutations and polymorphisms within the dystrophin gene of *mdx* mice can be approached via direct analysis of the 14 kb mRNA. We have used sequence data from the mouse cDNA clones to design PCR primers at intervals along the coding region of the mRNA. These primers can be used to amplify overlapping regions of the mRNA, enabling scanning methods to be used to identify sequence variations. The primary method we have employed involves chemical cleavage of base mismatches in heteroduplexes formed between a control and a test transcript. This approach has been used successfully to identify point mutations in the mRNA of a variety of genes (Cotton *et al.*, 1988).

In this method total RNA isolated from mouse skeletal muscle is reverse transcribed and regions of approximately 1 kb in size are amplified via PCR. Reverse transcription and PCR conditions have been described extensively elsewhere (Bies *et al.*, 1992). Briefly, 1 μg of total RNA is reverse transcribed with MMLV reverse transcriptase using random hexamer primers. The single-stranded cDNA is then precipitated with ethanol and dissolved in H_2O. One-tenth of the cDNA reaction is then added to a 50 μl PCR reaction and amplified for 35 cycles of PCR using primers specific for dystrophin cDNA. Amplification of dystrophin cDNA has worked well using primers between 20 and 25 bases in length with a G+C base composition near 50% (Chamberlain *et al.*, 1991a). Radioactive PCR products are produced by including 5 μCi of ^{32}P-dCTP in the PCR reaction. Amplified fragments are then excised from agarose gels, eluted and dissolved in H_2O. For chemical cleavage approximately 6 000 CPMs of radiolabelled *mdx* PCR product is annealed to a 10–15 fold excess of unlabelled wild-type PCR product for heteroduplex formation. These duplex molecules are then treated either with osmium tetroxide or hydroxylamine to modify base mismatches (Cotton *et al.*, 1988). The modified DNA is then cleaved with piperidine, and the products are analysed on a denaturing polyacrylamide gel to size any cleavage products that are generated. Identification of a cleavage product indicates that a base mismatch has been detected between the wild-type and mutant sequence. PCR products that produce mismatches can then be either subcloned for sequencing or directly sequenced using nested cDNA primers. Sequence analysis confirms the base mismatch and can reveal whether the change is a mutation or a polymorphism.

We have used the chemical cleavage method to scan the 5′ 6.5 kb of the dystrophin mRNA simultaneously in three strains of *mdx* mice: the original mutant (Bulfield *et al.*, 1984) and two of the mutant strains (467 and 551) (Chapman *et al.*, 1989). These results are illustrated in Figures 7.3–7.5. Initially a mismatch was detected when PCR product from either the 467 or

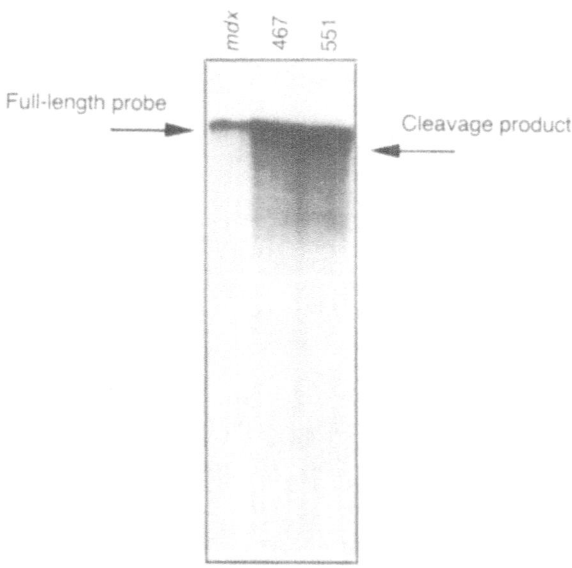

Figure 7.3 Chemical mismatch cleavage analysis of mouse dystrophin mRNA. PCR products for bases 727 to 1806 of the murine dystrophin mRNA were amplified from C57Bl/10, *mdx*, mdx^{467}, and mdx^{551} reverse transcribed RNA and purified via agarose gel electrophoresis. Amplification was as described in the text, with α-^{32}P dCTP included in the C57 reaction. The labelled PCR product was annealed with a 12 fold excess of cold template, and analysed as described (Chamberlain *et al.*, 1991a). Shown is an autoradiogram of the cleavage products obtained after treatment of the heteroduplexes with OsO_4 for 20 min.

the 551 mutant was annealed with the wild-type C57Bl/10 transcript (Figure 7.3). This experiment localized the mismatch to within 125 bases of one end of a PCR product that spanned mRNA bases 727 to 1806 (Chamberlain *et al.*, 1991a). Sequence analysis of the amplified product from the 467 mutant revealed the base change to be a C to T transition in exon 7 of the mouse gene at base 835 of the mRNA sequence (Figure 7.4). A second polymorphism was later found in the 467 and 551 mutants by direct cycle sequencing of PCR amplified mRNA. An A to G transition was observed at base 10 819 of the mRNA, which does not alter the proline codon at this position (data not shown). These polymorphisms clearly are not responsible for the mutant phenotype in the 467 and 551 mutants. Nonetheless these sequence differences enable the new mutants to be distinguished from the wild-type and the original *mdx* mouse, both of which are on a C57Bl/10 background. For this purpose mRNA from mouse skeletal muscle is amplified by PCR, dot or Southern blotted onto nylon

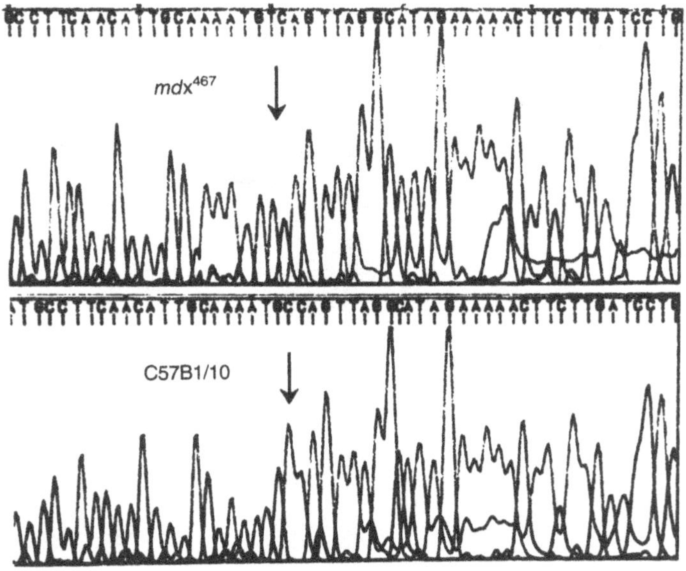

Figure 7.4 Sequence analysis of the mRNA region from Figure 7.3. Following PCR amplification of bases 727 to 1806, 5 units of the Klenow fragment of DNA polymerase were added to the reaction, which was then incubated for 10 min at 37°. The amplified fragments were purified on an agarose gel, and ligated into pTZ19r (Pharmacia) at a 10:1 insert:vector ratio for 4 hours at 22°. Following transformation of *E. coli* and preparation of single stranded DNA, the subcloned PCR fragment was sequenced on an Applied Biosystems model 370A automatic DNA sequencer. Shown are representative sequences from the mdx^{467} product, and from a wild-type (C57Bl/10) product. The mutant mouse displays a polymorphic base substitution when compared with the C57 sequence. The sequence of the original *mdx* mutant matched the C57 sequence, while the mdx^{551} sequence matched that of the mdx^{467} sequence.

membranes, and hybridized with allele-specific oligonucleotides (ASO) for either the control or 467/551 sequences (an example of ASO hybridization is described below). The ability to distinguish these strains is not otherwise trivial, as all the *mdx* mice display a mild phenotype and can not be discerned from the wild-type mice or each other by simple observation of the animals.

Chemical cleavage analysis also identified a base mismatch between the wild-type mouse and the original *mdx* mutant in a fragment spanning mRNA bases 2618 and 3433 (Figure 7.5). While this work was in progress Sicinski *et al.* independently PCR amplified, subcloned, and sequenced a variety of regions of the murine cDNA and identified a premature stop

Figure 7.5 Chemical mismatch cleavage analysis of bases 2618 and 3433 of the mouse dystrophin mRNA. PCR and chemical cleavage were performed as described in the legend to Figure 7.3. Shown is an autoradiogram of the cleavage products obtained from hydroxylamine treatment of heteroduplexes formed between wild-type probe and PCR products obtained from an *mdx* mouse (original isolate), and from a murine dystrophin cDNA clone. A mismatch was detected in the heteroduplex formed between the wild-type and *mdx* DNAs (arrows indicate the positions of the full-length and cleaved probes).

codon at base 3203 of the *mdx* mRNA (Sicinski *et al*., 1989). The cleaved cDNA fragments of approximately 590 and 230 base pairs (Figure 7.5) indicated that we had identified the same base change, and sequence analysis of the amplified fragment confirmed the C to T transition that creates a premature stop codon in the mRNA of the *mdx* mouse (Sicinski *et al*., 1989; Chamberlain *et al*., 1991a).

The mutation in the original *mdx* mouse can also be analysed via ASO hybridization of PCR amplified RNA (Chamberlain *et al*., 1991a). We sought to develop a genotyping assay for the *mdx* mouse that could be performed without sacrificing the animal so that colonies of *mdx* mice could be unambiguously distinguished from the wild-type mice. This required an assay that could be performed with amplified genomic DNA, which can be isolated from either a mouse tail piece or from blood obtained by retro orbital bleeding. As the gene structure for this region of dystrophin was not available we chose to directly isolate the relevant exon by using inverse PCR (Ochman *et al*., 1990) of mouse genomic DNA.

Inverse PCR was performed by digesting mouse genomic DNA with a variety of restriction enzymes, diluting the DNA and ligating to form circles. The circularized DNA was PCR amplified with primers flanking the *mdx* mutation after linearizing the circles with an enzyme, *Hae*III, which cut between the PCR primers. The greatest yield of product was obtained when *Alu*I was used to digest the genomic DNA prior to circularization. This inverse PCR reaction yielded a 713 bp fragment that was sequenced and found to contain the 3′ end of the intron immediately 5′ of the exon with the mutation (Figure 7.6). The flanking sequences 3′ of the exon were not obtained, as an *Alu*I site was found to be located further 3′ within the same exon (Figure 7.6). These results provided the 3′ splice site of mouse dystrophin exon 23, and enabled the design of an ASO assay that could be performed with genomic DNA.

To genotype the *mdx* mutant we designed two PCR primers and two ASO probes. One primer corresponds to a sequence within intron 22, while the second is complementary to a sequence within exon 23 (Figure 7.6). Amplification with these primers produces a 150 bp fragment that can be transferred to nylon membranes and hybridized with ASO probes for the wild-type or *mdx* mutation (Figure 7.7). This assay enables the original mutant to be distinguished from the wild-type, and the 467, 551 or 2019 alleles. Another advantage of this assay is that genomic DNA can be amplified without co-amplification of any dystrophin mini-gene vectors used to generate transgenic animals in tests of the feasibility of gene therapy for DMD. We have previously described an ASO assay for dystrophin RNA that distinguishes the wild-type and mutant sequences (Chamberlain *et al*., 1991a). That assay will amplify dystrophin cDNAs in the mini-gene vectors that several groups are currently testing for the ability to correct the mutant phenotype in *mdx* mice. We have determined that a combination of the gene amplification primers and the cDNA amplification primers enables rapid testing of potential transgenic mice for wild-type versus mutant genotype

```
AGCTTAGGTA AAATCAATGG ATTTATATAA TTAGTTAGAA TTTAAATATT AACAAACTAT
AACACTATGA TTAAATGCTT GATATTGAGT AGTTATTTTA ATAGCCTAAG TCTGGAAATT
AAATACTAGT AAGAGAAACT TCTGTGATGT GAGGACATAT AAAGACTAAT TTTTTTGTTG
ATTCTAAAAA TCCCATGTTG TATACTTATT CTTTTTAAAT CTGAAAATAT ATTAATCATA
TATTGCCTAA ATGTCTTAAT AATGTTTCAC TGTAGGTAAG TAAATGTATC ACATATATTA
ATAAACATAG TTATTAATGC ATAGATATTC AGTAAAATTA TGACTTCTAA ATTTCTGTCT
AAATATAATA TGCCCTGTAA TATAATAGAA ATTATTCATA AGAATACATA TATATTGCTT
TATCAGATAT TCTACTTTGT TTAGATCTCT AAATTACATA AACTTTTATT TACCTTCTTC

              |---------------a----------------->
TTGATATGAA TGAAACTCAT CAAATATGCG TGTTAGTGTA AATGAACTTC TATTTAATTT
   <-------------b---------------|      t=mdx
TGAGGCTCTG CAAAGTTCTT TGAAAGAGCA ACAAAATGGC TTCAACTATC TGAGTGACAC
     A  L   Q  S  S   L  K  E  Q   Q  N  G   F  N  Y   L  S  D  T
                        <-------------c-------------->
TGTGAAGGAG ATGGCCAAGA AAGCACCTTC AGAAATATGC CAGAAATATC TGTCAGAATT
  V  K  E   M  A  K   K  A  P  S   E  I  C   Q  K  Y   L  S  E  F
TGAAGAGATT GAGGGGCACT GGAAGAAACT TTCCTCCCAG TTGGTGGAAA GCT
  E  E  I   E  G  H   W  K  K  L   S  S  Q   L  V  E   S
```

Figure 7.6 Sequence of the 5′ flanking region of mouse dystrophin exon 23. This sequence was determined from the inverse PCR products described in the text. The sequence begins and ends at *Alu*I sites. The 3′ end of the sequence is in exon 23, and the amino-acids encoded are displayed below the DNA sequence. The first base of the exon is underlined. Also indicated are the location of various PCR primers used for analysis of this region. Primer 'a' is used with primer 'c' to amplify DNA for the ASO hybridization described in the legend to Figure 7.7. Two separate primers were made to the 'c' region, for amplification in both directions. The primer in the 3′–5′ orientation was used for the experiment in Figure 7.7, while the opposite primer was used for inverse PCR with primer 'b'. The C to T transition found in the original *mdx* mouse is indicated, and the location of the 15 base ASO probes is underlined flanking the *mdx* mutation. Immediately 5′ of primer 'c' is the *Hae*III site (in bold) used in the inverse PCR reaction.

and also reveals the presence of a dystrophin mini-gene in the genomic DNA of transgenic animals.

Chemical cleavage analysis has also proved useful in scanning cDNA clones for reverse transcriptase errors. Efforts to construct a full-length murine dystrophin cDNA clone resulted in the generation of four separate 14 kb cDNA clones, only one of which was able to produce full-size dystrophin following transfection into tissue culture cells (Lee *et al.*, 1991). Chemical cleavage analysis was performed on end-labelled *Eco*RI restriction fragments isolated from the four large clones used to construct the full-length clones (Figure 7.1). A single mutation was found in each of the non-functional cDNA clones by this analysis. Clone F5 contained an insertion of a single base at position 3718, resulting in a frameshift mutation. Clone C1 was found to have a two nucleotide deletion accompanied by the insertion of a single base at position 9726 of the cDNA

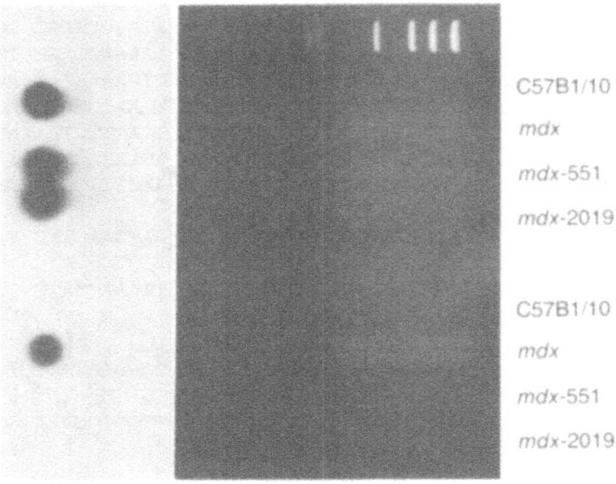

Figure 7.7 Allele-specific oligonucleotide hybridization analysis of amplified mouse genomic DNAs. A 150 bp stretch of the DNA spanning the exon 22 to intron 23 junction was PCR amplified from the indicated strains of mice using the primers shown in Figure 7.6. 10 µl of 50 µl reactions were electrophoresed on a 1.5% agarose gel, and the gel was photographed, transferred to a nylon membrane, cut in half, and each half was hybridized with one of two ASOs that had been end labelled with γ-^{32}P-dATP. The sequence of the ASOs is provided in Figure 7.6. Right: ethidium bromide stained agarose gel of the PCR products. Left, autoradiogram of the hybridized blots; Top, wild-type ASO; bottom, *mdx* ASO. Only the amplified *mdx* DNA hybridizes with the *mdx* ASO. Hybridization conditions have been described elsewhere (Chamberlain *et al.*, 1991a).

sequence, which also resulted in a frameshift mutation. Each of these reverse transcriptase errors was identified by chemical cleavage of cDNA fragments between 0.8 and 4 kb in size, and this method enabled sequence analysis of the defective 14 kb cDNA clones to be limited to the region of the mismatch.

7.3.3 A murine dystrophin dinucleotide repeat can be typed with PCR

Sequence analysis of mouse dystrophin cDNA clones resulted in the identification of a stretch of alternating dC-dA nucleotides in the 3′ non-translated region of the mouse dystrophin mRNA (Maichele and Chamberlain, 1992). Hundreds of examples have been reported where this class of dinucleotide repeat sequence has been found to exhibit length polymorphisms (Weber and May, 1989). In addition, a similar sequence at the identical position in the human dystrophin gene was found to display a

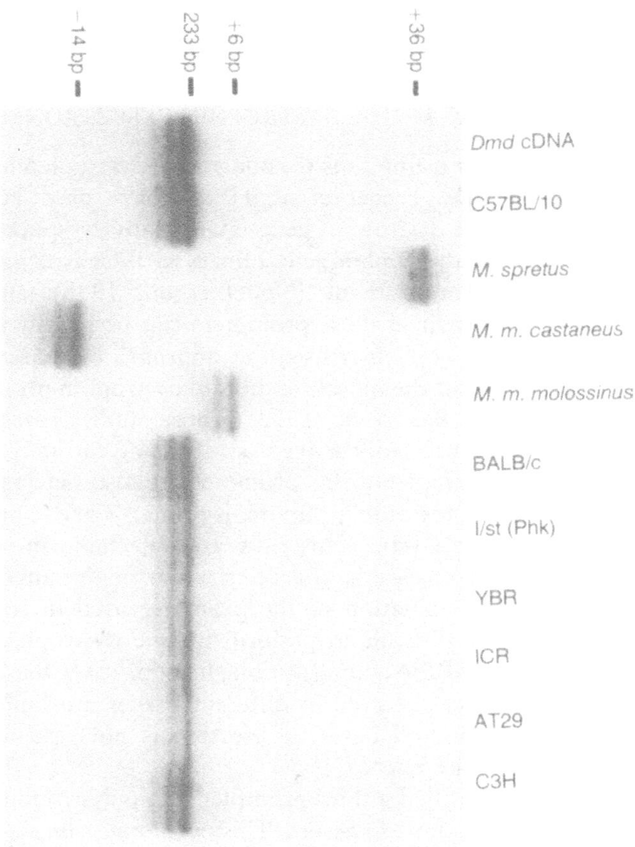

Figure 7.8 Analysis of a polymorphic dinucleotide repeat in the 3′ non-translated region of the mouse dystrophin gene. Genomic DNA was PCR amplified with primers flanking the $[CA]_n$ stretch in the 3′ non-translated region of the mouse dystrophin gene. PCRs included radioactive dCTP in the reaction. Amplified products were separated on a denaturing polyacrylamide gel and exposed to X-ray film. As shown, different sized products are obtained from several different species of mice. The 233 bp allele is the size found in cloned C57Bl/10 cDNAs. The relative size differences of the other alleles are indicated at the top of the photograph.

number of alleles and to be useful for haplotype and linkage analysis of DMD families (e.g. Clemens *et al.*, 1991). PCR analysis reveals that this mouse dystrophin $[dC\text{-}dA]_n$ repeat is present at a variety of different lengths in separate mouse strains and species (Figure 7.8). This dinucleotide repeat polymorphism in the 3′ terminal exon of the dystrophin gene may prove to be a useful X-linked marker for interspecific mouse crosses (Maichele and Chamberlain, 1992).

7.4 PCR ANALYSIS OF MOUSE DYSTROPHIN EXPRESSION

PCR is an extremely sensitive method for the analysis of dystrophin mRNA expression in various tissues. Feener *et al.* (1989) have used PCR to demonstrate that the human dystrophin gene is alternatively spliced in a variety of fetal tissues. The dystrophin gene utilizes at least two separate promoters in muscle and brain tissue (Nudel *et al.*, 1988), and the transcripts that arise from each of these promoters can be readily distinguished with PCR. We have previously used this approach to characterize the tissue expression pattern of the muscle and brain dystrophin promoters (Chamberlain *et al.*, 1991a; Bies *et al.*, 1992). Those studies reveal that while the muscle promoter is primarily active in skeletal and cardiac muscle, low levels of transcripts arising from this promoter are also detectable in mouse brain. Similarly, while the brain promoter is primarily active in brain tissue, low levels of the brain-type transcripts are detectable in muscle tissues. Each of these types of assays, characterization of the alternative splicing of the gene and examination of the promoter used in separate tissues, would be extremely difficult to perform for the dystrophin gene without PCR methods for mRNA amplification. In both cases the differences between the transcripts observed in different tissues are limited to short stretches of a few hundred bases, a size that is not resolved via Northern analysis of this 14 kb mRNA.

Bar *et al.* (1990) have identified a third promoter in the dystrophin gene that is primarily active in non-muscle tissues. This promoter is located near the 3′ end of the gene and drives the expression of a truncated transcript that encodes only the C-terminus of dystrophin. Northern analysis is able to resolve this shorter transcript from the full size 14 kb mRNA, and reveals that it is expressed either as a 4.8 or a 6.5 kb transcript (Figure 7.2). The precise organization of this transcript is not yet clear, and a complete analysis of the expression pattern of this non-muscle transcript will probably require isolation of unique portions of the sequence to enable specific PCR reactions that will not co-amplify the muscle and brain-type 14 kb mRNAs.

7.4.1 The 3′ portion of dystrophin mRNA displays extensive alternative splicing

The observation that human dystrophin mRNA is alternatively spliced in fetal tissues (Feener *et al.*, 1989) led us to examine dystrophin mRNA splicing during mouse development. Our results indicated that the overall pattern of splicing is conserved between mouse and human tissues, and that dystrophin isoforms are differentially expressed during development (Bies *et al.*, 1992). Alternative splicing of the dystrophin gene appears to be limited to a 1.2 kb stretch of the mRNA, from approximately base 10 000 to the C-terminal portion of the message. The 3′ terminal 2.7 kb of the mRNA is encoded on a single exon that contains the stop codons for the gene (J.S.C., unpublished). Models of the dystrophin protein structure divide the protein into four domains, the first three of which display sequence conservation with spectrin and α-actinin (Koenig *et al.*, 1988; Koenig and Kunkel, 1990). The fourth, C-terminal domain shares sequence similarity only with the dystrophin related protein, and it is this region that is alternatively spliced (Feener *et al.*, 1989). The function of this domain of dystrophin is not known; however, virtually all patients deleted for the carboxy-terminal domain display a severe phenotype. These phenotypic observations have been cited to suggest that the C-terminal domain is critical for the function of dystrophin (Koenig *et al.*, 1989; Hoffman *et al.*, 1991).

Alternative splicing of the C-terminal domain can be examined by PCR amplification of reverse transcribed RNA obtained from separate tissues at a variety of stages of development. Figure 7.9 displays the results obtained by PCR amplification of skeletal muscle, heart and brain RNA isolated from adult C57B1/10 mice. This approach has led to the identification of ten spliced forms of the mRNA in this region of the transcript. Sequence analysis of these ten isoforms reveals that they are generated by differential splicing of four exons of the gene (Bies *et al.*, 1992). Analysis of the transcripts isolated from fetal mouse tissues displays a less complex pattern of isoform diversity (Figure 7.10). Embryonic mouse tissues accumulate a subset of the adult transcripts, and an increase in the number of detectable isoforms occurs during development. Analysis of the alternative splicing pattern in *mdx* skeletal muscle reveals that the adult *mdx* mice display an isoform expression pattern more similar to embryonic than to adult wild-type muscle (Figure 7.10). These results are likely to reflect the pathology of *mdx* skeletal muscle. Dystrophic muscle fibres undergo more frequent degeneration and regeneration than do the corresponding wild-type muscles and it has been observed that regenerating muscle fibres frequently display an increased accumulation of embryonic isoforms of a variety of muscle proteins (Keller and Emerson, 1980). The embryonic pattern of alternative splicing observed in *mdx* muscle suggests that a greater degree of fibre degeneration and regeneration occurs in the adult *mdx* muscle relative to that of control animals.

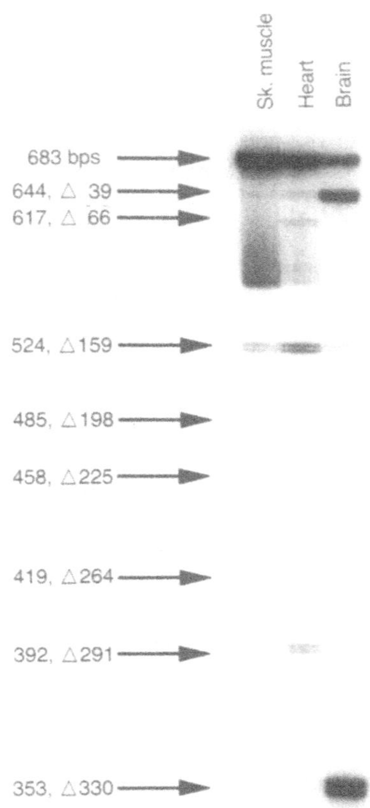

Figure 7.9 Analysis of dystrophin alternative splicing in adult *mdx* and C57Bl/10 mice. Bases 10 278 to 10 960 of the mouse transcript were amplified by PCR from various tissues and developmental stages of C57Bl/10 and *mdx* mice. Reactions included radioactive dCTP and were analysed on a denaturing polyacrylamide gel to eliminate heteroduplex formation. Ten isoforms are apparent on this and a longer exposure of the gel.

7.4.2 The dystrophin C-terminus is encoded by two spliced forms of the mRNA

PCR amplification of the C-terminal portion of the mRNA has identified two alternate forms of the mRNA that differ by the presence or absence of the penultimate exon of the gene (Feener *et al.*, 1989; Bies *et al.*, 1992).

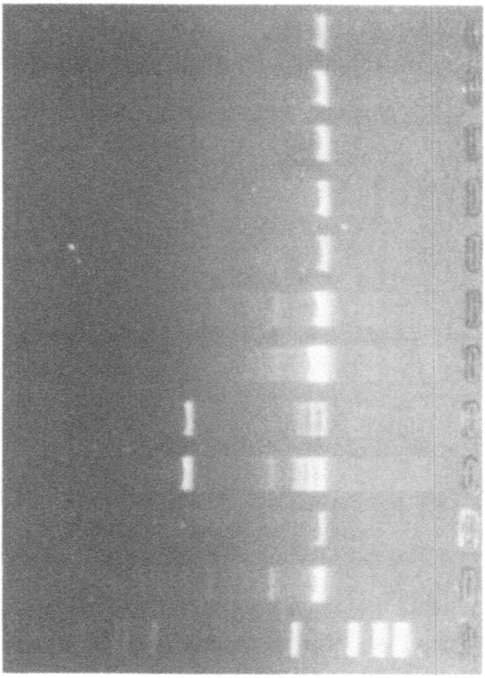

Figure 7.10 Differential splicing of the mouse dystrophin mRNA during development. The same region as displayed in Figure 7.9 was amplified in the absence of radioactivity for 35 cycles. 10 μl of a 50 μl reaction was electrophoresed on a 3% NuSieve agarose gel and photographed. Tissue and developmental stage specific isoforms are apparent. Note that the *mdx* skeletal muscle displays a pattern more closely resembling embryonic than adult normal muscle.

Amplification of transcripts from embryonic human or mouse tissues reveals that the majority of the mRNAs in skeletal muscle and brain lack this penultimate exon. An isoform transition is observed during development resulting in the retention of the penultimate exon in the majority of transcripts isolated from adult muscle and brain tissues (Bies *et al.*, 1992). Figure 7.11A displays an example of the isoform transition that occurs in mouse brain. The smaller spliced form of the mRNA contains an altered reading frame when compared with the larger form. This embryonic transcript encodes a longer and more hydrophilic C-terminus than the corresponding adult transcript (Bies *et al.*, 1992).

When skeletal muscle RNA from adult *mdx* mice is amplified by PCR, a greater percentage of the embryonic transcript is observed than in the adult wild-type mice. This relative ratio of the adult to the embryonic transcript in

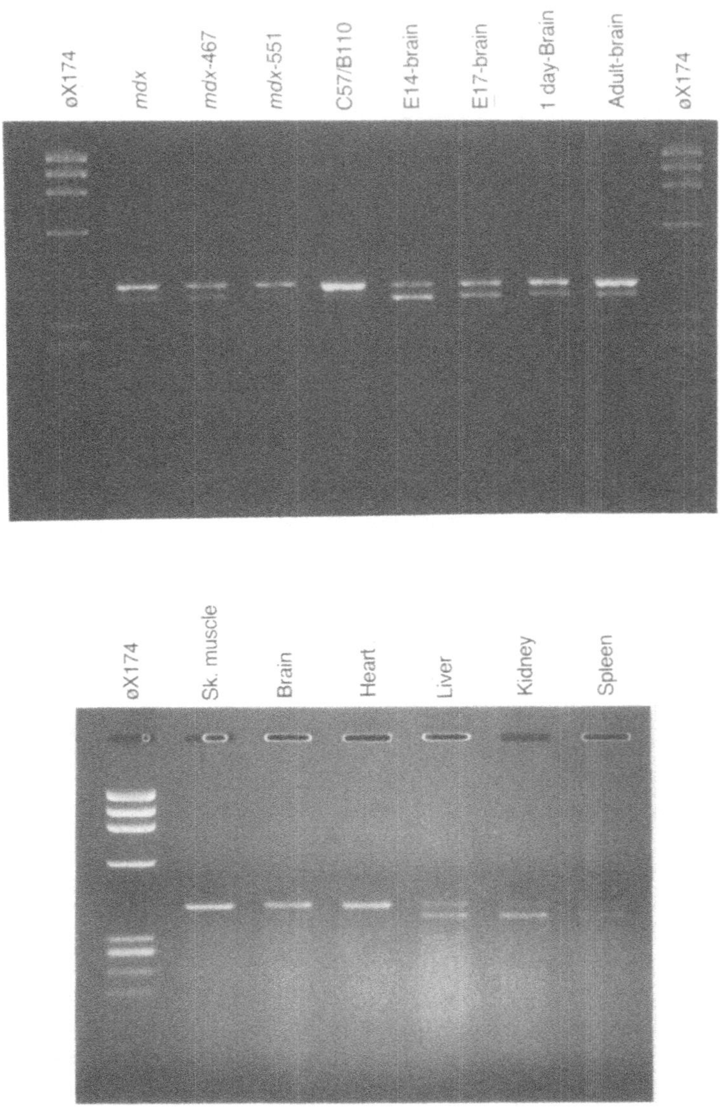

Figure 7.11 Alternative splicing of dystrophin mRNA leads to mRNAs that encode alternate C-termini of the protein. Bases 10 937 to 11 336 of the mouse transcript were amplified for 35 cycles as described in the legend to Figure 7.10 (A) Alternate splicing of the C-terminus in wild-type brain and in *mdx* muscle. (B) C-terminal isoforms in separate tissues of wild-type mice. The smaller spliced form of the transcript lacks the penultimate exon of the gene as determined by sequence analysis (Bies *et al.*, 1992).

mdx mice is more similar to late stage embryonic skeletal muscle from wild-type mice than it is to the wild-type adult muscle [Figure 7.11A; (Bies *et al.*, 1992)]. These observations are similar to those displayed in Figure 7.10, and support the assertion that adult *mdx* skeletal muscle is in a more regenerative developmental stage than the corresponding wild-type tissue.

The observation (Bar *et al.*, 1990) that dystrophin is expressed as a truncated C-terminal isoform in non-muscle tissues led us to examine the pattern of splicing of this truncated transcript in tissues other than muscle and brain. As displayed in Figure 7.11B, liver, kidney and spleen each contain dystrophin transcripts that can be amplified via PCR. These transcripts presumably arise from the truncated mRNAs observed via Northern analysis of these tissues (Figure 7.2). In contrast to the relative isoform ratios observed in muscle and brain, the non-muscle tissues display predominantly the shorter spliced form of the C-terminal mRNA. These results are in agreement with those of Bar *et al.* (1990) who reported that two cDNA clones isolated from liver lacked the penultimate exon of the dystrophin gene.

7.5 SUMMARY

PCR amplification has enabled a variety of studies to be performed on the murine dystrophin transcripts. Figure 7.12 displays a summary of the

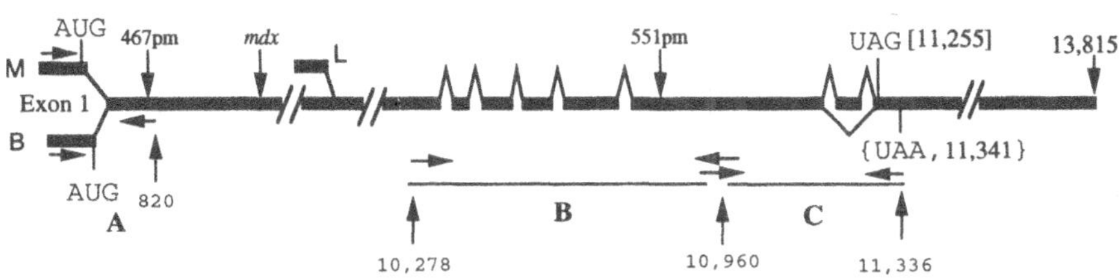

Figure 7.12 Summary of dystrophin isoforms identified in mouse tissues. Shown is a schematic illustration of features of the murine dystrophin mRNA described in the text. The gene uses two alternate promoters at the 5′ end, designated M (muscle) or B (brain), each of which is linked to a unique first exon. Expression from these promoters can be analysed using the PCR primer pairs in area 'A' (Bies *et al.*, 1992). A third promoter used in a variety of non-muscle cells is indicated by L (liver). The precise location of this promoter and the organization of transcripts produced through its use are not clear. Transcripts from the non-muscle promoter appear to become co-linear with the muscle transcript at approximately base 9200 of the muscle cDNA sequence (Bar *et al.*, 1990). The position of the *mdx* mutation and the 467/551 polymorphisms are indicated. Areas B and C are the alternatively spliced regions. Shown is the location and relative orientation of the PCR primers used to analyse this portion of the transcript (Bies *et al.*, 1992).

features of the murine dystrophin mRNA that have been described in this article. The location of the mutation in the original *mdx* mouse is indicated, as are the different spliced forms of the dystrophin transcript. Also shown are the location of various PCR primer binding sites that were used to deduce the alternative splicing pattern of the gene. It is likely that conventional cloning efforts aimed at identifying the variety of dystrophin spliced forms would have taken years to perform, particularly since several of the isoforms are expressed at levels significantly below the estimated 0.02% of total mRNA that dystrophin represents in skeletal muscle (Hoffman *et al.*, 1987a, b). Amplification of dystrophin mRNA simplifies scanning methods for the identification of DNA sequence variations. Attempts to re-isolate and sequence the 14 kb cDNA to determine the mutation in separate strains of *mdx* mice are not likely to be time or cost effective. PCR enables these types of questions to be answered in a relatively short period of time, and similar types of analyses can be applied to human DMD tissues.

Knowledge of the transcript diversity displayed by the dystrophin gene will enable the role of these separate isoforms to be addressed. Despite considerable effort by a variety of laboratories over the last five years, the precise functional role played by dystrophin remains unclear, and it can only be assumed that the separate isoforms act to modulate the functional role of dystrophin in separate tissues or in response to differing physiological states. PCR amplification of the dystrophin isoforms has enabled the variable regions of the transcript to be subcloned (Bies *et al.*, 1992). These clones have been used to reintroduce the variable regions into full-length mini-gene expression vectors, which are currently being tested for functional activity through the generation of transgenic *mdx* mice. The transgenic mice can be easily identified through the PCR–ASO assays described in this article, and the reverse transcriptase PCR assays will enable a detailed analysis of the expression pattern of the introduced mini-genes. It is hoped that such analyses will further attempts to determine the feasibility of using gene therapy as a treatment for DMD/BMD.

ACKNOWLEDGEMENTS

Supported by grants from the Muscular Dystrophy Association, and the March of Dimes Birth Defects Foundation. Also supported by grant NIH RO1-AR40864 (to J.S.C.). GAC was supported by a National Science Foundation Graduate Fellowship. A.J.M. was supported by a University of Michigan Rackham Predoctoral Fellowship and by grant NIH PO1-DK42718.

REFERENCES

Anderson, J.E., Bressler, B.H. and Ovalle, W.K. (1988) Functional regeneration in the hindlimb skeletal muscle of the mdx mouse. *J. Muscle Res. Cell Motil.*, **9**, 499–515.

Bar, S., Barnea, E., Levy, Z. *et al.* (1990) A novel product of the Duchenne muscular dystrophy gene which greatly differs from the known isoforms in its structure and tissue distribution. *Biochem. J.*, **272**, 557–60.

Bies, R.D., Phelps, S.F., Cortez, M.D. *et al.* (1992) Human and murine dystrophin mRNA transcripts are differentially expressed during skeletal muscle, heart, and brain development. *Nucleic Acids Res.*, 7, 1725–31.

Bridges, L.R. (1986) The association of cardiac muscle necrosis and inflammation with the degenerative and persistent myopathy of MDX mice. *J. Neurol. Sci.*, **72**, 147–57.

Bulfield, G., Siller, W.G., Wight, P.A. and Moore, K.J. (1984) X chromosome-linked muscular dystrophy (mdx) in the mouse. *Proc. Natl. Acad. Sci. U.S.A.*, **81**, 1189–92.

Chamberlain, J.S., Pearlman, J.A., Muzny, D.M. *et al.* (1988) Expression of the murine Duchenne muscular dystrophy gene in muscle and brain. *Science*, **239**, 1416–8.

Chamberlain, J.S., Ranier, J.E., Pearlman, J.A. *et al.* (1989) Analysis of Duchenne muscular dystrophy mutations in mice and humans, in Cellular and Molecular Biology of Muscle Development, *UCLA Symposia on Molecular and Cellular Biology*, New Series Vol. 93, (eds F. Stockdale, F. and L. Kedes), pp. 951–62.

Chamberlain, J.S., Farwell, N.J., Ranier, J.E. *et al.* (1991a) PCR analysis of dystrophin gene mutation and expression. *J. Cell. Biol.*, **46**, 334–6.

Chamberlain, J.S., Pearlman, J.A., Muzny, D.M. *et al.* (1991b) Mouse dystrophin cDNA Sequence, Genbank accession No. M68859. (Unpublished).

Chapman, V.M., Miller, D.M., Armstrong, D. and Caskey, C.T. (1989) Recovery of induced mutations for X chromosome-linked muscular dystrophy in mice. *Proc. Natl. Acad. Sci. U.S.A.*, **96**, 1292–6.

Chelly, J., Gilgenkrantz, H., Lambert, M. *et al.* (1990) Effect of dystrophin gene deletions on mRNA levels and processing in Duchenne and Becker muscular dystrophies. *Cell*, **63**, 1239–48.

Clemens, P.R., Fenwick, R.G., Chamberlain, J.S. *et al.* (1991) Carrier detection and prenatal diagnosis in Duchenne and Becker muscular dystrophy families, using dinucleotide repeat polymorphisms. *Am. J. Hum. Genet.*, **49**, 951–60.

Cotton, R.G.H., Rodrigues, N.R. and Campbell, R.D. (1988) Reactivity of cytosine and thymine in single-base pair mismatches with hydroxylamine and osmium tetroxide and its application to the study of mutations. *Proc. Natl. Acad. Sci. U.S.A.*, **85**, 4397–401.

den Dunnen, J.T., Grootscholten, P.M., Blonden, L.A. *et al.* (1989) Topography of the Duchenne muscular dystrophy (DMD) gene: FIGE and cDNA analysis of 194 cases reveals 115 deletions and 13 duplications. *Am. J. Hum. Genet.*, **45**, 835–47.

Feener, C.A., Koenig, M. and Kunkel, L.M. (1989) Alternative splicing of human dystrophin mRNA generates isoforms at the carboxy terminus. *Nature*, **338**, 509–11.

Gibbs, R.A., Chamberlain, J.S. and Caskey, C.T. (1989) Diagnosis of new mutation diseases using the polymerase chain reaction, in *PCR Technology: Principles and Applications of DNA Amplification*, (ed H.A. Erlich) Stockton Press, New York, pp. 171–91.

Hoffman, E.P., Brown, R.H., Jr. and Kunkel, L.M. (1987a) Dystrophin: the protein product of the Duchenne muscular dystrophy locus. *Cell*, **51**, 919–28.

Hoffman, E.P., Monaco, A.P., Feener, C.C. and Kunkel, L.M. (1987b) Conservation of the Duchenne muscular dystrophy gene in mice and humans. *Science*, **238**, 347–50.

Hoffman, E.P., Fischbeck, K.H., Brown, R.H. *et al.* (1988) Characterization of dystrophin in muscle-biopsy specimens from patients with Duchenne's or Becker's muscular dystrophy. *N. Engl. J. Med.*, **318**, 1363–8.

Hoffman, E.P., Morgan, J.E., Watkins, S.C. and Partridge, T.A. (1990) Somatic reversion/suppression of the mouse mdx phenotype in vivo. *J. Neurol. Sci.*, **99**, 9–25.

Hoffman, E.P., Garcia, C.A., Chamberlain, J.S. *et al.* (1991) Is the carboxyl-terminus of dystrophin required for membrane association? A novel, severe case of Duchenne muscular dystrophy. *Ann. Neurol.*, **30**, 605–10.

Hoffman, E.P. (1992) The animal models of Duchenne muscular dystrophy: Windows on the pathophysiological consequences of dystrophin deficiency, in *Ordering the Membrane-cytoskeleton Trilayer*, (eds M. Mooseker and J. Morrow), Academic Press, New York, in press.

Keller, E.R. and Emerson, C.P. (1980) Synthesis of adult myosin light chains by embryonic muscle cultures. *Proc. Natl. Acad. Sci. U.S.A.*, **77**, 1020–4.

Koenig, M., Monaco, A.P. and Kunkel, L.M. (1988) The complete sequence of dystrophin predicts a rod-shaped cytoskeletal protein. *Cell*, **53**, 219–26.

Koenig, M., Beggs, A.H., Moyer, M. *et al.* (1989) The molecular basis for Duchenne versus Becker muscular dystrophy: correlation of severity with type of deletion. *Am. J. Hum. Genet.*, **45**, 498–506.

Koenig, M. and Kunkel, L.M. (1990) Detailed analysis of the repeat domain of dystrophin reveals four potential hinge segments that may confer flexibility. *J. Biol. Chem.*, **265**, 4560–6.

Lee, C.C., Pearlman, J.A., Chamberlain, J.S. and Caskey, C.T. (1991) Expression of recombinant dystrophin and its localization to the cell membrane. *Nature*, **349**, 334–6.

Lemaire, C., Heilig, R. and Mandel, J.L. (1988) The chicken dystrophin cDNA: striking conservation of the C-terminal coding and 3′ untranslated regions between man and chicken. *EMBO J.*, **7**, 4157–62.

Maichele, A.J. and Chamberlain, J.S. (1992) Cross-species conservation of a polymorphic dinucleotide repeat in the dystrophin gene. *Mammalian Genome*, **3**, 290–2.

Moser, H. (1984) Review of studies on the proportion and origin of new mutants in Duchenne muscular dystrophy, in *Current Clinical Practices, Series 20*, (eds

L.P. Ten Tate, P.L. Pearson and A.M. Stadhouders) Excerpta Medica, Amsterdam, pp. 41–52.

Nudel, U., Zuk, D., Einet, P. *et al.* (1988) Duchenne muscular dystrophy gene product is not identical in muscle and brain. *Nature*, **337**, 76–8.

Ochman, H., Medhora, M.M., Garza, D. and Hartl, D.L. (1990) Amplification of flanking sequences by inverse PCR, in *PCR Protocols. A Guide to Methods and Applications*, (eds M.A. Innis, D.H. Gelfand, J.T. Sninsky and T.J. White), Academic Press, San Diego, pp. 219–27.

Scott, M.O., Sylvester, J.E., Heiman-Patterson, T. *et al.* (1988) Duchenne muscular dystrophy gene expression in normal and diseased human muscle. *Science*, **239**, 1418–20.

Sicinski, P., Geng, Y., Ryder-Cook, A.S. *et al.* (1989) The molecular basis of muscular dystrophy in the mdx mouse: a point mutation. *Science*, **244**, 1578–80.

Weber, J.L. and May, P.E. (1989) Abundant class of human DNA polymorphisms which can be typed using the polymerase chain reaction. *Am. J. Hum. Genet.*, **44**, 388–96.

8

Cell biology of the satellite cell

EDWARD SCHULTZ and KATHLEEN M. McCORMICK

8.1 INTRODUCTION

Over the past several years the satellite cell in skeletal muscle has been the subject of renewed interest because of its potential importance in therapy for human muscle diseases due to genetic defects such as Duchenne muscular dystrophy (Griggs and Karpati, 1990; Partridge, 1991). The importance of this cell lies in the fact that it can be isolated with relative ease, can be grown in culture and will survive when returned to the *in vivo* environment by implantation. The implanted cells are capable of not only surviving in the host muscle, but also of gaining access to myofibres by passing through their basement membranes and eventually fusing with them. The nucleus of the fused cell integrates into the fibre syncytium and carries with it that portion of the normal genome that may be altered or missing in the host muscle. This chapter provides some general background information concerning the functions and behaviour of satellite cells in normal and pathological muscles in order to illustrate some of the characteristics of the cells which might influence the behaviour of myogenic cells following implant therapy.

8.2 SATELLITE CELLS IN NORMAL MUSCLE

Embryonic muscle histogenesis gives rise to two types of muscle cell in mature skeletal muscle: myofibres and satellite cells. The well studied events of myofibre formation have illustrated that fibre formation is dependent upon populations of distinct embryonic myoblasts, characterized by their temporal and/or phenotypic appearance (White *et al.*, 1975; Bonner and Hauschka, 1974; Seed and Hauschka, 1984). Presumably, during the time that embryonic myoblasts enter terminal differentiation, align and fuse to form myotubes, presumptive satellite cells, admixed among these cells, also

Molecular and Cell Biology of Muscular Dystrophy
Edited by Terence Partridge
Published in 1993 by Chapman & Hall, London. ISBN 0 412 43440 7

align, but unlike embryonic myoblasts they do not fuse. In this manner a population of cells is created on the surface of each myotube that is carried forth on the surface of mature myofibres. Elaboration of basement membrane material brings the cells to their morphologically identifiable position between the myofibre membrane and the newly formed basement membrane. It was Alex Mauro's description in 1961 of the morphological relationship between mononucleated satellite cells and their myofibres in mature frog muscle and the insight of his speculation regarding the potential role of the cells during muscle regeneration that gave birth to the field of satellite cell studies.

The fact that presumptive satellite cells do not fuse is not because myofibres accept no more cells, but rather that presumptive satellite cells are different from embryonic myoblasts. Initial reports demonstrating such differences came from Cossu's laboratory by showing that satellite cells in culture were unaffected by the presence of a phorbol ester (TPA) whereas the differentiation of embryonic mouse myoblasts was reversibly inhibited, suggesting fundamental differences in phospholipid metabolism (Cossu *et al.*, 1983, 1985, 1988). In addition, it was shown that cultured satellite cells express functional acetylcholine receptors, whereas embryonic myoblasts do not (Cossu *et al.*, 1987; Eusebi and Molinaro, 1984). Finally, reports from other laboratories have provided additional evidence of behavioural differences between satellite cells and embryonic myoblasts in culture (Allen and Boxhorn, 1989; Senni *et al.*, 1987; Yablonka-Reuveni *et al.*, 1987; Le Moigne *et al.*, 1990). Taken together, these studies illustrate structural and functional differences that set satellite cells apart from the embryonic myoblast pool and, further, suggest that satellite cells are a distinct class or lineage of myogenic cells.

Satellite cells play an important role in postnatal skeletal muscle. Under normal circumstances they function to provide myonuclei at a pace that is closely coupled to myofibre growth (Moss and Leblond, 1971). Well before the discovery of satellite cells, it was clearly shown that the ability to divide is lost once a myoblast nucleus is incorporated into the syncytium of a myofibre (Stockdale and Holtzer, 1961). Yet concurrent studies clearly showed that the hallmark of the postnatal growth period is an enormous increase in the DNA content of the myofibres (Enesco and Puddy, 1964; MacConnachie *et al.*, 1964). The mechanism whereby the increase in the number of myonuclei is accomplished in postnatal muscle was demonstrated by the radioautographic work of Moss and Leblond (1971) using immature growing rats. Within 48 hours of a single injection of ^{3}H-thymidine, approximately half of the cohort of initially labelled satellite cells was incorporated into myofibres by fusion, showing that continued division of satellite cells, and the fusion of their daughter cells following a mitotic division, led to the increase in the total myonuclear DNA within the myofibre syncytia.

Quantitative studies by a number of laboratories using a wide variety of species, demonstrated that satellite cells are a prominent feature of young muscle, where approximately 30% of myofibre nuclei are actually within satellite cells (reviewed by Campion, 1984). With increasing age, the relative number of nuclei in satellite cells decreases so that in mature muscle they represent only 2–5% of myofibre nuclei (Snow, 1977b; Schmalbruch and Hellhammer, 1976). This age-related decrease in myofibre satellite cell nuclei is a combination of an increase in the number of myonuclei and of a loss of satellite cells. When the absolute number of satellite cells was determined, it was evident that the population of cells associated with predominantly glycolytic muscles is not as great as that in predominantly oxidative muscles (Gibson and Schultz, 1983). These differences in population sizes could not be accounted for on the basis of fibre number or fibre length in these muscles. Attrition in the number of satellite cells as a function of age is greater in glycolytic muscles than in oxidative muscles. The differences observed in whole muscles are also reflected at the fibre level, where satellite cells are not distributed equally between fibre types (Gibson and Schultz, 1982). That is, in a muscle of mixed fibre types, oxidative fibres, the fibres first and most frequently recruited during functional activity, contain the largest population of satellite cells. The exact reasons why there are differences in the distribution of satellite cells as a function of muscle and fibre type remain unknown. It has been suggested that the differences may be related to the growth or functional characteristics of the muscles. For example, in the former case, the oxidative soleus muscle in the rat accumulates nuclei at a higher rate than the glycolytic EDL over the same growth period (Gibson and Schultz, 1983). The difference in the rate of nuclear accretion between the two muscles is not the result of the cells in the soleus replicating more rapidly, but rather a result of a larger pool of cells producing myonuclei, because the calculated cell cycle time *in vivo*, in the order of 30 hours, is the same in both muscles (Schultz, unpublished observations). Another possible reason for the greater population of satellite cells in oxidative muscles is that tonic activity may simply produce more injury over the course of the lifetime and the larger cell population is an evolved higher repair reserve. Related to this is the possibility that fibre turnover occurs in oxidative muscles during the lifetime and the rate of turnover is related to the activity levels of the fibres (Giddings *et al.*, 1985). It must also be emphasized, however, that despite differences in the absolute numbers of satellite cells between muscles, the efficacy of regeneration of damaged muscle appears not to be affected by fibre type composition. For example, the EDL, which contains a relatively small number of satellite cells, and the soleus, which contains a relatively larger number of cells, exhibit no discernible difference in the rate or final outcome of the regeneration response following a free-graft procedure (Schultz, 1984; Clark and White, 1985).

Associated with age-related population changes of satellite cells are intracellular morphological alterations (Schultz, 1976). Satellite cells in immature muscles have a much greater array of intracellular organelles in a greater volume of cytoplasm than those in mature muscle. The well developed rough endoplasmic reticulum and Golgi in satellite cells of young muscle suggests an active role in elaboration and secretion of some product. Attempts to determine the destination of any satellite cell secreted products have been largely unsuccessful. Autoradiographic studies after injection of ^{3}H-tyrosine demonstrated that the prominent rough endoplasmic reticulum and Golgi apparatus in satellite cells associated with immature muscle are actively secreting products into the extracellular space and into the interspace with the adjacent myofibre (Figure 8.1). Parallel studies using ^{3}H-fucose as a tracer indicated that secreted products are glycosylated. The fact that none of the products elaborated by satellite cells is accumulated at any site in the vicinity of the secreting cells, has prevented any inference as to the function or potential importance of the secreted materials and to the organellar changes that occur in the cells as a function of age. The reduction in organelles and cytoplasm that occurs with age suggests that older satellite cells are considerably less metabolically active. The significance of these changes remains unknown, and has not been well investigated. However, the interesting observation that many perturbations of muscle result in an increase in the size of satellite cells as well as their organellar complement (Teravainen, 1970; Schultz, 1978) suggests that the cells may have functions in normal skeletal muscle other than the addition of nuclei to myofibres. Recent reports suggesting that satellite cells may be capable of producing known growth factors such as a-FGF (Alterio *et al.*, 1990; Groux-Muscatelli *et al.*, 1990) or insulin-like growth factor (Jennische *et al.*, 1987) during the period that they are actively proliferating make this an interesting area for future study.

In addition to satellite cell population differences among muscles and morphological changes in the cells as a function of age, there is also functional heterogeneity within satellite cell populations. This heterogeneity is related to the replicative potential of the cells. For example, when rat cells are placed in culture at clonal density, the size of the colony that each individual cell produces is not the same; some cells give rise to colonies of only a few myoblasts, whereas some give rise to colonies several hundred in number (Schultz and Lipton, 1982). This heterogeneity in colony size is evident despite the age of the source tissue, although the average size of the colonies decreases as a function of increasing age, leading to an overall reduction in the proliferative potential of the satellite cell population. These

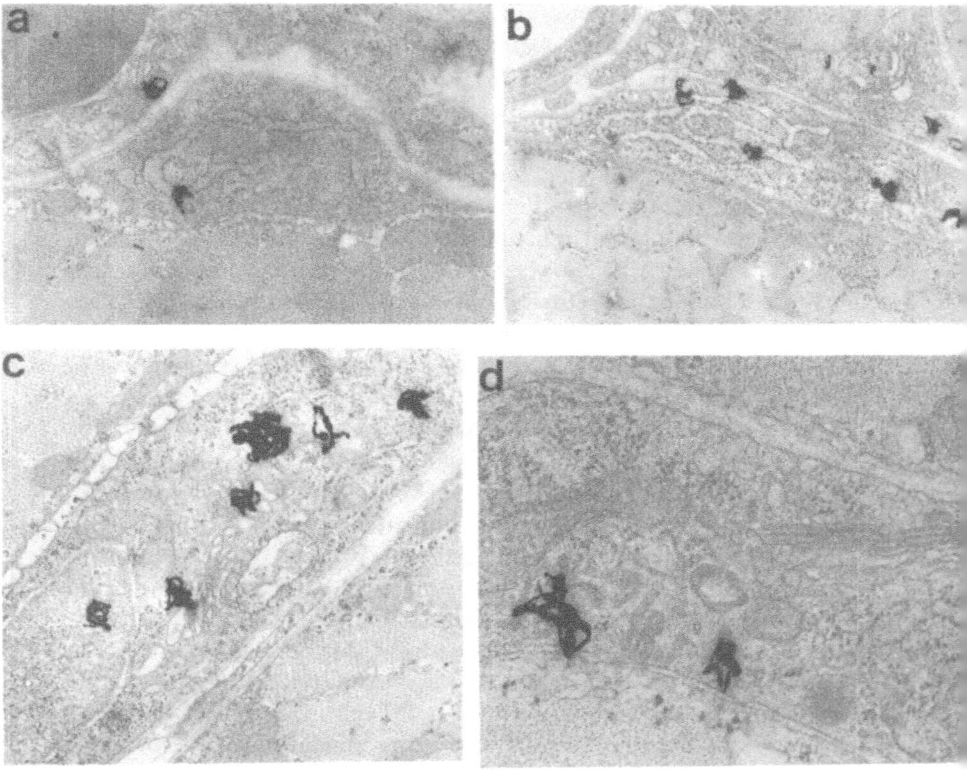

Figure 8.1 Autoradiographs of portions of satellite cells from tibialis anterior muscles of 30-day-old growing rats at various time periods after injection of ^{3}H-tyrosine illustrating synthesis of a product. (a) Two minutes following injection of ^{3}H-tyrosine a silver grain is located over a large cisterna of the rough endoplasmic reticulum located in the polar region of a satellite cell. The reticulum is typically well developed in many satellite cells at this age. Labelling is preferentially found over cells with lower nuclear:cytoplasmic ratios and with well developed organelles. (b) At ten minutes following injection of ^{3}H-tyrosine the number of grains per cell is increased and remains principally over areas rich in rough endoplasmic reticulum. Some grains are also over the Golgi (not shown). (c) At thirty minutes following injection of ^{3}H-tyrosine the majority of grains are located over the Golgi. (d) At thirty minutes some grains overlie vesicles in the region of the Golgi and vesicles at the periphery of the cell. Labelled vesicles are found at the margin of the cell facing the interstitial space as well as the margin facing the myofibre suggesting there are no preferential sites of secretion. The pattern of localization of grains illustrated at thirty minutes remained the same for up to four hours, after which the number of grains per cell declined. There were no sites near the cells that exhibited a build-up of the secretion products.

differences in the proliferative potential of individual satellite cells and age-related changes in the population as a whole are probably related to the previous proliferative history of the cells. Considering the large increases in myonuclei during the postnatal growth period (Enesco and Puddy, 1964), there is considerable replicative demand on the cells as they function to produce myonuclei. Since normal cells are considered to have a limited proliferative capacity (Hayflick, 1973) the number of mitotic divisions an individual cell has undergone during the growth period determines the proliferative potential of that cell at the time of culture. In rats, the proliferative potential of satellite cells continues to decrease over the first three months of age, a period during which muscle growth, expressed as a continued increase in the number of myonuclei, takes place. Beyond three months of age, when growth stops, there is little continued decrease in the colony forming potential of the cells. A direct relationship between proliferative activity of satellite cells *in vivo* and their colony-forming ability *in vitro* was demonstrated by showing that muscle injury and the ensuing regeneration response induced a premature reduction in the proliferative potential of satellite cells (Schultz and Jaryszak, 1985). More recent studies of pathological muscles have also shown that the proliferative demands of continual regeneration leads to a premature proliferative senescence of the cells (Wright, 1985; Webster and Blau, 1990). Taken together these studies have added further support to the notion that the observed decline in the proliferative capacity of satellite cells as a function of age is directly related to their previous mitotic activity *in vivo* and that mitotic divisions required to supply additional myonuclei during growth decreased the overall proliferation potential of the satellite cell population.

An observation of particular interest from the proliferation studies was that a small population of satellite cells retained the ability to form large colonies, even when derived from aged donors and from regenerated muscles that had been extensively damaged (Schultz and Lipton, 1982; Schultz and Jaryszak, 1985). In mature muscle donors this population constitutes approximately 10-15% of the total population of cultured cells. These observations suggested a mechanism whereby satellite cell proliferative heterogeneity is maintained during growth and even during a regeneration response. That is, satellite cells *in vivo* do not divide in an equivalent manner; some cells appear to divide more than others. However, it has been generally assumed that all satellite cells share equally in the production of myonuclei during the growth process since the *in vivo* studies of Moss and Leblond (1971), although if such were the case, the entire satellite cell population would be expected to be homogeneous with respect to their proliferation potential. We carried out a series of experiments in growing animals in order to determine whether all satellite cells contribute nuclei to muscle fibres in an equivalent manner. Miniosmotic pumps that contained

5-bromo-2-deoxyuridine (BrdU) were implanted subcutaneously in growing rats for periods of up to 14 days. The rationale was that if labelled DNA precursor were present on a continuous basis for extended periods of time and if all cells were dividing at an equal rate, then the entire satellite cell population would be rapidly labelled. The alternative situation, in which not all satellite cells were dividing, or some cells were dividing at a different rate to the general population, would result in a subpopulation of unlabelled cells that persisted for longer durations of the infusion period or simply remained unlabelled. Following the infusion period, the location of BrdU positive satellite cell nuclei was determined by immunogold labelling and electron microscopy (Figure 8.2). Labelling of satellite cells increased at a constant rate during the first 5 days of continuous labelling at which point approximately 80% were labelled (Figure 8.3). The rate of increase in labelling between 5 and 14 days of continuous labelling was reduced so that at the conclusion of this period approximately 10% of the cells remained unlabelled.

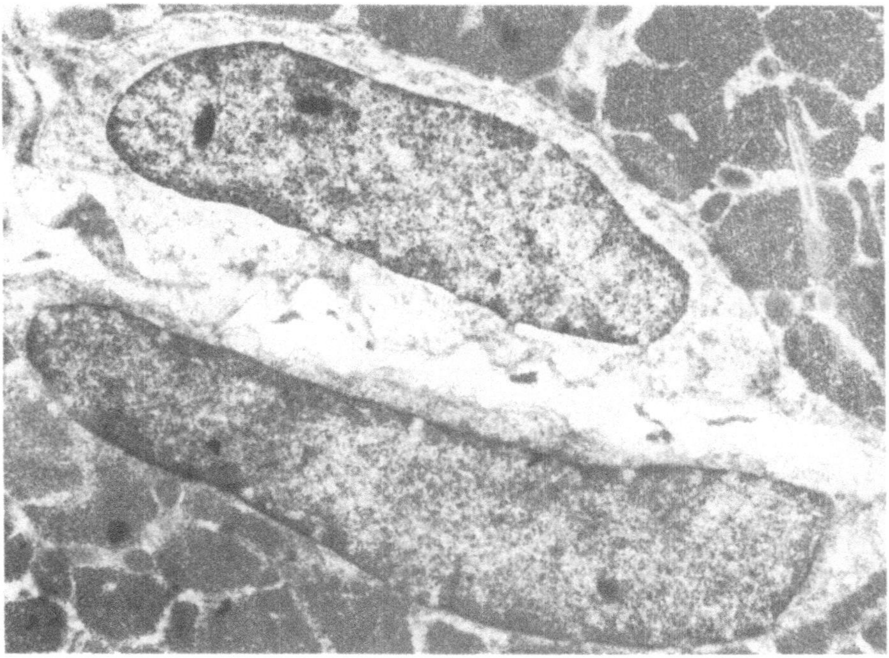

Figure 8.2 Immunogold labelling of a satellite cell nucleus following injection of 5-bromo-2-deoxyuridine. The myonucleus on the adjacent fibre is unlabelled whereas the satellite cell nucleus is covered with gold particles. The data in Figure 8.3 were obtained by counting labelled and unlabelled satellite cell nuclei.

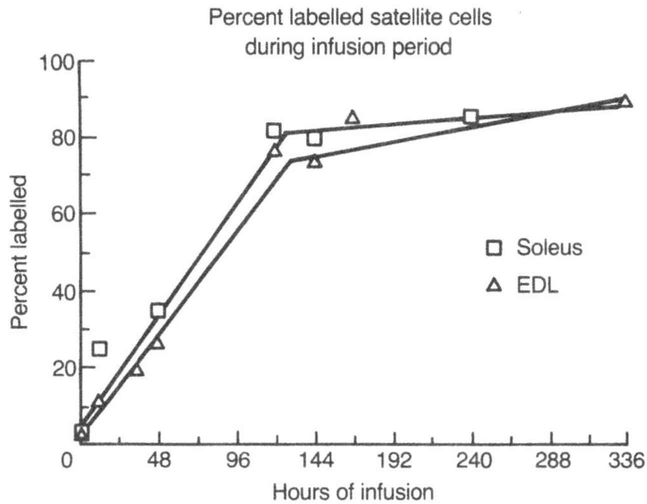

Figure 8.3 Plot showing the percentage of the satellite cell population labelled in 30-day-old rat soleus and extensor digitorum longus muscles after increasing periods of continuous infusion with 5-bromo-2-deoxyuridine administered by an osmotic pump. Both muscles show a rapid increase in the percentage of labelled cells over the initial 5 days. After this period the percentage of labelled cells continues to increase, but at a much slower rate. These data suggest that the satellite cell population in immature muscle is not homogeneous and can be subdivided into three subgroups on the basis of their proliferative behaviour. Group 1: cells that cycled within the first five days of infusion. Group 2: Cells that cycled during the next 9 days of infusion. Group 3: Cells that did not cycle during the 14-day infusion period.

The results of these experiments suggest that the satellite cell population in rat skeletal muscle comprises at least two sub-populations. One is relatively rapidly dividing and labels during the initial 5-day infusion period while a second relatively slowly dividing subpopulation is still not fully labelled even after 14 days of infusion. These *in vivo* findings are consistent with previous culture studies in mammalian muscle showing the heterogeneity in satellite cell colony forming potential (Schultz and Lipton, 1982; Schultz and Jaryszak, 1985). Moreover, these studies also suggest that the makeup of the satellite cell population in rat muscle exhibits many similarities to myogenic populations in embryonic avian muscles. Based upon *in vitro* analyses, Quinn *et al.*, (1984) proposed that myogenic cells in avian muscle are comprised of two populations: stem cells and cells committed to terminal differentiation. Stem cells can undergo either symmetric or self-renewing divisions, or asymmetric divisions that give rise to cohorts of cells that continue to divide before committing to terminal differentiation. Based upon our EM-BrdU labelling studies, satellite cells in mammalian postnatal muscle appear to be comprised of at least two populations which bear similarities to the myogenic cells described in the chicken. One population that does not label readily may function as progenitor or stem cells and through relatively slow or infrequent asymmetric mitotic divisions give rise to a renewing daughter progenitor cell and a second cell that is induced to divide or fuse depending upon environmental conditions or factors ultimately created by growth and/or functional demands upon the muscle (Konigsberg and Pfister, 1986). These latter cells make up a population that can divide an unknown number of times before entering terminal differentiation and fusing. In the growing rat, a significant proportion of the satellite cell population during growth appears to be capable of fusion following a single mitotic division (Moss and Leblond, 1971) because 50% of the initially labelled cells after a single injection of ^{3}H-thymidine had fused by 48 hours after injection, a duration that allowed only a single mitotic division. The minority of cells, constituting approximately 15–20% of the total cells, would be analogous to the progenitor population or an admixture of progenitor cells and cells that can divide multiple times before fusing depending upon local environmental factors.

Because the satellite cell population was heterogeneous, we also examined fusion and mitotic activity of satellite cells after continuous labelling of the satellite cell population, in order to determine if there is a preferential location where cells might divide and fuse along the myofibre length. Entire lengths of myofibre bundles of 5–10 fibres were examined for the location of labelled nuclei following a 5-day infusion period. Labelled nuclei would have included both myonuclei and satellite cell nuclei as well as nuclei of interstitial cells. There were no locations along the length of the myofibre that exhibited increased labelled nuclei. These preliminary observations suggest that satellite cell mitoses and fusion probably occurs at any point along the length of a myofibre. This lack of clustering of labelled nuclei was evident despite the fact that satellite cells are reportedly found at higher densities at neuromuscular junctions (Kelly, 1978; Wokke *et al.*, 1989), and that myofibres are elongating at their ends during growth (Williams and Goldspink, 1971).

8.3 BEHAVIOUR OF SATELLITE CELLS IN DAMAGED MUSCLE

The second major function of satellite cells in skeletal muscle is to repair damage incurred by myofibres (Snow, 1977a; Bischoff, 1975, 1989). Following damage, satellite cells are activated into a regeneration response that ultimately repairs or replaces necrotic myofibres. The mechanism of activation of the cells and the factors that guide their behaviour through regeneration are not clearly understood. Satellite cells do not have to be located immediately in an injured area of the muscle in order to exhibit signs of activation and participate in the regeneration response. Focal injury to muscle fibres was shown to induce satellite cell activation in areas of the muscle free of injury (Klein-Ogus and Harris, 1983; Schultz *et al.*, 1985). Likewise, satellite cells exhibit a pronounced activation/proliferation response even after bouts of exercise have produced microlesions throughout the muscle (Appell *et al.*, 1988; Darr and Schultz, 1989). Activation of the cells, as monitored by autoradiography after injection of ^{3}H-thymidine, occurs within 24–30 hours of insult (Schultz *et al.*, 1985; McGeachie and Grounds, 1987; Grounds and McGeachie, 1987). More recent studies using expression of myogenin or *MyoD* transcripts as markers of activation suggest that satelite cells actually respond to muscle injury in a matter of hours and that even this early response is widespread (Grounds *et al.*, 1992). Activation of the cells is thought to occur through the release of factors that are present in normal muscle but immobilized or compartmentalized and consequently unavailable to the cells (Bischoff, 1990; DiMario *et al.*, 1989, Kardami *et al.*, 1988; Yamada *et al.*, 1989). Damage to the muscle results in liberation of the activating factors leading to activation and proliferation of satellite cells which in adult muscle are mitotically quiescent (Schultz *et al.*, 1978).

Migration is an important facet of activated satellite cell behaviour. Cells in normal muscle exhibit migratory behaviour which becomes apparent if they are labelled with ^{3}H-thymidine or BrdU and monitored at various times after they have divided. Immediately following a mitosis the daughter cells can be visualized as doublets, however, within hours of the mitosis the cells have dispersed along the surface of the myofibre so that no morphological indication of the previous mitotic event is seen (Schultz, unpublished observations). Thus, even in intact muscle there is evidence that satellite cells migrate along the length of their fibres to at least some degree. In fact, inspection of fibres following various periods of labelling infrequently shows labelled cells close to one another suggesting that there may actually be some mechanism whereby the cells evenly disperse on myofibres and clustering of the cells is prevented.

Migration is important in the survival of the cell during regeneration of damaged muscle and for participation of large numbers of cells in the repair or replacement of the myofibres. The ability of satellite cells to migrate has been examined in a number of systems (Lipton and Schultz, 1979; Grounds *et al.*, 1980; Grounds and Partridge, 1983; Roth and Oron, 1985; Schultz *et al.*, 1985, 1986; Ghins *et al.*, 1984; Morgan *et al.*, 1987; Watt *et al.*, 1987; Bischoff, 1990; Hughes and Blau, 1990; Phillips *et al.*, 1990). In whole muscle grafts, which result in large scale ischaemic necrosis of the muscle, the grafted muscles undergo a characteristic and predictable pattern of centripetal degeneration/regeneration that has been well documented (Hansen-Smith and Carlson, 1979; Hansen-Smith *et al.*, 1980). During the degeneration phase, it was suggested that satellite cells in the central regions of the muscle belly remained dormant until the centripetal wave of regeneration extended to their position, at which time the cells were activated. However, studies of the central regions of whole muscle free-grafts suggested that satellite cells were no better able to withstand the ischaemic conditions within the muscle than any other cell type (Phillips *et al.*, 1987); cultures of the central or core regions of the grafted muscle indicated there were no viable cells present by 48 hours after the graft (Schultz *et al.*, 1988). Based upon the absence of satellite cells in the necrotic regions of the muscle, it was suggested that the cells were able to survive the grafting procedure by migrating to more favourable conditions at the peripheral regions of the muscle (Schultz *et al.*, 1988). Thus, during the initial phases of muscle injury, particularly where widespread damage has produced significant areas of ischaemia, the ability of the cells to migrate is of utmost importance in their survival.

Whether satellite cells actually leave their fibres and migrate transversely in the muscle or remain on the surface of their fibres was not determined. However, studies from other laboratories suggest that transverse migration is possible (Maltin *et al.*, 1983) and recent reports by Phillips *et al.* (1990) and Hughes and Blau (1990) have also provided compelling evidence of transversely migrating satellite cells. The precise cues guiding such migratory behaviour in damaged muscles are unknown although the invasion of macrophages and the ingrowth of vasculature are potential sources (Venkatasubramanian and Solursh, 1984). An interesting implication of such widespread migration between fibres is that there may be an as yet unidentified population of myogenic cells in the interstitial space that are simply in transit or in permanent residence. The avian overload model of strength induced hypertrophy (Sola *et al.*, 1973) may illustrate the existence of interstitial myogenic cells. Within a week of initiation of a weight stimulus, small diameter ventricular-like embryonic myosin positive fibres appear in the interfascicular spaces of the ALD muscle (Kennedy *et al.*, 1988, 1989). There seems little doubt that these are nascent fibres since all nuclei in these fibres are labelled under the conditions of continuous labelling (McCormick and Schultz, 1992). The rapid formation of nascent myofibres in the overloaded chicken muscle may be a reflection of the appropriate stimulus for these cells to differentiate and form fibres. Although specific markers of the satellite cell population are becoming available (Wakshull *et al.*, 1983; Schubert *et al.*, 1989) they are not suitable for all species so that little work has been done to document unequivocally the existence of myogenic precursor cells in the interstitial spaces of a muscle.

Although cells can be readily recruited from distant regions within a muscle, can exogenous cells be recruited from another muscle? Independent studies from laboratories using rats as experimental animals have concluded that satellite cell migration between muscles does not occur (Ghins *et al.*, 1984; Schultz *et al.*, 1986). When a muscle in the rat is killed completely by freezing it is repopulated by cells, but there are no myogenic cells among them. The frozen muscle is initially invaded by nerve, blood vessels and non-muscle cells. Apparently, most nerve and blood vessels are subsequently resorbed since within several weeks of freezing all that remains is a well formed tendon. Myofibres are formed if a small portion of the muscle is not frozen and serves as a source of myogenic cells (Schultz *et al.*, 1986) or if a muscle is killed but cultured myogenic cells are implanted (Ghins *et al.*, 1985; Schultz *et al.*, 1986; Morgan *et al.*, 1987; Alameddine *et al.*, 1989). Likewise, if an adhesion is created between the killed muscle and an intact adjacent muscle then myogenic cells will invade the frozen muscle and form myofibres. Thus, it appears that a requirement for myogenic cell movement between muscles is that the connective tissue investments surrounding both the killed muscle and the adjacent intact muscle must be interrupted. In the experiments described, the additional requirement that the adjacent muscle must also be injured was not completely ruled out. In the case of the frozen muscle experiments, placing a small injury in the intact muscle may have induced a regeneration response that simply extended into the frozen muscle.

The findings that satellite cells do not migrate extensively between muscles are not universal and distinct species differences are evident. Experiments using mice as experimental models have demonstrated migration between muscles but it is at best limited (Grounds *et al.*, 1980; Morgan *et al.*, 1987; Watt *et al.*, 1987). For example, the results of Grounds *et al.* (1980) have demonstrated in the mouse through the use of GP6 isoforms that muscles transplanted into a host are in fact invaded by host myoblasts. The invading myoblasts participate in the regeneration process as indicated by the presence of hybrid enzyme. The invasion of the host myoblasts required a course through the same connective tissue investments that in the rat prevented movement of the cells. Although the connective tissue of the mouse is probably more delicate than in the rat, which to some extent explains the differences in the migratory potentials of myoblasts between species, a complete explanation of the factors restricting or promoting migratory capacities *in vivo* has not been developed.

8.4 SATELLITE CELLS, MYOBLASTS AND IMPLANT THERAPY

Considering the ease with which they can be harvested and grown in culture, satellite cells of mature muscle will probably be a major source of myoblasts for implant therapy in Duchenne muscular dystrophy. Certainly the age of the host, pathology in the muscle from which the cells are obtained as well as the duration of growth *in vitro* are factors to be considered prior to implantation because each of these influence the yield or proliferative history and, consequently, the proliferative potential of the implanted cells. The proliferative capacity of harvested satellite cells is directly related to the proliferative demands on the cells *in vivo* as illustrated by the observed premature senescence of cells from regenerated muscles (Schultz and Jaryszak, 1985) or from animal models of myopathy (Wright, 1985). Thus, it would be expected that the proliferation potential of human satellite cells derived from Duchenne dystrophic muscle is prematurely reduced because of the continued proliferative demand of an ongoing regeneration response (Webster and Blau, 1990). In fact, Webster and Blau (1990) found that proliferative potential was severely reduced and only cells derived from young dystrophy patients could be grown to the extent that would make therapeutic genetic engineering and subsequent reimplantation of the dystrophic cells a possibility. Thus, from the standpoint of simply generating a large enough population of cells for the implant, normal muscle appears to be the optimum source. Interestingly, even though satellite cells of normal human muscle exhibit the same pattern of reduced number and proliferative potential with age as observed in rat muscle, given their very large proliferative reserve, harvesting a suitable number of satellite cell-derived myoblasts for implant therapy seems not to be a limiting factor even when the cells are derived from adult donors. Finally, differences in the fibre type composition of the donor and recipient muscles may prove to be important, but based upon current evidence it is uncertain what impact these differences might have. That is, it is unclear whether the fibre type composition of the donor and the recipient muscle should be approximated. This would be an issue if satellite cell-derived myoblasts remained faithful to the fibre type from which the parent satellite cell was derived. However, recent evidence suggests that mammalian satellite cell myoblasts are indiscriminate with respect to the fibres they form (Dusterhoft *et al.*, 1990; Whalen *et al.*, 1990) as is the case with avian muscle (Feldman and Stockdale, 1991). On the basis of these findings it would not be expected that implanted myoblasts 'sort out' and become associated only with specific fibre types.

A feature observed in animal studies that perhaps deserves further consideration when preparing human cells for implant therapy is the heterogeneity in the colony-forming potential of the population. A goal of implant therapy has been to stimulate growth of the satellite cells in culture in order to maximize the total number of cells that can be derived from the source tissue. Growth of the cells in culture would stimulate all cells to grow to their maximum and, as a result, abrogate the natural proliferative heterogeneity that characterizes this cell population. When these cells are harvested and implanted into a host muscle, it would be expected that virtually all cells that gain access to myofibres would fuse with them in a relatively short period of time. On the other hand, cells harvested and implanted immediately or after culture periods of short duration would be expected to more accurately reflect the *in vivo* situation in that many cells might gain access to myofibres but not fuse with them and take up positions as satellite cells. More importantly, progenitor cells with high proliferative capacity may assume positions at the periphery of fibres where they would presumably continue to function as satellite cells as the host muscle continued to grow. In this manner, therapy of immature muscle would have the longer-lasting benefit of the prolonged incorporation of additional normal myonuclei as the fibres continue growing.

There are two lines of evidence to suggest that some of the implanted myoblasts might be expected to function as satellite cells. The results of an implant study in which marked cells were followed at the EM level suggested that while a large proportion of surviving cells eventually fused with myofibres, some cells associated with fibres did not fuse and retained a satellite cell position (Lipton and Schultz, 1979). Unfused cells probably explain the recent findings of Morgan *et al.* (1991) who cultured a host muscle after long-term implant and were able to retrieve a small number of the implanted cells. These cells obviously did not fuse, but the position they occupied is unknown and, just as in the previous study, it remains to be determined whether they functioned as satellite cells by providing additional myonuclei to myofibres.

Longitudinal spread of implanted cells within a muscle appears to be widespread, but these observations are largely based upon experiments using rodent muscles. The maximum capacity of cells to disperse longitudinally in larger human muscles is still untested and requires further study. There appears to be no specific site along the longitudinal axis of a muscle where cells should be introduced because there is no evidence of regions where fusion of satellite cells occurs preferentially along the length of myofibres. Consequently, wherever a myoblast makes contact with a fibre it should be able to fuse.

Although evidence suggests that implant sites along the longitudinal axis of a muscle do not need to be specific, a major concern is lateral spread of the cells from the implant site. The most important finding from these animal studies has been the demonstration of the limited ability of myoblasts to penetrate the connective tissue investments covering a muscle.

The epimysia of muscles appears to represent a significant barrier to movement of myogenic cells such that implants of cells must certainly be placed within the belly of recipient muscles and not between muscles. The best examples of transverse migration without any injury or damage imposed upon the muscle have been in immature muscle so the exact times at which effects of age and associated increases in connective tissue impact on the ability of the cells to move is unknown. The extent to which the perimysium acts as a barrier in human muscle remains to be established especially if the muscle has undergone some fibrosis. The simplest method to nullify the potential role of the perimysium in inhibiting transverse spread of the cells is to place multiple injections across the belly of the muscle and to account for the fusiform or penniform arrangement of the fibres.

In addition to the problem of increased fibrosis and the potential barriers to migration, the environment of pathological muscle is very different from normal muscle because there are numerous foci of degeneration and regeneration. Based upon the behaviour of satellite cells in focally damaged muscles the implanted cells may be preferentially drawn toward those foci of degeneration and regeneration at the expense of the undamaged myofibres. Under these conditions many of the 'intact' myofibres that were not undergoing some aspect of the degeneration/regeneration cycle might not serve as a target for the implanted cells and, as a consequence, would be lost. Moreover, the implanted cells must 'compete' with the endogenous population of satellite cells in adding myonuclei to the fibres. Thus, a better understanding of satellite cell or myoblast behaviour is needed in order to devise strategies to suppress the division and fusion activity of the endogenous satellite cell population while allowing the implanted cells to carry out these same functions.

REFERENCES

Alameddine, H.S., Dehaupas, M. and Fardeau, M. (1989) Regeneration of skeletal muscle fibers from autologous satellite cells multiplied in vitro. An experimental model for testing cultured cell myogenicity. *Muscle Nerve*, **12**, 544–55.

Allen, R.E. and Boxhorn, L.K. (1989) Regulation of skeletal muscle satellite cell proliferation and differentiation by transforming growth factor-beta, insulin-like growth factor I, and fibroblast growth factor. *J. Cell Physiol.*, **138**, 311–15.

Altereio, J., Courtois, Y., Robelin, J. *et al.* (1990) Acidic and basic fibroblast growth factor mRNAs are expressed by skeletal muscle satellite cells. *Biochem. Biophys. Res. Commun.*, **166**, 1205–12.

Appell, H.J., Forsberg, S. and Hollmann, W. (1988) Satellite cell activation in human skeletal muscle after training: evidence for muscle fiber neoformation. *Int. J. Sports Med.*, **9**, 297–9.

Bischoff, R. (1975) Regeneration of single skeletal muscle fibers in vitro. *Anat. Rec.*, **182**, 215–35.

Bischoff, R. (1989) Analysis of muscle regeneration using single myofibers in culture. *Med. Sci. Sports Exerc.*, **21**, S164–S172.

Bischoff, R. (1990) Cell cycle commitment of rat muscle satellite cells. *J. Cell Biol.*, **111**, 201–7.

Bonner, P.H. and Hauschka, S.D. (1974) Clonal analysis of vertebrate myogenesis. I. Early developmental events in the chick limb. *Dev. Biol.*, **37**, 317–28.

Campion, D.R. (1984) The muscle satellite cell: a review. *Int. Rev. Cytol.*, **87**, 225–51.

Clark, K.I. and White, T.P. (1985) Morphology of stable muscle grafts in rats: Effects of gender and muscle type. *Muscle Nerve*, **8**, 99–104.

Cossu, G., Molinaro, M. and Pacifici, M. (1983) Differential response of satellite cells and embryonic myoblasts to a tumor promoter. *Dev. Biol.*, **98**, 520–4.

Cossu, G., Cicinelli, P., Fieri, C. *et al.* (1985) Emergence of TPA-resistant 'satellite' cells during muscle histogenesis of human limb. *Exp. Cell Res.*, **160**, 403–11.

Cossu, G., Eusebi, F., Grassi, F. and Wanke, E. (1987) Acetylcholine receptor channels are present in undifferentiated satellite cells but not in embryonic myoblasts in culture. *Dev. Biol.*, **123**, 43–50.

Cossu, G., Ranaldi, G., Senni, M.I. *et al.* (1988) 'Early' mammalian myoblasts are resistant to phorbol ester-induced block of differentiation. *Development*, **102**, 65–9.

Darr, K.C. and Schultz, E. (1989) Hindlimb suspension suppresses muscle growth and satellite cell proliferation. *J. Appl. Physiol.*, **67**, 1827–34.

DiMario, J., Buffinger, N., Yamanda, S. and Strohman, R.C. (1989) Fibroblast growth factor in the extracellular matrix of dystrophic (mdx) mouse muscle. *Science*, **244**, 688–90.

Dusterhoft, S., Yablonka-Reuveni, Z. and Pette, D. (1990) Characterization of myosin isoforms in satellite cell cultures from adult rat diaphragm, soleus and tibialis anterior muscles. *Differen.*, **45** 185–91.

Enesco, M. and Puddy, D. (1964) Increase in the number of nuclei and weight in skeletal muscles of rats of various ages. *Am. J. Anat.*, **114**, 235–44.

Eusebi, F. and Molinaro, M. (1984) Acetylcholine sensitivity in replicating satellite cells. *Muscle Nerve*, **7**, 488–92.

Feldman, J.L. and Stockdale, F.E. (1991) Skeletal muscle satelite cell diversity: Satellite cells form fibers of different types in cell culture. *Dev. Biol.*, **143**, 320–34.

Ghins, E., Colson-Van Schoor, M. and Marechal, G. (1984) The origin of muscle stem cells in rat triceps surae regenerating after mincing. *J. Musc. Res. Cell Motility*, **5**, 711–22.

Ghins, E., Colson-Van Schloor, M., Maldague, P. and Marechal, G. (1985) Muscle regeneration induced by cells autografted in adult muscles. *Arch. Int. Physiol. Biochim.*, **93**, 143–53.

Gibson, M.C. and Schultz, E. (1982) The distribution of satellite cells and their relationship to specific fiber types in soleus and extensor digitorum longus muscles. *Anat. Rec.*, **202**, 329–37.

Gibson, M.C. and Schultz, E. (1983) Age-related differences in absolute numbers of skeletal muscle satellite cells. *Muscle Nerve*, **6**, 574–80.

Giddings, C.J., Neaves, W.B. and Gonyea, W.J. (1985) Muscle fiber necrosis and regeneration induced by prolonged weight-lifting exercise in the cat. *Anat. Rec.*, **211**, 133–41.

Griggs, R.C. and Karpati, G. (eds). (1990) Myoblast transfer therapy. *Adv. Exp. Med. Biol.*, **208**, Plenum Press, New York, N.Y.

Grounds, M.D. and Partridge, T.A. (1983) Isoenzyme studies of whole muscle grafts and movement of muscle precursor cells. *Cell Tissue Res.*, **230**, 677–88.

Grounds, M.D. and McGeachie, J.K. (1987) A comparison of muscle precursor replication in crush injured skeletal muscle of Swiss and BALBc mice. *Cell Tissue Res.*, **255**, 385–91.

Grounds, M., Partridge, T.A. and Sloper, J.C. (1980) The contribution of exogenous cells to regenerating skeletal muscle: an isoenzyme study of muscle allografts in mice. *J. Pathol.*, **132**, 325–41.

Grounds, M.D., Garrett, K.L. *et al.* (1992) Identification of skeletal muscle precursor cells *in vivo* by use of MyodI and myogenin probes. *Cell Tissue Res.*, **267**, 99–109.

Groux-Muscatelli, B., Bassaglia, Y., Barritault, D. *et al.* (1990) Proliferating satellite cells express acidic fibroblast growth factor during in vitro myogenesis. *Dev. Biol.*, **142**, 380–5.

Hansen-Smith, F.M. and Carlson, B.M. (1979) Cellular responses to free grafting of the extensor digitorum longus muscle of the rat. *J. Neurol. Sci.*, **41**, 149–73.

Hansen-Smith, F.M., Carlson, B.M. and Irwin, K.L. (1980) Revascularization of the freely grafted extensor digitorum longus muscle in the rat. *Am. J. Anat.*, **158**, 65–82.

Hayflick, L. (1973) The biology of human aging. *Am. J. Med. Sci.*, 265, 432–45.

Hughes, S.M. and Blau, H.M. (1990) Migration of myoblasts across basal lamina during skeletal muscle development. *Nature*, **345**, 350–3.

Jennische, E., Skottner, A. and Hansson, H.A. (1987) Satellite cells express the trophic factor IGF-I in regenerating skeletal muscles. *Acta Physiol. Scand.*, **129**, 9–15.

Kardami, E., Spector, D. and Strohman, R.C. (1988) Heparin inhibits skeletal muscle growth in vitro. *Dev. Biol.*, **126**, 19–28.

Kelly, A.M. (1978) Perisynaptic satellite cells in the developing and mature rat soleus muscle. *Anat. Rec.*, **190**, 891–903.

Kennedy, J.M., Eisenberg, B.R., Reid, S.K. *et al.* (1988) Nascent muscle fiber appearance in overloaded chicken slow-tonic muscle. *Am. J. Anat.*, **181**, 203–15.

Kennedy, J.M., Sweeney, L.J. and Gao, L.Z. (1989) Ventricular myosin expression in developing and regenerating muscle, cultured myotubes, and nascent myofibers of overloaded muscle in the chicken. *Med. Sci. Sports Exerc.*, **21**, S187–S197.

Klein-Ogus, C. and Harris, J.B. (1983) Preliminary observations of satellite cells in undamaged fibres of the rat soleus muscle assaulted by a snake-venom toxin. *Cell Tissue Res.*, 230, 671–6.

Konigsberg, I.R. and Pfister, K.K. (1986) Replicative and differentiative behavior in daughter pairs of myogenic stem cells. *Exp. Cell Res.*, **167**, 63–74.

Le Moigne, A., Martelly, I., Barlovatz-Meimon, G. *et al.* (1990) Characterization of myogenesis from adult satellite cells cultured in vitro. *Int. J. Dev. Biol.*, **34**, 171–80.

Lipton, B.H. and Schultz, E. (1979) Developmental fate of skeletal muscle satellite cells. *Science*, **205**, 1292–4.

MacConnachie, H.F., Enesco, M. and Leblond, C.P. (1964) The mode of increase in the number of skeletal muscle nuclei in the postnatal rat. *Am. J. Anat.*, **114**, 245–51.

Maltin, C.A., Harris, J.B. and Cullen, M.J. (1983) Regeneration of mammalian skeletal muscle following the injection of the snake-venom toxin, taipoxin. *Cell Tissue Res.*, **232**, 565–77.

Mauro, A. (1961) Satellite cell of skeletal muscle fibers. *J. Biophys. Biochem. Cytol.*, **9**, 493–8.

McCormick, K.M. and Schultz, E. (1992) Mechanisms of nascent fiber formation during avian skeletal muscle hypertrophy. *Dev. Biol.*, **150**, 319–34.

McGeachie, J.K. and Grounds, M.D. (1987) Initiation and duration of muscle precursor replication after mild and severe injury to skeletal muscle. *Cell Tissue Res.*, **248**, 125–30.

Morgan, J.E., Coulton, G.R. and Partridge, T.A. (1987) Muscle precursor cells invade and repopulate freeze-killed muscles. *J. Muscle Res. Cell Motil.*, **8**, 386–96.

Morgan, J.E., Pagel, C.N. and Partridge, T.A. (1991) The immediate and long-term fates of myogenic cells implanted into MDX mouse muscles. *J. Cell. Biochem.*, **15C**, 40.

Moss, F.P. and Leblond, C.P. (1971) Satellite cells as a source of nuclei of growing rats. *Anat. Rec.*, **170**, 421–36.

Partridge, T.A. (1991) Invited review: Myoblast transfer: A possible therapy for inherited myopathies? *Muscle Nerve*, **14**, 197–212.

Phillips, G.D., Lu, D., Mitashov, V.I. and Carlson, B.M. (1987) Survival of myogenic cells in freely grafted rat rectus femoris and extensor digitorum longus muscles. *Am. J. Anat.*, **180**, 365–72.

Phillips, G.D., Hoffman, J.R. and Knighton, D.R. (1990) Migration of myogenic cells in the rat extensor digitorum longus muscle studied with a split autograft model. *Cell Tissue Res.*, **262**, 81–8.

Quinn, L.S., Nameroff, M. and Holtzer, H. (1984) Age-dependent changes in myogenic precursor cell compartment sizes. Evidence for the existence of a stem cell. *Exp. Cell Res.*, **154**, 65–82.

Roth, D. and Oron, U. (1985) Repair mechanisms involved in muscle regeneration following partial excision of the rat gastrocnemius muscle. *Exp. Cell Res.*, **53**, 107–114.

Schmalbruch, H. and Hellhammer, U. (1976) The number of satellite cells in normal human muscle. *Anat. Rec.*, **185**, 279–87.

Schubert, W., Zimmernan, K., Cramer, M. and Starzinski-Powitz, A. (1989) Lymphocyte antigen Leu-19 as a molecular marker of regeneration in human skeletal muscle. *Proc. Natl. Acad. Sci. U.S.A.*, **86**, 307–11.

Schultz, E. (1976) Fine structure of satellite cells in growing skeletal muscle. *Am. J. Anat.*, **147**, 49–70.

Schultz, E. (1978) Changes in the satellite cells of growing muscle following denervation. *Anat. Rec.*, **190**, 299–312.

Schultz, E. (1984) A quantitative study of satellite cells in regenerated soleus and extensor digitorum longus muscles. *Anat. Rec.*, **208**, 501–6.

Schultz, E., Gibson, M.C. and Champion, T. (1978) Satellite cells are mitotically quiescent in mature mouse muscle: an EM and radioautographic study. *J. Exp. Zool.*, **206**, 451–6.

Schultz, E. and Lipton, B.H. (1982) Skeletal muscle satellite cells: changes in proliferation potential as a function of age. *Mech. Ageing Dev.*, **20**, 377–83.

Schultz, E. and Jaryszak, D.L. (1985) Effects of skeletal muscle regeneration on the proliferation potential of satellite cells. *Mech. Ageing Dev.*, **30**, 63–72.

Schultz, E., Jaryszak, D.L. and Valliere, C.R. (1985) Response of satellite cells to focal skeletal muscle injury. *Muscle Nerve*, **8**, 217–22.

Schultz, E., Jaryszak, D.L., Gibson, M.C. and Albright, D.J. (1986) Absence of exogenous satellite cell contribution to regeneration of frozen skeletal muscle. *J. Muscle Res. Cell Motil.*, **7**, 361–7.

Schultz, E., Albright, D.J., Jaryszak, D.L. and David, T.L. (1988) Survival of satellite cells in whole muscle transplants. *Anat. Rec.*, **222**, 12–17.

Seed, J. and Hauschka, S.D. (1984) Temporal separation of the migration of distinct myogenic precursor populations into the developing chick wing bud. *Dev. Biol.*, **106**, 389–93.

Senni, M.I., Castrignano, F., Poiana, G. *et al.* (1987) Expression of adult fast pattern of acetylcholinesterase molecular forms by mouse satellite cells in culture. *Differentiation*, **36**, 194–8.

Snow, M.H. (1977a) Myogenic cell formation in regenerating rat skeletal muscle injured by mincing. II. An autoradiographic study. *Anat. Rec.*, **188**, 201–17.

Snow, M.H. (1977b) The effects of aging on satellite cells in skeletal muscles of mice and rats. *Cell Tissue Res.*, **185**, 399–408.

Sola, O.M., Christensen, D.L. and Martin, A.W. (1973) Hypertrophy and hyperplasia of adult chicken anterior latissimus dorsi muscles following stretch with and without denervation. *Exp. Neurol.*, **41**, 76–100.

Stockdale, F.E. and Holtzer, H. (1961) DNA synthesis and myogenesis. *Exp. Cell Res.*, **24**, 508–20.

Teravainen, H. (1970) Satellite cells of striated muscle after compression injury so slight as not to cause degeneration of the muscle fibers. *Z. Zellforsch.*, **103**, 320–7.

Venkatasubramanian, K. and Solursh, M. (1984) Chemotactic behavior of myoblasts. *Dev. Biol.*, **104**, 428–33.

Wakshull, E., Bayne, E.K., Chiquet, M. and Fambrough, D.M. (1983) Characterization of a plasma membrane glycoprotein common to myoblasts, skeletal muscle satellite cells, and glia. *Dev. Biol.*, **100**, 464–77.

Watt, D.J., Morgan, J.E., Clifford, M.A. and Partridge, T.A. (1987) The movement of muscle precursor cells between adjacent regenerating muscles in the mouse. *Anat. Embryol. (Berl).*, **175**, 527–36.

Webster, C. and Blau, H.M. (1990) Accelerated age-related decline in replicative life-span of Duchenne muscular dystrophy myoblasts: implications for cell and gene therapy. *Somat. Cell Mol. Genet.*, **16**, 557–65.

Whalen, R.G., Harris, J.B., Butler-Browne, G.S. and Sesodia, S. (1990) Expression of myosin isoforms during notexin-induced regeneration of rat soleus muscles. *Dev. Biol.*, **141**, 24–40.

White, N.K., Bonner, P.H., Nelson, R. and Hauschka, S.D. (1975) Clonal analysis of vertebrate myogenesis. VI. Medium-dependent classification of colony forming cells. *Dev. Biol.*, **44**, 346–57.

Williams, P.E. and Goldspink, G. (1971) Longitudinal growth of striated muscle fibres. *J. Cell Sci.*, **9**, 751–67.

Wokke, J.H., Van den Oord, C.J., Leppink, G.J. and Jennekens, F.G. (1989) Perisynaptic satellite cells in human external intercostal muscle: a quantitative and qualitative study. *Anat. Rec.*, **223**, 174–80.

Wright, W.E. (1985) Myoblast senescence in muscular dystrophy. *Exp. Cell Res.*, **157**, 343–54.

Yablonka-Reuveni, Z., Quinn, L.S. and Nameroff, M. (1987) Isolation and clonal analysis of satellite cells from chicken pectoralis muscle. *Dev. Biol.*, **119**, 252–9.

Yamada, S., Buffinger, N., DiMario, J. and Strohman, R.C. (1989) Fibroblast growth factor is stored in fiber extracellular matrix and plays a role in regulating muscle hypertrophy. *Med. Sci. Sports Exerc.*, **21**, S173–S180.

9

Molecular and cell biology of skeletal muscle regeneration

MIRANDA D. GROUNDS and ZIPORA YABLONKA-REUVENI

9.1 INTRODUCTION

When skeletal muscle is damaged, it is repaired by the proliferation of mononuclear muscle precursor cells (mpc) which fuse either with one another to form young multinucleated muscle cells (myotubes) or with the ends of damaged myofibres (Robertson *et al.*, 1990). The success of new muscle formation is related to the size of the injury, as after major trauma and extensive disruption of the external lamina of muscle fibres there is often significant replacement by fibrous and cellular connective tissue. Impaired muscle regeneration and progressive replacement by fat and connective tissue is a feature of myopathies such as Duchenne muscular dystrophy (DMD), although this results from many small discrete lesions constantly recurring over a long period of time rather than from a single large injury. Failed regeneration can be seen in simplistic terms as a failure of muscle precursor replication. In this review we shall concentrate on the biology of muscle precursor cells. For coverage of other aspects of regeneration such as resealing of damaged myofibres, revascularization and reinnervation, see Grounds (1991).

It is of particular interest to compare the biology of muscle precursor cells from humans with those from mice and dogs as these animals provide valuable models for the human X-linked dystrophin myopathies like Duchenne and Becker muscular dystrophy (Partridge, 1991a). Skeletal muscles from patients with DMD and from X-linked dystrophic *mdx* mice and dystrophic golden retriever dogs, lack dystrophin and undergo cycles of necrosis and regeneration. The limb muscles remain relatively healthy in *mdx* mice, whereas they become progressively weaker and are replaced by fat and connective tissue in the dog model which clinically more closely

Molecular and Cell Biology of Muscular Dystrophy
Edited by Terence Partridge
Published in 1993 by Chapman & Hall, London. ISBN 0 412 43440 7

resembles DMD (see other chapters in this book). Although the limb muscles of *mdx* mice remain relatively healthy, there is a progressive degeneration of diaphragm muscles which are extensively replaced by fibrous tissue by one year of age (Stedman *et al.*, 1991). Before the identification of these animal models of X-linked dystrophy, other animals which were widely studied as models of DMD were the 129ReJ *dy/dy* mouse and dystrophic chickens, although both of these are due to factors other than a lack of dystrophin. It is clearly important to understand more about the pathology of the genetically closely related dystrophin myopathies and the factors controlling muscle regeneration in these different species.

In mature muscle fibres, the mpc that give rise to new muscle after trauma are considered to be derived from satellite cells located between the plasmalemma and external lamina (widely referred to as basal lamina) of myofibres (first described by Mauro, 1961: reviewed by Mazanet and Franzini-Armstrong, 1980; Campion, 1984; White and Esser, 1989). That satellite cells do give rise to mpc has been demonstrated elegantly *in vitro* (Bischoff, 1979, 1990a); however, one cannot rule out the possibility that other locally derived cells might also give rise to muscle precursors *in vivo*, particularly following injury (Kennedy *et al.*, 1988; reviewed by Grounds, 1990a, 1991). The relationship between satellite cells and mpc will be discussed later. Since the proliferative capacity of muscle precursor cells appears to be central to effective muscle regeneration we shall examine the behaviour of satellite cells with respect to: (1) the time of formation during development and their identification *in vivo*; (2) when replication ceases during postnatal growth; (3) the numbers and proliferative capacity of such mpc from animals of different ages; and (4) the influence of various growth factors and extracellular matrix components on proliferation. Data will be evaluated from tissue culture and *in vivo* studies in a range of species.

9.2 DEVELOPMENTAL ORIGIN OF SATELLITE CELLS AND THEIR IDENTIFICATION *IN VIVO*

9.2.1 Cell culture studies

Studies in chick/quail chimaeras indicate that satellite cells are part of the same somitically derived myogenic lineage as embryonic mpc (Armand *et al.*, 1983). The simple notion that satellite cells are merely embryonic* mpc trapped under the developing external lamina may be incorrect, as evidence is now accumulating from cell culture studies that satellite cells from adult

* For simplicity the term 'embryonic' is used throughout this discussion to describe all prenatal or pre-hatching stages of development: it includes both the early and later (often referred to as fetal) stages.

vertebrate muscle may represent a myogenic cell population distinct from embryonic mpc. Studies by Yablonka-Reuveni and colleagues have demonstrated that, compared to chicken embryonic mpc, cultured mpc from adult chickens fuse into myotubes later (Yablonka-Reuveni *et al.*, 1987), express desmin as cycling cells more frequently (Yablonka-Reuveni and Nameroff, 1990), have more receptors for platelet-derived growth factor (Yablonka-Reuveni *et al.*, 1990b; also Yablonka-Reuveni *et al.*, manuscript submitted for publication) and regulate differently the expression of myosin heavy chain isoforms (Hartley *et al.*, 1991). These two cell populations also have a different response to photosensitization by merocyanin 540 plus light, i.e. photosensitization destroyed mpc from 10-day embryos, but not from adult muscle (Nameroff and Rhodes, 1989). Additionally, quail-chicken mpc transfer experiments indicate that, in contrast to mpc from early stages of development, postnatal mpc do not participate in myogenesis in embryos (Chevallier *et al.*, 1986). Studies on cultured mammalian mpc (reviewed by Cossu and Molinaro, 1987) have shown differences between embryonic and adult mpc in their sensitivity to a tumour promotor (Cossu *et al.*, 1983), and in the expression of acetylcholine receptors (Cossu *et al.*, 1987) and isoforms of acetylcholinesterase (Senni *et al.*, 1987). Furthermore, the above studies in both avia and mammals and additional studies in avia (Hartley *et al.*, 1992) indicate that these satellite cell characteristics become apparent in myogenic cell cultures from late stages of embryogenesis.

The cell culture results suggest that adult mpc are different from embryonic mpc, and that satellite cells are not simply embryonic mpc trapped underneath the external lamina. However, we cannot rule out the possibility that adult and embryonic mpc are derived from a single cell type which alters its phenotype as the *in vivo* environment changes during development, growth and regeneration. Possible differences between embryonic and adult mpc (or their environ) may have important implications for understanding the severity of the pathological changes in muscular dystrophy in embryonic and postnatal muscles of dystrophic animals and humans. It is important to note that all mpc in the developing embryo may not be identical, and studies in both mammals and avia suggest the presence of different lineages (Hauschka, 1974a, b; White *et al.*, 1975; Mouly *et al.*, 1987; Cossu *et al.*, 1988; Vivarelli *et al.*, 1988; Stockdale and Miller, 1987; Miller and Stockdale, 1989). Hauschka and Stockdale and their colleagues subdivide avian mpc in developing embryos into those from earlier periods of development (when embryonic morphogenesis is occurring), and those from later periods (when growth is occurring) (summary in Stockdale and Miller, 1987; Miller, 1991). It is not even clear whether satellite cells in postnatal young muscle, which are still replicating (see below), represent the same cell population as satellite cells in mature muscle, which are primarily quiescent and enter the cell cycle only in response to the appropriate stress or injury.

9.2.2 External lamina and electron microscopy studies

The term 'satellite cells' is based on the physical location of the cell beneath the external lamina of the myofibre (Figure 9.1). An external lamina is physically apparent at the electron microscopic level around the time of secondary myotube formation (Table 9.1), although components of this membrane, such as laminin and agrin, are present much earlier (in pre-muscle masses) before the differentiation of mpc (Godfrey *et al.*, 1988). Immunohistochemical studies of embryonic chicken muscles with anti-laminin show relatively complete stain around myofibres at 16 days and some laminin in earlier developmental stages (Kieny and Mauger, 1984). At 18 and 19 days respectively, a full lamina was demonstrated using antibodies to collagen type IV (Mayne *et al.*, 1989), and to laminin and

Table 9.1 Time when satellite cells can first be identified in morphologically developing skeletal muscles

Species	*Approximate age at birth*	*Cells with the appearance of satellite cells (but external lamina not pronounced)*	*Definite external lamina present (overlying satellite cells)*	*Reference*
Chicken	21 days	14–18 days	not at 18 days	Nathanson (1979)
			21–22 days	Przyblska (1983)
			21–22 days	Armand *et al.*, (1983)
Mouse	19 days	–	14–15 days	A.J. Harris (pers. comm.)
			by 19 days	Cardasis (1979)
Rat	19 days	–	16–18 days (not at 16 days)	Kelly and Zacks (1969)
			16–17 days	Ontell (1979)
Dog	9 weeks	N/A	N/A	
Human	40 weeks	–	12 weeks	Schmalbruch (pers. comm.)
		10 weeks	15 weeks	Ishikawa (1970)
		10–14 weeks	–	Conen and Bell (1970)

N/A, not available

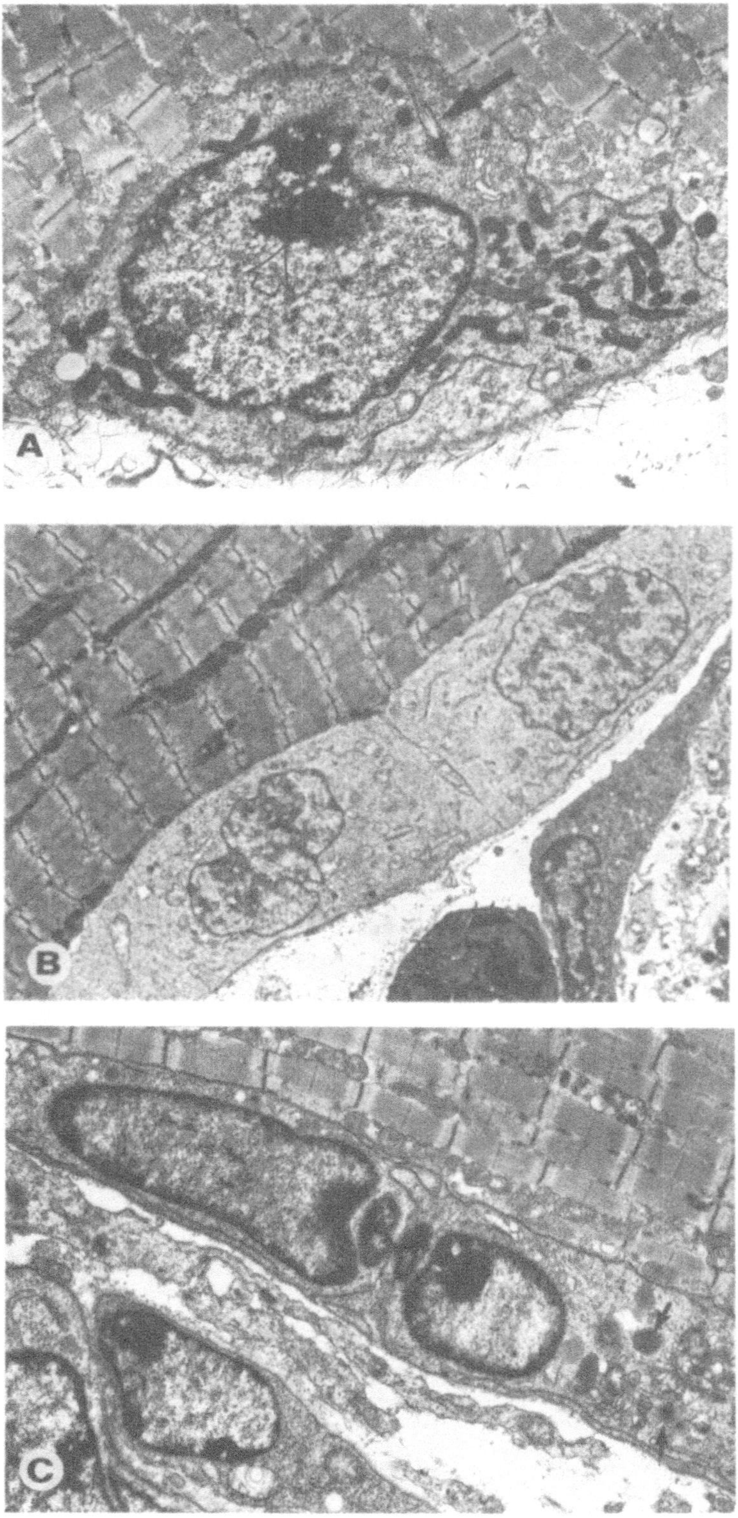
A
B
C

collagen type IV (Yablonka-Reuveni – unpublished). Examples of developing and adult chicken skeletal muscle stained with an antibody to laminin are shown in Figure 9.2. The staining around the developing fibres is similar to adult fibres, suggesting that the 18–19 day chicken embryonic muscle has a complete external lamina. However, this has not been verified ultrastructurally (see discussion below regarding Nathanson's work). The anti-laminin binding pattern of the developing muscle (Figure 9.2) is similar to that reported for other components of the external lamina in adult muscle from chicken and rat (Bayne *et al.*, 1984; Sanes and Cheney, 1982). It should be noted that components of the external lamina were not revealed by immunohistochemistry in muscle from embryonic day 15 rat (Sanes *et al.*, 1986).

The ultimate identification of satellite cells is at the electron microscopy level by their physical location and with reference to their appearance (Ontell, 1974; Ontell *et al.*, 1984; Kahn and Simpson, 1974; Cull-Candy *et al.*, 1980). However, the identification of satellite cells in mature muscle *in vivo* is not straightforward. Pericytes can resemble satellite cells (Venable and Lorenz, 1970), and macrophages (see Figure 9.1C) or other cells infiltrating beneath the external lamina may be morphologically indistinguishable from satellite cells at the light microscope level (Trupin *et al.*, 1979; Franzini-Armstrong, 1979). Furthermore, satellite cells may be completely encircled by one or more layers of external lamina (Snow 1977; Franzini-Armstrong, 1979). It has even been suggested that in the newt following metamorphosis, satellite cells become encircled with external lamina material and become pericytes (Popiela, 1976). These structures were later termed post-satellite cells and shown to be distinct from vascular pericytes and to be myogenic (discussion in Cameron *et al.*, 1986).

The identification of satellite cells in embryonic muscle is also rather arbitrary as there are proliferating mpc at many stages of differentiation, many in close proximity to, or lying within, the external lamina of developing myotubes. In serial sections of developing muscle Ontell (1979) showed that many cells which appeared to be satellite cells were in fact early myotubes and contained myofilaments. 'Satellite-like' cells can be identified

---

Figure 9.1 Electron micrographs of 'activated' cells lying between the plasmalemma and external lamina of injured mature mouse muscle fibres. (A) An activated satellite cell, 4 days after injury. Note the numerous mitochondria, polyribosomes and prominent Golgi: a cilium is also present (arrow) (× 9200). (B) Two daughter satellite cells following division, 2 days after injury (× 4700). (C) Two macrophages in the 'satellite cell position'. Note the numerous lysosomes in the cytoplasm (arrows) (× 9100). (Electron micrographs were provided courtesy of T. Robertson, Department of Pathology, University of Western Australia).

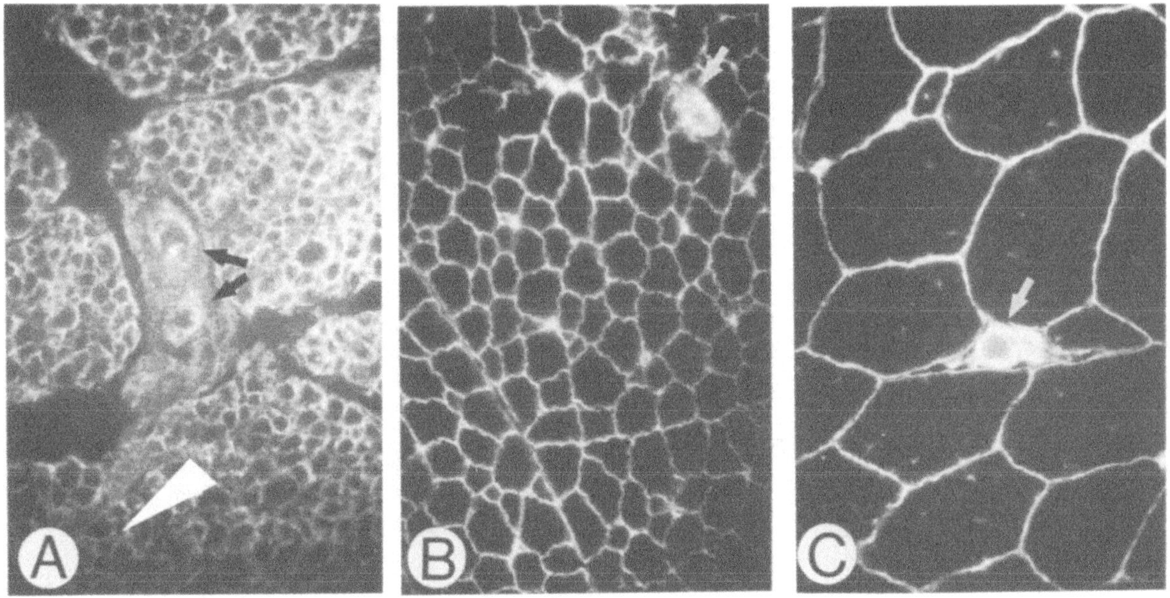

Figure 9.2 Demonstration by indirect immunofluorescence of the laminin component of the external lamina surrounding muscle fibres in the breast muscle from embryos, young and mature chickens. Unfixed frozen cross sections were prepared from (A) 19-day-old embryo; (B) 19-day-old chicken; and (C) 2-months-old chicken. Sections were reacted with a monoclonal antibody against chicken laminin (prepared by Bayne *et al.*, 1984) followed by a fluorescein-labelled goat anti-mouse IgG, arrows indicate blood vessels which are also positive for laminin (40 × objective was used for all micrographs) (Yablonka-Reuveni, unpublished).

before the time of birth in mice and rats, but relatively earlier, around mid-term, in fetal muscle of humans (Table 9.1). Studies on quail and chicken muscle identified satellite cells as early as the first day following hatching (Armand *et al.*, 1983; Przyblska, 1983). An important question is whether the appearance of satellite cells is tightly associated with the appearance of complete external lamina (i.e. if satellite cells do differ from mpc in the developing embryo they might be identifiable prior to the presence of external lamina). Attempts to identify satellite cells in embryonic chicken muscle (prior to the appearance of a complete external lamina) using the location, heterochromaticity of nuclei and nuclear/cytoplasm ratios as criteria (Nathanson, 1979) suggested that very low numbers of these satellite-like cells can first be detected in 14-day-old embryonic muscle, and that they increase in 18-day-old embryonic muscle. Although distinct external lamina was not seen at this time by Nathanson (1979), antibodies to components of the external lamina demonstrate that such constituents completely surround each individual fibre in chicken muscles from 18-day-old embryos (see discussion above). It is of interest that tissue culture studies using the differential expression of myosin heavy chain isoforms (ventricular and embryonic) showed that 'satellite cells' become predominant around day 18 of chicken embryogenesis (Hartley *et al.*, 1992). In earlier stages of development (day 10, day 14) the frequency of these cells is much lower. 'Satellite-like' cells were also first seen in developing muscles of the fruit-bat at a stage (before the midpoint of intrauterine development) when complete external lamina could not be demonstrated ultrastructurally (Church, 1969).

9.2.3 Immunohistochemical markers for satellite cells

Studies into the time of satellite cell appearance during development would be greatly facilitated if satellite cell specific markers, which are independent of the developmental stage of the muscle, the presence of external lamina, and the differentiative stage of the cell (i.e. markers that can identify mpc prior to expressing differentiation-specific characteristics) were available. Unfortunately, the very undifferentiated and quiescent nature of satellite cells in uninjured mature muscle implies that they may not express specific genes, and therefore it may not be possible to obtain specific molecular markers for these cells. Several antibodies (as discussed below) have been reported which recognize satellite cells in adult muscle, but they also recognize embryonic mpc and/or other cell types, and therefore are not suitable to address the issue of the emergence of satellite cells during development.

Isoforms of neural cell adhesion molecules (N-CAM) represent potential markers for satellite cells in humans (Schubert *et al.*, 1989), rats (Covault and Sanes, 1986) and dogs (Alameddine *et al.*, 1990), although the extent to which they can define all satellite cells (activated and quiescent) *in vivo* has not been demonstrated. The antibody 5.1H11 used by Webster *et al.*, (1988) to sort mpc from fibroblasts for cultures of human myogenic cells is now known to be an antibody to N-CAM (Walsh *et al.*, 1989). Similarly, monoclonal antibodies to N-CAM have been used to purify mpc from developing mouse muscle (Jones *et al.*, 1990). The use of antibodies against N-CAM to identify satellite cells during development and regeneration can be problematic since N-CAMs are expressed by developing and regenerating myofibres, by mpc, and by cells at or near the neuromuscular junction (Rieger *et al.*, 1985; Covault and Sanes, 1986; Gatchalian *et al.*, 1989; Tassin *et al.*, 1991; Yablonka-Reuveni unpublished results with chicken cells). A different monoclonal antibody, H36, recognizes an integral membrane glycoprotein on undifferentiated, replicating, and terminally differentiated cultured rat mpc (Kaufman and Foster, 1988). The appearance of H36+/desmin+ cells in embryonic rat muscle occurs between 12 and 16 days (Kaufman *et al.*, 1991). Whether the H36 or desmin antibodies can recognize quiescent satellite cells in adult rat muscle *in vivo* is under investigation. In chickens, the C3/1 antibody prepared against embryonic chicken mpc by Wakshull *et al.*, (1983) recognizes a plasma membrane glycoprotein with a molecular weight on SDS polyacrylamide gels of 38 000. Using isolated single fibres, it has been demonstrated that the antibody can recognize chicken satellite cells, but the usefulness of this antibody for identifying satellite cells *in vivo* without resorting to electron microscopy is questionable, as other cells are also recognized by this antibody (Wakshull *et al.*, 1983). Combinations of monoclonal antibodies can also identify undifferentiated mpc in the chicken embryo (George-Weinstein *et al.*, 1988), but their ability to identify adult satellite cells was not determined.

Whereas many of the above antibodies were prepared against embryonic cells, a monoclonal antibody against adult chicken mpc, named SAT2H10, has been prepared by Yablonka-Reuveni (1988a; also manuscript in preparation). The antibody recognizes an antigen associated with actin filaments which may be cytoplasmic myosin heavy chain, and gives a strong signal with cultured satellite cells and their dividing and fused progeny. Developing chick myofibres also stain intensely with the SAT2H10 antibody, but there is very weak, if any, reaction with adult myofibres. In addition, smooth muscle cells in both developing and adult vasculature, and cells at the periphery of adult myofibres which are associated with a circular, laminin-positive structure, are stained intensely by the antibody (Figure 9.3). In location and appearance, these laminin-positive structures resemble capillaries. It is possible that the SAT2H10-positive cells associated with the circular laminin positive structures are related to the mpc in

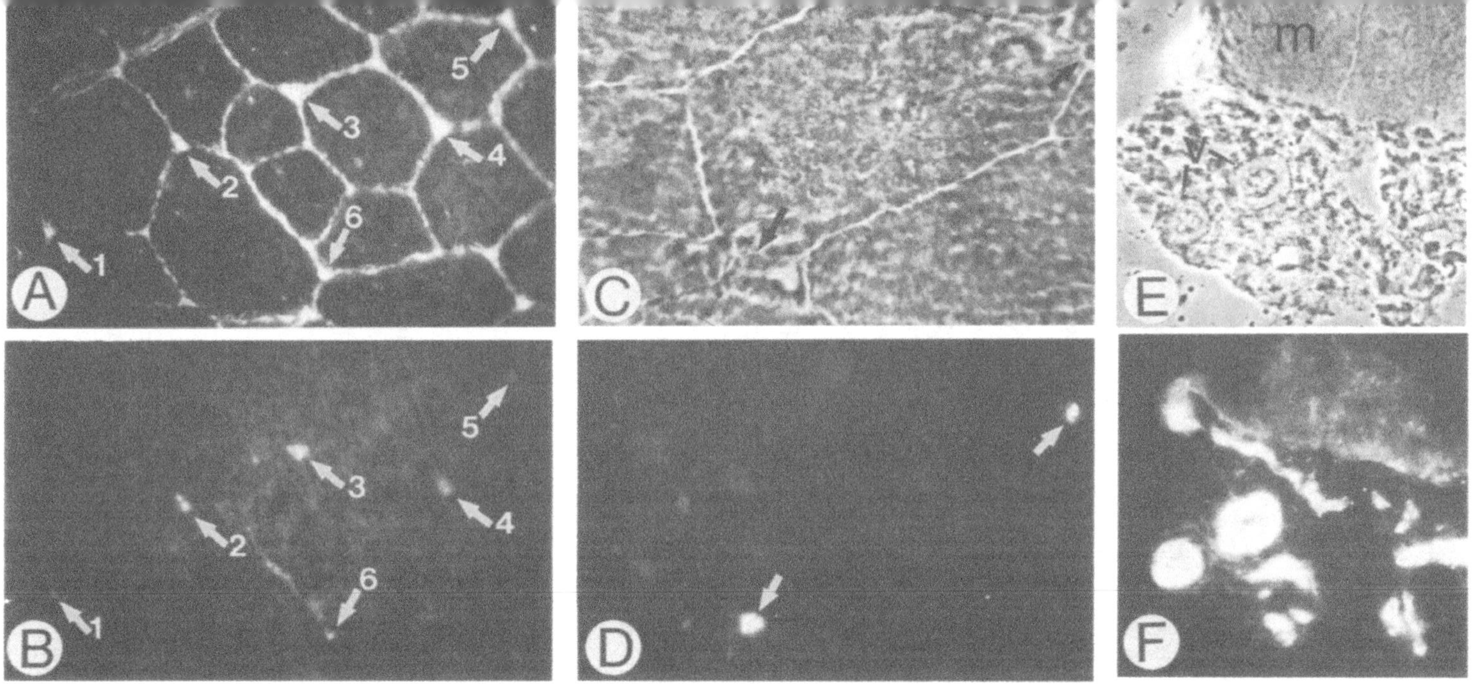

Figure 9.3 Indirect immunofluorescence analysis of adult chicken skeletal muscle reacted respectively with the monoclonal antibody SAT2H10 and anti-laminin. Unfixed frozen cross-sections were prepared from the breast muscle of 2-month-old chicken. Fluorescein-labelled rabbit anti-mouse IgG was used as a secondary antibody. (A) and (B), serial sections reacted with SAT2H10 and anti-laminin respectively, demonstrate that the majority of the SAT2H10-positive structures (potentially satellite cells) are surrounded by laminin-positive material (40 × objective). (C) and (D) higher magnification of a section reacted with SAT2H10 (100 × objective): the same field is shown in phase contrast (C) and by immunofluorescence (D). E and F show phase contrast and immunofluorescence respectively of a cross section reacted with SAT2H10, demonstrating that blood vessels are also positive: reactivity of myofibres with the antibody is very low, if any (40 × objective). Arrows indicate the position of SAT2H10 positive structures in the paired micrographs. In (E) 'm' and 'v' refer to muscle fibre region and blood vessels in connective tissue, respectively. (Yablonka-Reuveni, manuscript in preparation).

the newt which are enclosed within a basement membrane and resemble pericytes as discussed above (i.e., Popiela, 1976; Cameron *et al.*, 1986). The studies on localization of SAT2H10 positive entities suggest that skeletal muscle may contain mpc in locations other than the classically described satellite cells, although further work is required to substantiate such a notion. With respect to this last point it is noteworthy that new fibre formation is observed in extrafascicular spaces when chicken muscle undergoes hypertrophy in response to overloading (Kennedy *et al.*, 1988): such new muscle fibres might be derived from satellite cells emigrating from their original location beneath the external lamina of adjacent 'undamaged' myofibres, or they might result from mpc which were never in the satellite cell position.

Desmin, an intermediate filament protein, is present in proliferating mpc from embryonic and adult rats in culture (Foster *et al.*, 1987; Allen *et al.*, 1991; Kaufman *et al.*, 1991). It is also detected (although at lower frequency than in rat cultures) in dividing mpc from adult chicken and in diminishing numbers in mpc from progressively younger chicken embryos (Yablonka-Reuveni and Nameroff, 1990). Furthermore, desmin-positive cells are reported in regenerating rat skeletal muscle at 24 hours or more after injury (Helliwell, 1988; Allen *et al.*, 1991). These results suggest that desmin might be a potential marker for quiescent as well as activated satellite cells, but with limited usefulness since developing and mature myofibres as well as vascular smooth muscle express desmin. Furthermore, in our search for chicken satellite cell markers we could immunolocalize desmin expression to myofibres and blood vessels, but hardly ever to structures which are positive for N-CAM or SAT2H10 (Yablonka-Reuveni, unpublished). Desmin expression by satellite cells might be somewhat species-specific, since adult bovine mpc do not express desmin in their proliferative phase in contrast to proliferating mpc from adult rats (Allen *et al.*, 1991).

Although cell culture studies have identified molecular differences between adult-derived mpc and embryonic mpc, no specific immunohistochemical markers are available at present that can reliably identify satellite cells (particularly quiescent satellite cells) and distinguish them from embryonic mpc or other cells *in vivo*. The method of identifying quiescent and activated satellite cells in frog muscle at the light microscope level by the use of vital fluorescent dyes (Herrera and Banner, 1990) or at the electron microscope level by exclusion of horseradish peroxidase (Cull-Candy *et al.*, 1980) may offer an alternative approach to the study of satellite cells in adult muscle. In addition, with selective immunohistochemistry it may be possible to clearly distinguish the plasmalemma (using antibodies to dystrophin) and the external lamina (using antibodies to collagen IV or laminin), with the result that satellite cells located between these two membranes can be definitively identified at the light microscope level (Zhang and McLennan, personal communication).

9.3 REPLICATION DURING POSTNATAL GROWTH

9.3.1 Cessation of satellite cell replication in developing muscle

Satellite cells proliferate during embryonic and early postnatal growth of muscle fibres, but eventually become quiescent. During the period of rapid muscle growth, daughter cells from dividing satellite cells (Figure 9.1B) presumably fuse with young myofibres to maintain the myonuclear/sarcoplasmic ratio. The increase in myofibre length and associated increase in number of muscle nuclei in postnatal muscles of mice has been documented by Cardasis (1979), but the extent of growth and final number of muscle nuclei can vary widely between different types of muscles (Enesco and Puddy, 1964; Kelly, 1978; Gibson and Schultz, 1983). Autoradiographic studies, which rely on the uptake of tritiated thymidine (^{3}H-Tdr) by cells synthesizing DNA, indicate that within weeks of birth in rodents, the proliferation of satellite cells and fusion with myofibres is greatly reduced and effectively ceases (Table 9.2). The subsequent growth of muscles is largely due to hypertrophy (increase in sarcoplasmic volume). Cessation of mpc replication is relatively rapid in postnatal muscles of mice (Table 9.2): in 15-day-old mice, samples removed one hour after injection of ^{3}H-Tdr have essentially no labelling of (premitotic) 'satellite' cells, neither is labelling seen in (post-mitotic) myotube nuclei when samples were removed more than 7 days after ^{3}H-Tdr injection (McGeachie *et al.*, unpublished observations). Due to sampling problems, autoradiographic studies cannot exclude the possibility that some mpc are still replicating at this time, but they do indicate that proliferation is minimal compared with that seen at birth or in regenerating muscle. That the replication of mpc in mice does decline rapidly after birth is supported by a postnatal decrease in *MyoD* and myogenin gene transcription, which is reduced to adult levels by 3 weeks of age (discussed in the next section). In rats, however, significant satellite cells replication is seen at 2 weeks after birth, decreases only slightly by 3 weeks (Table 9.2: see also Kelly, 1978), and is still significant (14% of satellite cells nuclei labelled) at 30 days of age (Schultz *et al.*, 1985). Little DNA synthesis was detected in satellite cells of 5–6-week-old rats (Snow, 1979), although arrested mitosis of satellite cells was demonstrated in colchicine-treated muscles of 5-week-old rats (MacConnachie *et al.*, 1964), and numbers of muscle nuclei appeared to roughly double between 5 and 11 weeks of age in rats (Enesco and Puddy, 1964), indicating that some satellite cells were still proliferating during this time. The longer period of mpc replication in rats as compared with mice is presumably related to the

Table 9.2 Cessation of satellite cell replication in postnatal muscle: data from autoradiographic studies

Species	*Strain*	*Time after birth*	*% of (premitotic) satellite cells labelled*[a]	*Reference*
Chicken		(around 21 days)		(see text)
Mouse	Charles River	9 and 10 days	2.3%	Venable and Lorenz (1970)
	C57Bl/10sn, *mdx* mice	15–17 days	≈ 0	McGeachie *et al.*, (unpublished)
	BALB/c, SJL/J	30 days	0	McGeachie and Grounds (unpublished)
Rat	Sherman	14–17 days	≈ 2.5%	Moss and Leblond (1971)
	Sprague Dawley	15–17 days	≈ 3.5%	Snow (1979)
	Sprague Dawley	7–21 days	≈ 2.3%	Schultz (1979)
	Sprague Dawley	35–42 days	0	Snow (1979)
Dog		N/A	N/A	
Human		N/A	N/A	

[a]The % of labelled satellite cells is calculated as a proportion of total muscle nuclei (i.e. myonuclei and satellite cells)

increased length of muscle fibres associated with the difference in body size of these animals, and hence the longer period of growth required in rats (see Table 9.3).

There appears to be no equivalent autoradiographic data for developing

Table 9.3 Comparison of growth parameters[a] in different species

	Mouse	*Rat*	*Chickens (white leghorn)*	*Dog (golden retriever)*	*Human*
Body weight at birth[a]	2 g	5 g	0.04 kg	0.5 kg	3 kg
Adult body weight	30 g	300 g	1.5 kg	30 kg	70 kg
Relative postnatal increase in weight (adult/birth weight)	15	60	40	60	24
Life span (years)[a]	2	2½	6	12	70

[a]Body weights and life spans represent only approximate average values

muscles of chickens. Autoradiographic studies do show that mpc in adult white Leghorn hens (weighing 600–800 g) are quiescent, as labelling of mpc was not seen in (uninjured) pectoralis, anterior and posterior latissimus dorsi muscles (Grounds and McGeachie, 1989a). An electron microscopic study of post-hatched hens aged 1–60 days characterized the morphology (cytoplasm, nucleus, ribosome, etc.) of satellite cells, and identified two distinct populations (Przyblska, 1983). The populations of active ('dark') and quiescent ('light') satellite cells both exist initially: the active (proliferating) cells are abundant during the first 21 days following hatching which correlates well with the time of most intensive body growth of chickens, but eventually the quiescent type become predominant. The cessation of mpc replication around 3 weeks post-hatching is supported by data on mpc extracted from chicken muscles of different ages. The proportion of mpc which are proliferating at the time of cell isolation declines drastically in preparations from post-hatch chicken muscle and from embryos closer to hatching, compared to embryonic day 10 and 14 preparations (Hartley and Yablonka-Reuveni, unpublished; Feldman and Stockdale, 1991a). To the best of our knowledge, there are no studies which measured directly the number of proliferating satellite cells in the chicken muscle *in vivo*.

There is now evidence that in larger animals such as dogs, humans or sheep (Wilson *et al.*, 1992) muscle fibre formation is more complex than in small laboratory rodents (Ross *et al.*, 1987). In these larger species it is not feasible to carry out autoradiographic studies, and data comparing the total numbers of myonuclei in young and old muscle fibres (which would indicate the extent of mpc proliferation required during growth) do not appear to be available. Because much greater muscle growth is required to attain adult size in these larger species, it is anticipated that mpc might continue to proliferate for a much greater time after birth, and into late childhood and adolescence in humans (see Cheek, 1985). This implies (depending on the size of the initial population and the pattern of replication versus differentiation) that satellite cells may divide many more times in humans and dogs than in rodents before they become quiescent. Since it is often proposed that the capacity of satellite cells to replicate is exhausted in X-linked muscular dystrophies, such an explanation might, in part, underlie the lesser regenerative capacity of dystrophic humans and dogs, compared with smaller animals such as mice. Since the absolute number of satellite cells (reviewed Campion, 1984) and the extent of postnatal proliferation (Schultz, 1979) is known to vary between different muscles (see also Chapter 8 by Schutz and McCormick in this volume), it is of great interest to determine the relative extent of mpc replication in different muscles of growing and mature animals as this might influence the pathology of the disease. As indicated below it may now be possible to gather such data for human and dog muscles, by examining the expression of the skeletal muscle specific regulatory genes *MyoD* or myogenin (or their homologues) which serve as markers for activated mpc *in vivo* (see next section).

9.3.2 Identification of activated muscle precursors in mature muscle

It has already been emphasized that the only undifferentiated mpc that can be identified *in vivo* are satellite cells since these are essentially defined geographically (Figure 9.1A, B). Thus replicating cells (labelled by ^{3}H-Tdr uptake) or dividing cells (showing mitotic figures) which are lying between the external lamina and plasma membrane of myofibres are assumed to be activated mpc. If autoradiographic labelling or mitotic figures (or the presence of a particular growth factor, etc.) is seen in undifferentiated cells located in the extracellular space (i.e. not in the satellite cell position), these cells cannot be distinguished from other mononuclear cells and identified as mpc. Since many proliferating mpc are undifferentiated and do not synthesize the distinctive thick and thin (myosin and actin) filaments, it has not been possible to identify such anonymous cells as mpc unless they are lying in the satellite cell position. However, where a mononuclear cell does contain thick and thin filaments it can be readily identified as a mpc and such differentiated mpc are widely referred to as myoblasts. In regenerating muscle where the external lamina has been damaged, and even in hypertrophing muscles where there is minimal myofibre damage (Kennedy *et al.*, 1988), myoblasts are often conspicuous in the extracellular space: it cannot be proven that these myoblasts are the progeny of satellite cells (although they are widely assumed to be so), and it may be that they are derived from mpc originating in interstitial tissues.

The absence of markers for undifferentiated mpc *in vivo* has limited the interpretation of events in regenerating muscle. However, the analysis of mRNA for the recently described skeletal muscle-specific regulatory genes *MyoD* (reviewed Weintraub *et al.*, 1991) and myogenin (Wright *et al.*, 1989) and their homologues (e.g. *Myf3* and *Myf4* in humans [Braun *et al.*, 1989]) provide powerful markers for identifying activated mpc *in vivo*. *In situ* hybridization studies on tissue sections do not detect mRNA for *MyoD* or myogenin in quiescent mpc of mature skeletal mouse muscle (Grounds *et al.*, 1991b), and this is supported by Northern blot analysis which shows very low levels of *MyoD* and myogenin mRNAs in uninjured adult mouse muscle (Beilharz *et al.*, 1992). In developing muscles, Northern blot analysis shows that mRNAs for *MyoD* and myogenin in normal mice decrease to adult levels between 1 and 3 weeks after birth (Eftimie *et al.*, 1991; Beilharz *et al.*, 1992) and this corresponds to the cessation of postnatal mpc replication based on autoradiographic data (Table 9.2). In contrast to the lack of transcription in quiescent mpc of mature muscle, transcription of *MyoD* and myogenin is seen in mononuclear cells by 6 hours after muscle injury, peaks at 24 to 48 hours (Figure 9.4) and declines to pre-injury levels by about 8 days (Grounds *et al.*, 1991b). Furthermore, the striking demonstration of mRNAs for *MyoD* and myogenin in what corresponds to myoid cells (skeletal mpc) in mouse thymus tissue (Grounds *et al.*, 1991a), confirms the use of these transcripts as markers for undifferentiated 'activated' mpc. *MyoD* and myogenin expression can be studied using either nucleic acid probes (to detect mRNA) or antibodies (to detect the gene product). There is evidence from studies of mouse muscles regenerating *in vivo* (Fuchtauer and Westphal, 1992) and from developing mouse muscles (Cusella-De Angelis *et al.*, 1992) that, at least in some situations, there may be a delay of several days between the appearance of these mRNA transcripts and their protein products.

Northern analysis showed a close correlation between the presence of the mRNAs for *MyoD* or myogenin and populations of activated/replicating mpc in the crush injury model of muscle regeneration in mice (Beilharz *et al.*, 1992). This demonstrated that this technique can quantitate 'regenerative activity' in whole muscle, and it was then used to quantitate the extent

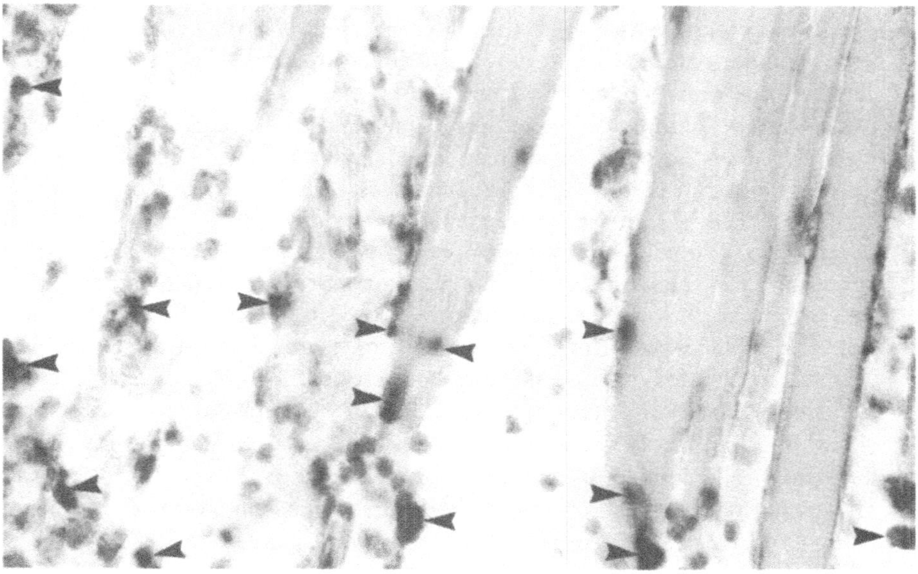

Figure 9.4 Identification of activated mpc by hybridization of digoxigen-labelled *MyoD* riboprobes to paraffin sections of injured mature mouse muscle. At 36 hours after crush injury the dark staining *MyoD* positive nuclei (arrows) are closely associated with damaged myofibres and also with cells lying in the extracellular space (× 40 objective) (Garrett, Grounds and Beilharz, unpublished observations).

of skeletal muscle regeneration (and by implication myonecrosis) in dystrophic muscles. Rapid down-regulation of the expression of these genes was seen initially in postnatal muscles of *mdx* and control mice. However, three weeks after birth mRNA levels for both genes increased in *mdx* (but not control) mice and remained higher than control levels in muscles sampled up to 420 days of age (Beilharz *et al.*, 1992). This shows that the process of muscle necrosis and regeneration is ongoing even in older *mdx* mice.

9.4 NUMBERS AND PROLIFERATIVE CAPACITY OF MPC FROM ANIMALS OF DIFFERENT AGES

Studies have been carried out both *in vitro* and *in vivo* to determine the absolute numbers of mpc in muscles of different ages, and whether the proliferative capacity of mpc diminishes with the age of the animal, or after repeated cycles of necrosis and regeneration resulting from either experimental injury or muscular dystrophy. Data from such experiments will be discussed, particularly with respect to the interpretation of tissue culture experiments.

In tissue culture studies, mononuclear cells extracted from skeletal muscle divide and subsequently fuse to form myotubes. Early studies used heterogeneous muscle cultures, where interpretation of data on mpc was complicated by the presence of fibroblasts and other cells (see Witowski, 1986; discussion in Yablonka-Reuveni and Nameroff, 1987). Several approaches were introduced to purify mpc, including differential plating and the use of different substratum to enrich for mpc (discussed in Yablonka-Reuveni and Nameroff, 1987). More recently, methods were developed for the physical separation of mpc from non-myogenic cells by density centrifugation (Yablonka-Reuveni *et al.*, 1987, 1988), or by more elaborate techniques of cell sorting according to light scattering of the different cells (Yablonka-Reuveni, 1988b, 1989), or after labelling mpc with fluorescently labelled antibodies directed against cell surface antigens (Hurko *et al.*, 1986; Webster *et al.*, 1988; Schweitzer *et al.*, 1987). Several laboratories used clonal mpc cultures where cells are cultured at very low density and individual cells divide and give rise to colonies which can be identified as myogenic by the appearance of differentiated mpc or myotubes (Konigsberg, 1963; Yaffe, 1973; Hauschka and Konigsberg, 1966; Hauschka, 1974a, b; Hauschka *et al.*, 1979). Such clones can be expanded, leading to many cells all derived from the same progenitor. More recently, mpc from humans (Webster *et al.*, 1988) and mice (Jones *et al.*, 1990; Coleman *et al.*, 1991) have been purified for clonal tissue culture by their specific binding to N-CAM antibodies. Clonal studies can examine (i) the replicative capacity of the mpc by assessing the number of clones that arise, and (ii) how many cell divisions each mpc is capable of by assessing the final size of each clonal colony.

9.4.1 Numbers of available mpc

Assessment of the size of the *in vivo* mpc population is clearly influenced by the ability to extract all mpc from the muscle tissue. For example, although it was initially reported that about $1–4 \times 10^5$ cells per gram of pectoral muscle (primarily mpc) could be isolated from adult chickens (Yablonka-Reuveni *et al.*, 1987) this yield has been increased to 10^6 cells/gram tissue (Hartley *et al.*, 1991). These yields are far higher than that reported by Feldman and Stockdale (1991b) for the pectoralis, but are comparable with those recently reported for satellite cell preparations from rat muscles (Düsterhöft *et al.*, 1990). It is also possible that only a particular selected subpopulation of mpc is successfully liberated from muscle tissue and survives in tissue culture. Furthermore, if a population of mpc can be recruited *in vivo* in response to damage, e.g. from pericytes in new blood vessels, then the traditional procedure of extracting cells from uninjured muscles for tissue culture would not include this additional population of mpc.

It is almost certainly more difficult to extract mpc from older muscle where connective tissue is thicker and may be more resistant to enzymatic digestion, and this could lead to an apparent decrease in numbers of mpc. A doubling in the thickness of external lamina was reported in old rats by Snow (1977), and complete enclosure of satellite cells by external lamina and their separation from the muscle fibre was observed also. Multiple layers of external lamina were also noted in muscles of 8-month-old *mdx* mice (Anderson *et al.*, 1987). Increased connective tissue is seen in older muscles from normal mice (Marshall *et al.*, 1989) and markedly increased collagen deposition is a feature of older dystrophic muscles in humans (Lipton, 1979; Duance *et al.*, 1980), *mdx* and 129/ReJ *dy*/*dy* mice (Marshall *et al.*, 1989; Ontell *et al.*, 1984) and chickens (Feit *et al.*, 1989).

Determining relative numbers of mpc by the analysis of myogenic clones obtained from muscles of different ages is also problematic, since the efficiency of clonal plating (i.e., the number of clones obtained) is highly dependent on the medium used. It might also be considered that mpc which have undergone many replicative cycles *in vivo* (either due to dystrophy or multiple cycles of regeneration induced experimentally) might have subtly different growth requirements from ‘younger’ mpc, and this might be reflected by relatively poor proliferation under standard tissue culture conditions designed for maximal growth of ‘younger’ mpc. Similarly, clones from different ages may require different nutrients for differentiation, and ‘non-myogenic’ clones might differentiate and fuse given optimal nutrients

(Hauschka, 1974a; White *et al.*, 1975; Seed and Hauschka, 1988). Additionally, some clones may contain very narrow myotubes which are difficult to identify without the aid of an immunocytochemical marker for differentiated cells (Yablonka-Reuveni, unpublished). Another complication is that some clones might consist of only a small number of differentiated cells which do not necessarily fuse into myotubes (Quinn *et al.*, 1984, 1985) and such clones can be easily overlooked when analysing myogenic clones after 1–2 weeks in culture. Because of the problems outlined above, caution is required when interpreting such tissue culture data. For these reasons it is difficult to assess the physiological significance of tissue culture studies which suggest a reduction in yields of mpc extracted from muscles of older animals.

9.4.2 Proliferative capacity

One application of clonal tissue culture studies has been for determining the number of cell divisions that an individual mpc is capable of. It is widely considered that all cells have a limited replicative capacity (reviewed Walton, 1982), and it has been proposed that progressive shortening at the ends of chromosomes, called telomeres, may account for such cellular senescence (Wright and Shay, 1992). Several authors have observed a decreased proliferative capacity of cloned mpc isolated from patients with DMD (Blau *et al.*, 1983a, b) or from the 129/ReJ dystrophic mouse (Hauschka *et al.*, 1979; Summers and Parsons, 1981), and they and others have speculated that the regenerative demands of muscular dystrophy might exhaust the potential for mpc cell division (Blau *et al.*, 1983a, b; Hauschka *et al.*, 1979; Mastaglia *et al.*, 1970; Lipton, 1979). Hurko *et al.* (1986) provide data indicating no intrinsic proliferative disadvantage to cells carrying the *DMD+* mutation on their active X chromosome. Of particular interest are elegant experiments by Ontell on the 129/ReJ mouse, which shows progressive muscle dystrophy (although this myopathy is autosomally linked and is not the result of dystrophin deficiency). The denervation of muscles in this model rescues the muscle from progressive cycles of necrosis and regeneration (Moschella and Ontell, 1987). Tissue culture studies on mpc extracted from innervated and long-term denervated 129/ReJ muscles strongly support the proposal that the replicative capacity of mpc is limited: where mpc were cultured from innervated muscles which had undergone repeated cycles of regeneration the mpc had a reduced proliferative capacity compared with those from the protected denervated muscle (Ontell *et al.*, 1991).

Wright (1985) studied the number of cell divisions that mpc from normal and dystrophic chickens of different ages can undergo in culture (the chicken model of dystrophy is also not due to a lack of dystrophin). Prior to two months of age, dystrophic mpc exhibited relatively normal proliferative capacity: however, as the disease progressed the proliferation of mpc was severely affected. This supports the hypothesis that the decline in proliferative capacity is a secondary response rather than an intrinsic property of dystrophic mpc, and agrees with conclusions from the study on DMD muscle by Webster *et al.*, (1986) that the limited proliferative potential of *DMD*+ mpc from DMD muscle is secondary to the expression of the disease.

Interest has focused recently on whether the capacity for regeneration of normal muscle decreases with age. Cross transplantation studies of muscles isografted between young and old rats (5 to 26 months) (Gutman and Carlson, 1976; Carlson and Faulkner, 1989), and mice (20 to 140 days) (Grounds, 1987) show that the regenerative capacity of young and old muscles is essentially the same when the transplants are compared in hosts of the same age (i.e. both into young, or both into old hosts). The observation by Zacks and Sheff (1982) that regeneration is impaired in old male Swiss mice is not seen with old female Swiss mice nor with other strains (Grounds, 1987). These *in vivo* results do not indicate an impaired proliferative capacity of mpc from old muscles (at least after one cycle of regeneration) and are supported by tissue culture analysis of mpc from old rats (more than 30 months) (Allen *et al.*, 1980). In contrast, Schultz and Lipton (1982) found a reduced proliferative rate and number of progeny in cultured satellite cells from old rats, and the proliferative capacity of mpc after 80 days in culture was reduced with increasing age in humans (Webster *et al.*, 1986). It is difficult to know to what extent these tissue culture results are a function of the *in vitro* environment (for the reasons outlined earlier) and to what extent they might represent the situation *in vivo*. It is of interest to try and reconcile the *in vitro* and the *in vivo* results since *in vivo* cross-transplantation experiments are unlikely to be carried out in humans.

At this point it is relevant to compare the situation in animal models with that in humans. As outlined in Section 9.3, the growth phase, size and life span of the inbred laboratory animal models (mice and rats) is extremely short compared with that in humans. The principles established from animal experiments are very important, but it is almost impossible to know whether they are overridden by the above differences between the species. The life span of humans is not particularly relevant to the situation in DMD, where failed muscle regeneration is apparent in very young children, and is pronounced by adolescence. However, the relative size and extent of growth may be important, and for these reasons it is of great interest to carry out similar cross transplantation and injury experiments at different ages in a larger animal model such as the dog, in order to assess the influence of these growth parameters on mpc proliferation and the host environment.

Studies of muscle regeneration in old host animals consistently report impaired regeneration compared with that seen in younger host animals (Table 9.4). These differences between regeneration in young and old hosts might be attributed, not to a reduced intrinsic replicative capacity of mpc, but rather to a less favourable environment in old host animals often leading to increased fibrous tissue formation (Ullman *et al.*, 1990), as was concluded by Studitsky (1988) from numerous studies in the Russian laboratories. Enhanced connective tissue formation in tissues of older animals probably also accounts for the increased proportion of branched fibres in regenerating old muscle grafts in rats (Blaivas and Carlson, 1991). It would be of great interest to determine precisely what the factors are that contribute to the unfavourable host environment, and whether they can be overcome.

Since dystrophic muscle undergoes repeated cycles of necrosis and regeneration, it is also relevant to examine the regenerative capacity of normal skeletal muscle subjected to repeated trauma (either multiple injuries or transplantations). The general concensus of opinion from such studies in experimental animals is that mpc have a sustained capacity for regeneration (Manda and Kakulas, 1986; reviewed Morlet *et al.*, 1989): however, connective tissue tends to increase with each regenerative phase and the persistent accumulation of fibrous tissue may result in slightly reduced new muscle formation after each cycle of regeneration. In this

Table 9.4 *In vivo* investigations into the influence of host age on skeletal muscle regeneration

Animal	*Ages studied (months)*	*Injury type*	*Time of analysis after injury*	*Reference*
Rats (Wistar)	5–26	Muscle grafts	60 days	Gutman and Carlson (1976)
Rats (Long Evans)	3–24	Bupivacaine	up to 26 days	Sadeh (1988)
Rats (Wistar)	2–24	Intract muscle grafts	60 days	Carlson and Faulkner (1989)
Rats (Sprague-Dawley)	6–27	Ischaemia	10 weeks	Ullman *et al.*, (1990)
Mice (C57Bl/6)	2–27	Contraction induced	up to 28 days	Brooks and Faulkner (1990)
Mice (C57Bl/6)	2–27	Contraction plus free radical	3 days	Zerba *et al.* (1990)
Mice (*mdx*)	1–2	Devascularization and denervation	up to 6 weeks	Zacharias and Anderson (1991)

context it is of interest that connective tissue collagen is higher at all ages in muscles of *mdx* than control mice, and is 2 to 3 times higher in one-year-old *mdx* mice (Marshall *et al.*, 1989).

These animal models show that the host environment for muscle regeneration becomes less favourable with age, and that connective tissue accumulates slightly with each cycle of regeneration. It would seem likely that the combination of these factors, in conjunction with the influence of components of the extracellular matrix on myogenesis (see Section 9.5), contributes to the pathology in DMD. However, other factors such as mechanical activity (Valentine and Cooper, 1991) must also be involved, as evidenced by the progressive degeneration of diaphragm but not other muscles in the *mdx* mouse model (Stedman *et al.*, 1991).

9.5 THE INFLUENCE OF GROWTH FACTORS AND EXTRACELLULAR MATRIX COMPONENTS ON THE PROLIFERATION AND DIFFERENTIATION OF MPC

9.5.1 Relevance to muscular dystrophy

With respect to potential therapies for muscular dystrophies there are two major reasons why there is an urgent need to understand the effects of various factors on the replication and proliferation of mpc. The first concerns myoblast transfer therapy (Partridge, 1991b). This represents a potential gene replacement treatment which utilizes the transplantation of genetically competent mpc and their fusion with dystrophic myofibres. Before such a treatment could be put into large scale use, a number of problems will need to be overcome, including the ability to produce sufficient numbers of myoblasts from primary human muscle sources (Karpati, 1991). To this end it is of interest to understand which factors facilitate mpc replication in tissue culture. Furthermore, enhancing the proliferation and fusion of such implanted mpc *in vivo* may be fundamental to the success of this proposed therapy (Grounds, 1990b).

The second potential treatment for muscular dystrophy focuses on improving the regeneration of muscles which lack dystrophin. If the factors underlying the difference between the apparently successful muscle regeneration seen in the *mdx* mouse model and the impaired regeneration of dystrophic muscles of dogs and humans could be defined, it might be possible to manipulate conditions *in vivo* in order to enhance new muscle formation. It is important to mention that the apparently sustained regeneration in *mdx* mice might also result from an acquired resistance to necrosis, such that in older mice myofibres no longer need to regenerate to the same extent (discussed in Grounds and McGeachie, 1991). With respect to this point, Northern analysis of *MyoD* and myogenin indicates that while the process of necrosis and regeneration peaks around 3 to 5 weeks of age, it is ongoing throughout the life of *mdx* mice (Beilharz *et al.*, 1992). It has been demonstrated that muscle regeneration after experimental injury is similar in *mdx* and the control parental mouse strain, which emphasizes that the different regenerative capacity is species-specific (Zacharias and Anderson, 1991; Grounds and McGeachie, 1992). To date there is no conclusive evidence as to whether such species-specific differences might be due to, (i) a fundamental difference in the proliferative capacity of satellite cells related to the size and relative postnatal growth of muscles (Section 9.3), (ii) the different life spans and an ageing phenomenon (Section 9.4), (iii) the result of the lower metabolic rate in dogs and humans as compared with mice (as discussed by Studitsky, 1988), (iv) more vigorous connective tissue (fibrous and adipose) formation in the larger species (Studitsky, 1988; Sweeney and Brown, 1981) which may also directly inhibit myogenesis (Lipton, 1979), or (v) differences in the availability of factors (produced either by mpc or their environ) which favour mpc replication and new muscle formation. The last point is the subject of the following discussion.

9.5.2 Growth factors

A growth factor can be defined as a peptide signalling molecule that stimulates, inhibits or regulates cellular function. Its action may be autocrine (secreted by and acting on the same cell), paracrine (acting on other cells in the local environment), juxtacrine (membrane bound and requiring cell–cell contact) or endocrine (acting on distant target cells). Although the focus of this review is on adult mpc (satellite cells), we will summarize the literature pertaining to embryonic mpc and to myogenic cell lines as these have been used to determine the effect of so many agents (see Table 9.5).

The availability in recent years of many highly purified agents that can modulate myogenesis in culture has enabled serum-free media to be developed for culturing myogenic cells (Askanas and Gallez-Hawkins, 1985; Allen *et al.*, 1985; Ham *et al.*, 1988; McFarland *et al.*, 1991), and has dramatically increased the understanding of the role of such agents during proliferation and differentiation of mpc. There have been several reviews in recent years regarding the effects of growth factors and hormones during myogenesis *in vitro* (Florini, 1987; Florini and Magri, 1989; White and Esser, 1989; Allen and Rankin, 1990; Ewton and Florini, 1990; Florini *et al.*, 1991a) and *in vivo* (Grounds, 1991). These reviews discuss at length the role of the most extensively studied agents which are basic (bFGF) and acidic (aFGF) fibroblast growth factor (FGF), the insulin-like growth factors (IGFs), and the transforming growth factor-beta (TGF-beta) family. We shall briefly discuss some aspects of these factors plus platelet-derived growth factor (PDGF). The effects of these and less studied agents on the

Table 9.5 Factors affecting the behaviour of skeletal muscle precursor cells in tissue culture

	Proliferation	*Differentiation and fusion*	*Cell type*	*Reference*[a]
Growth factors (GF)				
Basic fibroblast GF	↑	↓	Satellite cells on intact myofibres (rat)	Bischoff (1986a)
			mpc (rat)	Allen and Boxhorn (1989)
			mpc (cow)	Greene and Allen (1991)
			mpc (chick)	Kardami *et al.* (1988)
			C2 (mouse)	Yablonka-Reuveni *et al.* (1990c)
			mpc (mouse, human)	Austin *et al.* (1992)
Platelet-derived GF (BB)	↑	↓	C2 (mouse)	Yablonka-Reuveni *et al.* (1990)
			L6J and L8 (rat)	Jin *et al.* (1991a, b)
Platelet-derived GF (BB and AB)	↑	↓	mpc (chick)	Yablonka-Reuveni *et al.* (1992)
Bischoff muscle GF	↑	–	Satellite cells on intact myofibres (rat)	Bischoff (1986a; 1990c)
Insulin GF-I and II, insulin	↑	↑	L6 (rat)	Florini *et al.* (1986)
Insulin GF-I	↑	↑	mpc (rat)	Allen and Boxhorn (1989)
Transforming GF-β	–	↓	C2 (mouse)	Olson *et al.* (1986)
	↓	↓	L6Ea and L8 (rat), mpc (chick)	Massague *et al.* (1986)
			mpc (rat)	Allen and Boxham (1989)
Transforming GF-α	↑	–	mpc (mouse), mpc (human)	Austin *et al.* (1992)
Epidermal GF	↑	–	DD-I (mouse)	Lim and Hauschka (1984)
			mpc (human)	Ham *et al.* (1988)
Interferon	-	↑	mpc (human)	Fisher *et al.* (1983)
	-	↓	mpc (chick), MM14DZ (mouse)	Multhauf and Lough (1986)
Cytokines				
Leukaemia inhibitory factor	↑	-	mpc (mouse)	Austin and Burgess, (1991)
			mpc (human)	Austin *et al.* (1992)

Table 9.5 Factors affecting the behaviour of skeletal muscle precursor cells in tissue culture

	Proliferation	*Differentiation and fusion*	*Cell type*	*Reference*[a]
Interleukin-6	↑	–	mpc (mouse)	Austin and Burgess (1991)
Macrophage colony stimulating factor–1	↑	–	C2, mpc (mouse)	Jones *et al.* (1991)
Neuropeptides				
Adrenocorticotropin	↑	–	mpc (mouse)	Cossu *et al.* (1989)
Enkephalins	↓	–	mpc (mouse)	Cossu *et al.* (1990)
Other				
Prostaglandin E1	–	↑	mpc (chick, human)	Zalin (1979, 1987)
Hemin	↓	↑	mpc (rat)	Funanage *et al.* (1989)
Transferrin	↑	–	mpc (chick), L6 (rat)	Ozawa and Hagiwara (1981) reviewed Ozawa (1989)
Beta-adrenergic agonists	↑	–	mpc (chick)	Grant *et al.* (1990)
Bone morphogenic protein-2	–	↓	L6 (rat)	Yamaguchi *et al.* (1991)
Extracellular matrix components				
Laminin	↑	–	mpc (rat)	Kuhl *et al.* (1986)
			mpc (rat)	Foster *et al.* (1987)
			MM14Dy (mouse)	Ocalen *et al.* (1988)
Fibronectin	↑	↓	L6 (rat)	Podleski *et al.* (1979)
			MM14Dy, mpc (mouse)	Von de Mark and Ocalan (1989) reviewed Grounds (1991)
Hyaluronic acid	–	↓	mpc (chick)	Yoshimura (1985)
			mpc (chick)	Kujawa *et al.* (1986)
Heparin, or Heparin sulphate	↓	–	mpc (chick)	Kardami *et al.* (1988)
Collagen	↑	↑	mpc (chick)	Hauschka and Konigsberg (1966)
	–	↑	L6 (rat)	Nandan *et al.* (1990)
N-acetylglycosamine	–	↑	L6 (rat)	Cates *et al.* (1984)

[a]The few references given indicate a review or generally a more recent paper: due to space limitations many excellent papers are not cited

proliferation and differentiation in tissue culture of myogenic cells from a range of species are indicated in Table 9.5. There are reports of species-related differences in mpc response to growth factors (discussed in Florini and Magri, 1989; Ham *et al.*, 1988), and human mpc may be less responsive to bFGF but also responsive to epidermal growth factor, in contrast with non-human species (Askanas and Gallez-Hawkins, 1985; Ham *et al.*, 1988). (The mechanisms of growth factors action and the increasing literature on the role of various oncogenes and of the transcription factors such as MyoD and myogenin [Olson *et al.*, 1991] are beyond the scope of this review).

It is generally accepted that bFGF stimulates the proliferation but depresses differentiation of mpc (Allen and Boxhorn, 1989). In addition to its mitogenic stimulation of mpc, the angiogenic effects of FGF are important during muscle regeneration (reviewed by Gospodarowicz, 1990). IGFs stimulate both the proliferation and differentiation of mpc and so does insulin although at much higher concentrations than the IGFs (Florini *et al.*, 1986; Ham *et al.*, 1988). While IGFs mediate the effects of growth hormone there is now evidence that in some instances growth hormone may also act directly on mpc (Adamafio *et al.*, 1991; Jennische and Andersson, 1991). As indicated later in this section, there is very strong evidence that IGFs are produced by mpc and act in an autocrine or paracrine manner (see also Grounds, 1991; Ullman *et al.*, 1990). In contrast with bFGF and IGFs, TGF-beta has no effect or slightly suppresses mpc proliferation, but potently inhibits differentiation (Olson *et al.*, 1986; Allen and Boxhorn, 1989; Bischoff, 1990b; Greene and Allen, 1991; Hathaway *et al.*, 1991). There are various exceptions to these generalizations: for example the rat-derived myogenic cell line L6 is not affected by FGF and does not express FGF receptors (Florini *et al.*, 1986; Moore *et al.*, 1991). It has been further proposed that the right combinations of bFGF, IGF and TGFb-beta can elicit proliferation only, proliferation and differentiation, or a state of no proliferation and no differentiation (Allen and Rankin, 1990). Using the mouse-derived myogenic cell line MM14, Clegg *et al.* (1987) and Olwin and Hauschka (1988) indicated that under conditions of low serum and high FGF concentration, FGF no longer supports proliferation of myogenic cells but keeps them in an undifferentiated stage which could resemble the quiescent stage of satellite cells.

The negative growth regulator TGF-beta (bound to the extracellular matrix) might also contribute to the quiescent state of satellite cells (Bischoff, 1990b). An important role for TGF-beta *in vivo* may relate to its potent inhibition of mpc differentiation (Allen and Boxhorn, 1989; Allen and Rankin, 1990), which might prevent activated satellite cells from leaving the proliferative phase and thereby maintain mpc replication. Of additional interest with respect to muscle regeneration is the fact that TGF-beta is chemotactic for monocytes, and plays an extensive role in enhancing the formation and stabilization of components of the extracellular matrix (reviewed Sporn *et al.*, 1987; Massague, 1990).

Recent studies have indicated that PDGF is also a modulator of myogenesis. Using the myogenic cell line C2 (derived from adult mouse muscle, presumably from satellite cells [Yaffe and Saxel, 1977]), Yablonka-Reuveni *et al.*, (1990a) demonstrated that of the three isoforms of PDGF (AA, AB, BB) it is PDGF-BB that significantly promotes proliferation and depresses differentiation of mpc. The effect on proliferation of C2 cells mediated by PDGF-BB and bFGF together is not additive, i.e. cells do not proliferate beyond their proliferation in response to each growth factor alone (Yablonka-Reuveni *et al.*, 1990c). Jin *et al.*, (1990; 1991a, b) observed similar effects of PDGF-BB on proliferation and differentiation of the rat-derived myogenic cell lines L6J1 and L8. Both PDGF-AB and PDGF-BB are also potent promoters of proliferation of clonally-derived chicken myoblasts (Yablonka-Reuveni *et al.*, 1990b; Yablonka-Reuveni and Seifert, 1993). As PDGF is also a chemoattractant for many mesenchymal cells (see references in Yablonka-Reuveni *et al.*, 1990a) it could potentially enhance cell proliferation and guide migration of mpc from areas distant to the injury in muscle regenerating *in vivo*. Whether PDGF isoforms can promote the initial activation of quiescent satellite cells is currently being investigated.

9.5.2.1 Activation of satellite cells

Two crucial questions are: what factors maintain satellite cells in their quiescent state, and what agents cause the initial activation of satellite cells in response to trauma? It seems likely that these situations result from a delicate balance of electrical activity, and growth factor and extracellular matrix composition, such that a perturbation to this environment results in mpc activation (discussed in Bischoff, 1990a, b; Grounds, 1991). The importance of interactions between the extracellular matrix environ and mpc is emphasized by a report that a factor which is mitogenic for mpc but not fibroblasts is produced by proliferating (but not quiescent) fibroblast-like cells in the connective tissue of skeletal muscles (Quinn *et al.*, 1990). Can there be just one growth factor that mediates the initial activation of these quiescent cells? Bischoff (1986a, 1990c) using cultured intact rat myofibres with satellite cells attached demonstrated that bFGF might be such an activator (or competence factor as this is sometimes referred to). Bischoff further showed that similar competence activity is present in rat muscle extracts collected from mildly injured muscles, and suggested that this activity in the muscle extract is different, although closely related, to bFGF (Bischoff, 1986b, 1989, 1990c). The elegant experiments by Bischoff using isolated muscle fibres in culture, have provided some of the best indications as to what the situation might actually be *in vivo*. He concludes that a factor produced by injured myofibres can activate satellite cells but that a serum factor is required for cells to move through the cell cycle and replicate (Bischoff, 1990c). This accords with observations discussed below with respect to the possible role of infiltrating macrophages.

9.5.2.2 Detection of growth factors

What is the relevance of the growth factor studies in culture to the actual situation in muscle tissue? *In vivo* studies have been facilitated by the recent availability of specific antibodies to various growth factors: however, the localization of a growth factor to a cell alone does not confirm that the particular cell produced this factor, as exogenous growth factor could have been taken up into, or bound by, the cell. Furthermore, results between laboratories may vary due to different epitopes recognized by different antibody preparations. Antibody studies show bFGF located in the extracellular matrix around muscle fibres in mice (DiMario *et al.*, 1989; Yamada *et al.*, 1989), in the cytoplasm of embryonic muscles and mpc (Joseph-Silverstein *et al.*, 1989), and another antibody (Kardami *et al.*, 1990) localized bFGF to the periphery, cytoplasm and nuclei of myogenic cells in regenerating muscles of *mdx* mice (Anderson *et al.*, 1991).

Immunofluorescent studies with antibodies to IGF-I demonstrate that rat muscles regenerating after ischaemic necrosis, produce IGF-I in satellite cells, nerves and blood vessels within 24 hours, and IGF levels remained elevated for at least 2 weeks (Jennische and Hansson, 1987; Jennische *et al.*, 1987, Jennische, 1989). To our knowledge there are no published antibody studies with TGF-beta and regenerating muscle *in vivo*. The status of PDGF in normal and regenerating muscle tissue has not yet been resolved, although PDGF is likely to be present in injured muscle as it is released from injured vessels, platelets and macrophages.

The demonstration that mRNAs for various growth factors can be expressed by mpc further supports an autocrine or paracrine role for such factors during myogenesis. For example, mRNAs for both bFGF and aFGF were detected in cultured satellite cells (Alterio *et al.*, 1990) and rodent cell lines (Moore *et al.*, 1991), and proliferating rodent primary satellite cells express mRNA for aFGF (Groux-Muscatelli *et al.*, 1990). *In situ* hybridization studies using primary cell cultures of rat mpc showed mRNA for bFGF in myotubes and mpc aligned for fusion, but not in earlier mpc (Gutheridge *et al.*, 1992). The extension of this study to rat muscle regenerating *in vivo*, showed bFGF mRNA transcripts in damaged myofibres up to three days after injury, and in myotubes and mononuclear cells after this time: however the relative distribution of bFGF in mpc, macrophages, fibroblasts and other mononuclear cells could not be determined (Gutheridge *et al.*, 1992).

Other studies with regenerating rat muscle show that mRNA for IGF-I increased at 24 hours, peaked strongly at 3 days and returned to normal by 10 days (Edwall *et al.*, 1989). This corresponds to the pattern of mpc replication in regenerating rodent muscles (Grounds and McGeachie, 1989b). Similarly, high mRNA levels for IGF-I were seen in cultured replicating C2 cells (Brunetti *et al.*, 1989). To date there appears to be no relevant data for TGF-beta, and relatively few studies with PDGF although mRNA for PDGF A-chain was detected in the rat myogenic cell line L6J1 (Jin *et al.*, 1990).

9.5.2.3 Detection of receptors for growth factors

Studies on expression of cell surface receptors for the various growth factors have utilized radioisotopically-labelled growth factor binding, and specific antibody and mRNA detection. Studies of FGF receptors in several adult mouse cell lines support a correlation between FGF receptor loss and acquisition of a post-mitotic phenotype (Olwin and Hauschka, 1988); however, there appears to be essentially no information available for FGF receptors on satellite cells *in vivo*. The close relationship between replication, differentiation, and production of IGF-I by mpc *in vivo*, is supported by observations with tissue cultured C2 cells where IGF-I receptors are expressed at their highest level in replicating mpc and decrease by 50% after differentiation (Brunetti *et al.*, 1989). Receptors for IGFs (unlike other growth factors) are not down-regulated during differentiation (Florini and Magri, 1989). The coordinated production of IGF-I and IGF-II, and their receptors, and of IGF binding proteins during myogenesis (reviewed by Grounds, 1991) strongly supports an autocrine (Florini *et al.*, 1991b), or paracrine role for these growth factors on myogenic cells *in vitro* and *in vivo*. Similarly, the virtual disappearance of TGF-beta binding sites on mpc during differentiation and fusion in some cultured cell lines (Ewton *et al.*, 1988) supports the idea that TGF-beta levels must decrease to allow differentiation to occur. Receptor binding studies demonstrate that both C2 cells and chicken mpc express receptors to PDGF, predominantly to PDGF-BB (Yablonka-Reuveni *et al.*, 1990a; Yablonka-Reuveni and Seifert, 1993). Studies by Jin *et al.* (1990) indicate that mRNA for the beta receptor of PDGF decreases upon differentiation in the rat-derived myogenic cell line L6RJ, which suggests that these receptors are reduced or eliminated during terminal differentiation.

9.5.2.4 Growth factors and muscular dystrophy

With respect to the X-linked muscular dystrophies, the only growth factor that has been studied in any detail to date is bFGF. In muscles of *mdx* mice, it is difficult to know whether the increased bFGF associated with the extracellular matrix (DiMario *et al.*, 1989) and myogenic cells (Anderson *et al.*, 1991) is a function of the regenerative process, or reflects a direct response to the dystrophic condition. The fact that increased extracellular bFGF is reported at 2 weeks of age before foci of necrosis and regeneration are apparent (DiMario *et al.*, 1989), and that mpc from *mdx* mice respond to lower concentrations of bFGF than do mpc from control mice (DiMario and Strohman, 1988) suggested that this might be related directly to the dystrophic condition in *mdx* mice and that the increased sensitivity of mpc in *mdx* muscle might favour mpc proliferation rather than fibroblast proliferation, and hence minimize fibrosis (DiMario and Strohman, 1988).

To test the proposal that high levels of bFGF in mouse muscle might underlie the effective muscle regeneration seen in *mdx* mice, immunofluorescent studies with the same bFGF antibodies (Anderson *et al.*, 1991) were carried out on muscles from DMD, other myopathies and normal human muscles, and on muscles from normal and dystrophic dogs (Anderson *et al.*, 1992). There was a striking difference between the species in the binding of bFGF antibodies to extracellular matrix, particularly at the periphery of myofibres; this was pronounced in mouse, but weak or absent in human and dog muscle. Binding to muscle nuclei and sarcoplasm was also stronger in mice than in humans and dogs, and in all species was more pronounced in fetal than adult muscle. Increased binding of bFGF antibodies was seen in damaged and regenerating muscle cells in all myopathic specimens where these were present. This was associated with the regenerative process rather than with myopathy, as a similar pattern of bFGF expression was seen in normal mouse muscle regenerating after experimental crush injury. The higher extracellular/peripheral staining for bFGF around mouse myofibres correlates with the successful muscle regeneration in dystrophic mice, and because of the sensitivity of rodent mpc to bFGF it is suggested that bFGF at the fibre periphery might increase locally the numbers of muscle precursor cells available to respond to muscle injury in this species. It is not known whether this apparent species difference can alone account for the markedly different regenerative capacity seen between dystrophic muscles of *mdx* mice, dogs and DMD patients, and it is clearly of interest to evaluate other growth factors such as IGFs in such dystrophic muscles.

9.5.3 Role of other cells and serum factors

At this point it is relevant to refer to the role that macrophages, other inflammatory cells, and fibroblasts might play in muscle regeneration *in vivo*. It is well documented that polymorphonuclear leukocytes and macrophages rapidly accumulate at a site of injury (Papadimitriou *et al.*, 1991) and that macrophages (and fibroblasts) are intimately associated with mpc and myotubes (Robertson *et al.*, manuscript in preparation). The physical proximity of macrophages during the early stages of regeneration suggests that these cells, or rather one or more of their secretory products (Nathan, 1987; also Clark and Henson, 1988), could be ideal candidates for activating mpc after trauma. There is strong evidence from both tissue culture (Bischoff, 1986a; Allen and Rankin, 1990) and *in vivo* studies (Roberts *et al.*, 1989; Grounds and McGeachie, 1991) that, in response to trauma, mpc require an exogenous factor before they can start to replicate. It was shown in regenerating muscle grafts in mice, that the onset of mpc replication occurs in the absence of intimate association with macrophages, although it is closely associated with revascularization and macrophage infiltration (Mastaglia *et al.*, 1975; Roberts *et al.*, 1989; Roberts and McGeachie, 1990): it was concluded that diffusible factors, either from macrophages or from the circulation, were responsible for the mpc response. A role for a serum-derived factor is strongly supported by results from irradiation studies in mice (Robertson *et al.*, 1992) where mpc proliferation and fusion occurred in the absence of infiltrating macrophages or other inflammatory cells. In these experiments mice received whole body irradiation (1600 rads), muscles were crush injured 24 hours later (at a time when levels of circulating white blood cells are severely depleted) and 4 days after injury many myoblasts and myotubes were present among the persisting necrotic tissue. A central role for a serum derived factor is also strongly indicated by the experiments of Bischoff (1990c) with isolated muscle fibres in culture.

Although the evidence argues against a vital mitogenic role for macrophages, it is well recognized that macrophages are the cells primarily responsible for phagocytosing necrotic tissue (Grounds, 1987), and they are of central importance for successful muscle regeneration. It has also been proposed that excessive macrophage activity may result in an accumulation of fibrous tissue (Lang *et al.*, 1987; Krieg and Heckmann, 1989) and, as discussed below, this might impair mpc proliferation.

9.5.4 Extracellular matrix

The different components of the extracellular matrix are known to affect diverse aspects of cell behaviour, such as motility, adhesion, proliferation and differentiation (for examples, see Table 9.5). In addition, the binding of extracellular matrix components to various growth factors can regulate the amounts of growth factors available to mpc: this is particularly relevant for bFGF which is known to bind to the extracellular matrix of muscle fibres

(DiMario *et al.*, 1989; Yamada *et al.*, 1989; Flaumenhaft *et al.*, 1989), and to TGF-beta (reviewed Massague, 1990). Relative changes in the extracellular matrix components, collagen, laminin, fibronectin, hyaluronic acid, heparin and heparan sulphate, chondroitin/dermatan sulphate proteoglycans, and various glucoconjugates are a feature of general wound repair (see Clark and Henson, 1988) and have been documented in regenerating muscles (reviewed by Grounds, 1991). However, it is difficult to determine precisely the relevance and outcome of these complex interactions within the dynamic and three-dimensional *in vivo* environment.

Valuable information regarding the interaction between extracellular components and myogenic cells has been gained from tissue culture studies. Mpc have been grown routinely on gelatin (denatured type I collagen) (Hauschka, 1974a, b), and different laboratories have reported the culturing of mpc on other extracellular matrix components (see Table 9.5). Studies regarding the culture of mpc on components of the external lamina are of interest as it has been suggested that *in vivo* the old basement membrane serves as a scaffolding for new myofibre formation. Kuhl *et al.*, (1986) demonstrated that mpc adhere faster to matrices of laminin/type IV collagen (i.e. external lamina components) whereas fibroblasts adhere faster to matrices of fibronectin/type I collagen. Furthermore, laminin stimulates mpc proliferation (Foster *et al.*, 1987; Ocalan *et al.*, 1988), supports the differentiation of murine mpc (von der Mark and Ocalan, 1989), and stimulates mpc motility (Ocalan *et al.*, 1988; Goodman *et al.*, 1989). Proliferation of mpc is inhibited on a matrix of heparin sulphate proteoglycan (Kardami *et al.*, 1988): it was considered that this might have been an indirect affect since the mitogen bFGF binds to these matrices and its availability for cell proliferation is reduced. However, it has been shown that heparin and heparin sulphate bind directly to BC3H-1 cells (a mouse derived muscle cell line), and possibly interfere directly with growth factor binding or signalling (Vannucchi *et al.*, 1990). Extracellular components can also increase the longevity of myotubes in culture. A three-dimensional gel of collagen was shown to support cultures of highly contractile avian myotubes (Vandenburgh *et al.*, 1988). Matrigel, a matrix of reconstituted basement membrane (composed primarily of type IV collagen, laminin and heparin sulphate proteoglycan) was shown to increase the initial adherence of myogenic cell preparations; it did not effect cell proliferation significantly, but delayed differentiation by 1–2 days (Hartley and Yablonka-Reuveni, 1990). This matrigel matrix supported very long-term, highly contractile avian myotubes which expressed the more advanced isoforms of fast myosin heavy chains (Hartley and Yablonka-Reuveni, 1990) and were narrower, more uniformly orientated and far less branched compared with control myotubes on gelatin (Yablonka-Reuveni *et al.*, 1990b).

As mentioned above, changes in the composition of the extracellular matrix can clearly influence many aspects of myogenesis and muscle regeneration (reviewed by Grounds, 1991), and it has been proposed that an alteration in the balance of connective tissue components might inhibit new muscle formation and account for the pathological changes seen in DMD (Lipton, 1979; Duance *et al.*, 1980; Sweeney and Brown, 1981). In DMD and animal dystrophies, the main change that has been documented is an increase in the amount of extracellular collagen (Ionasescu and Ionasescu, 1982; Ontell *et al.*, 1984; Marshall *et al.*, 1989), and altered forms of collagen (Sweeney and Brown, 1981; Feit *et al.*, 1989). Tissue culture studies of clonal DMD myogenic cells (Ionasescu and Ionasescu, 1982) indicate that an increased collagen synthesis is related to the primary defect of the disease and is not due to secondary replacement fibrosis. It is important to determine precisely what the implications are of the increased collagen in dystrophic muscle (see Sweeney and Brown, 1981; Ionasescu and Ionasescu, 1982). In addition, it would be valuable to document changes in the other extracellular matrix components in DMD muscles, and in muscles of *mdx* mice and dystrophic dogs.

The range of experiments that can be designed to investigate the factors which control various aspects of myogenesis *in vivo* has been greatly advanced by the availability of highly purified growth factors and other substances, specific antibodies, the capacity to localize mRNAs to individual cells, a more sophisticated understanding of gene regulation at the DNA level, and the identification of skeletal muscle specific genes. With the advent of such tools there has been a renaissance of interest in the biology of skeletal muscle regeneration.

ACKNOWLEDGEMENTS

The use of unpublished observations and constructive criticisms of this review by various colleagues is greatly appreciated. We thank Margaret Seats in the Department of Pathology, University of Western Australia for her excellent typing. Consistent support for M.D. Grounds from the National Health and Medical Research Council of Australia, and support to Z. Yablonka-Reuveni from the National Institutes of Health (AR39677), the Muscular Dystrophy Association, the American Heart Association and the American Heart Association, Washington Affiliate is gratefully acknowledged.

REFERENCES

Adamafio, N.A., Towns, R.J. and Kostyo, J.L. (1991) Growth hormone receptors and action in BC3H-1 myocytes. *Growth Regul.*, **1**, 17–22.

Alameddine, H., Sharp, N., Dehaupas, S. *et al.* (1990) Lymphocyte Leu-19 antigens expression in regenerating canine muscles. *J. Neurol. Sci.*, **98S**, 296.

Allen, R.E. and Boxhorn, L.A. (1989) Regulation of skeletal muscle satellite cell proliferation by transforming growth factor-beta, insulin-like growth factor 1, and fibroblast growth factor. *J. Cell Physiol.*, **138**, 311-15.

Allen, R. and Rankin, L.L. (1990) Regulation of satellite cells during skeletal muscle growth and development. *Proc. Soc. Exp. Biol. Med.*, **194**, 81–6.

Allen, R.E., McAllister, P.K. and Masak, K.C. (1980) Myogenic potential of satellite cells in skeletal muscle of old rats. A brief note. *Mech. Age. Dev.*, **13**, 105–9.

Allen, R.E., Dodson, M.V., Luiten, L.S. and Boxhorn, L.K. (1985) A serum free medium that supports the growth of cultured skeletal satellite cells. *In vitro Cell Dev. Biol.*, **21**, 636–40.

Allen, R.E., Rankin, L.L., Greene, E.A. *et al.* (1991) Desmin is present in proliferating rat muscle satellite cells but not in bovine muscle satellite cells. *J. Cell Physiol*, **149**, 525–35.

Alterio, J., Courtois, Y., Robelin, J. *et al.* (1990) Acid and basic fibroblast growth factor mRNAs are expressed by skeletal muscle satellite cells. *Biochem. Biophys. Res. Comm.*, **166**, 1205–12.

Anderson, K.E., Ovalle, W.K. and Bressler, B.M. (1987) Electron microscopic and autoradiographic characterisation of hindlimb muscle regeneration in the mdx mouse. *Anat. Rec.*, **219**, 243–57.

Anderson, J.E., Liu, L. and Kardami, E. (1991) Distinctive patterns of bFGF distribution in degenerating and regenerating areas of dystrophic (mdx) striated muscles. *Dev. Biol.*, **147**, 96–109.

Anderson, J.E., Kakulas, B.A., Jacobsen, P.F. *et al.* (1992) Comparison of basic fibroblast growth factor in X-linked dystrophin deficient myopathies of human, dog and mouse. Submitted for publication.

Armand, O., Boutineau, A.-M., Manger, A. *et al.* (1983) Origin of satellite cells in avian skeletal muscles. *Arch. d'Anat. Micr.*, 72, 163–81.

Askanas, V. and Gallez-Hawkins, G. (1985) Synergistic influence of polypeptide growth factors on cultured human muscle. *Arch. Neurol.*, **42**, 749–52.

Austin, L. and Burgess, A.W. (1991) Stimulation of myoblast proliferation in culture by leukaemia inhibitory factor and other cytokines. *J. Neurol. Sci.*, **101**, 193–7.

Austin, L., Bower, L., Kurek, J. and Vakakis, N. (1992) The effects of leukaemia inhibitory factor and other cytokines on murine and human myoblast proliferation. *J. Neurol. Sci.*, **112**, 185–91.

Bayne, E.K., Anderson, M.J. and Fambrough, D.M. (1984) Extracellular matrix organization in developing muscle: correlation with acetylcholine receptor aggregates. *J. Cell Biol.*, **99**, 1486–501.

Beilharz, M.W., Lareu, R., Garrett, K.L. *et al.* (1992) Quantitation of muscle precursor cell activity in skeletal muscle by Northern analysis of MyoD and myogenin: Application to dystrophic (mdx) mouse muscle. *Mol. Cell Neurosci.*, 3, 326–33.

Bischoff, R. (1979) Tissue culture studies on the origin of myogenic cells during muscle regeneration in the rat, in *Muscle Regeneration* (eds A. Mauro *et al.*), Raven Press, New York, pp. 13–29.

Bischoff, R. (1986a) Proliferation of muscle satellite cells on intact myofibers in culture. *Dev. Biol.*, **115**, 129–39.

Bischoff, R. (1986b) A satellite cell mitogen from crushed adult muscle. *Dev. Biol.*, **115**, 140–7.

Bischoff, R. (1989) Analysis of muscle regeneration using single myofibers in culture. *Med. Sci. Sports Exerc.*, **21**, S164–S172.

Bischoff, R. (1990a) Interaction between satellite cells and skeletal muscle fibres. *Development*, **109**, 943–52.

Bischoff, R. (1990b) Control of satellite cell proliferation. *Adv. Exp. Biol. Med.*, **280**, 147–58.

Bischoff, R. (1990c) Cell cycle commitment of rat muscle satellite cells. *J. Cell Biol.*, **111**, 201–7.

Blaivas, M. and Carlson, B.M. (1991) Muscle fibre branching – differences between grafts in old and young rats. *Mech. Ageing Dev.*, **60**, 43–53.

Blau, H.M., Webster, C., Chiu, C-P., Guttman, S. and Chandler, F. (1983a) Differentiation properties of pure populations of human dystrophic muscle cells. *Exp. Cell Res.*, **144**, 495–503.

Blau, H.M., Webster, C. and Pavlath, G.K. (1983b) Defective myoblasts identified in Duchenne muscular dystrophy. *Proc. Natl. Acad. Sci. USA*, **80**, 4856–60.

Braun, T., Bober, E., Buschhausen-Denker, G. *et al.* (1989) Differential expression of myogenic determination genes in muscle cells: possible autoactivation by the Myf gene products. *EMBO J.*, **8**, 3617–25.

Brooks, S.V. and Faulkner, J.A. (1990) Contraction-induced injury: recovery of skeletal muscles in young and old mice. *Am. J. Physiol.*, **258** (Cell Physiol 27) C436–C442.

Brunetti, A., Maddux, B.A., Wong, K.Y. and Goldfine, I.D. (1989) Muscle cell differentiation is associated with increased insulin receptor biosynthesis and messenger RNA levels. *J. Clin. Invest.*, **83**, 192–8.

Cameron, J.A., Hillgers, A.R. and Hinterberger, T.J. (1986) Evidence that reserve cells are a source of regenerated adult newt muscle in vitro. *Nature*, **321**, 607–10.

Campion, D.R. (1984) The muscle satellite cell: a review. *Int. Rev. Cytol.*, **87**, 225–51.

Cardasis, C.A. (1979) Isolated single mammalian muscle fibres aid in the study of satellite cells and myonuclear populations, in *Muscle Regeneration* (eds A. Mauro *et al.*), Raven Press, New York, pp. 155–66.

Carlson, B.M. and Faulkner, J.A. (1989) Muscle transplantation between young and old rats: age of host determines recovery. *Am. J. Physiol.*, **256**, C1262–C1266.

Cates, G.A., Kaur, H. and Sanwal, B.D. (1984) Inhibition of fusion of skeletal myoblasts by tunicamycin and its reversal by N-acetyl glucosamine. *Can. J. Biochem. Cell. Biol.*, **62**, 28–35.

Cheek, D.B. (1985) The control of cell mass and replication. The DNA unit – a personal 20-year study. *Early Hum. Dev.*, **12**, 211–39.

Chevallier, A., Pautou, M.P., Harris, A.J. and Kieny, M. (1986) On the non-equivalence of skeletal muscle satellite cells and embryonic myoblasts. *Arch d'Anat. Microscop.*, **75**, 161–6.

Church, J.C.T. (1969) Satellite cells and myogenesis; a study in the fruit-bat web. *J. Anat.*, **105**, 419–38.

Clark, R.A.F. and Henson, P.M. (1988) *The Molecular and Cell Biology of Wound Repair*, Plenum Press, New York, 597 pp.

Clegg, C.H., Linkhart, T.A., Olwin, B.B. and Hauschka, S.D. (1987) Growth factor control of skeletal muscle differentiation: Commitment to terminal differentiation occurs in G1 phase and is repressed by fibroblast growth factor. *J. Cell Biol.*, **105**, 949–56.

Coleman, M.G., Prattis, S., Kornegay, J.N. *et al.* (1991) Affinity purification of myogenic cells for myoblast transfer. *J. Cell. Biochem.*, **15C**, 36.

Conen, P.E. and Bell, C.A. (1970) Study of satellite cells in mature and fetal human muscle and rhabdomyosarcoma, in *Regeneration of Striated Muscle and Myogenesis*, (eds A. Mauro, S.A. Shafiq and A.T. Milhorat), Exerpta Medica, Amsterdam, 194–211.

Cossu, G. and Molinaro, M. (1987) Cell heterogeneity in the myogenic lineage. *Cur. Top. Dev. Biol.*, **23**, 185–208.

Cossu, G., Molinaro, M. and Pacific, M. (1983) Differential response of satellite cells and embryonic myoblasts to a tumor promoter. *Dev. Biol.*, **98**, 520–4.

Cossu, G., Eusebi, F., Grassi, F. and Wanke, E. (1987) Acetylcholine receptors are present in undifferentiated satellite cells but not in embryonic myoblasts in culture. *Dev. Biol.*, **98**, 520–4.

Cossu, G., Ranaldi, G., Senni, M.I. *et al.* (1988) Early mammalian myoblasts are resistant to phorbol ester-induced block of differentiation. *Development*, **102**, 65–9.

Cossu, G., Cusella-De Angeles, M.G., Senni, M.I. *et al.* (1989) Adrenocorticotropin is a specific mitogen for mammalian myogenic cells. *Dev. Biol.*, **131**, 331–6.

Cossu, G., De Angelis, L., Cusella-De Angelis, M.G. and Molinaro, M. (1990) A role for neuropeptides in the proliferation of different classes of myogenic cells during muscle histogenesis. *Proc. XIXth European Conference on Muscle Contraction and Cell Motility*, (Sept, 1990, Brussels), 43.

Covault, J. and Sanes, J.R. (1986) Distribution of N-CAM in synaptic and extrasynaptic portions of developing and adult skeletal muscle. *J. Cell Biol.*, **102**, 716–30.

Cull-Candy, S.G., Miledi, R., Nakajima, Y. and Uchitel, O.D. (1980) Visualisation of satellite cells in living muscle fibres of the frog. *Proc. Roy. Soc. Lond. B.*, **209**, 563–8.

Cusella-De Angelis, M.C., Lyons, G., Sonnino, C. *et al.* (1992) MyoD1, myogenin independent differentiation of primordial myoblasts in mouse somites. *J. Cell. Biol.*, **116**, 1243–55.

DiMario, J. and Strohman, R.C. (1988) Satellite cells from dystrophic (mdx) mouse muscle are stimulated by fibroblast growth factor in vitro. *Differentiation*, **39**, 42–9.

DiMario, J., Buffinger, N., Yamada, S. and Strohman, R.C. (1989) Fibroblast growth factor in the extracellular matrix of dystrophic (mdx) mouse muscle. *Science*, **244**, 688–90.

Duance, C.V., Stephens, H.R., Dunn, M. *et al.* (1980) A role for collagen in the pathogenesis of muscular dystrophy. *Nature*, **248**, 470–2.

Düsterhöft, S., Yablonka-Reuveni, Z. and Pette, D. (1990) Characterization of myosin isoforms in satellite cell cultures from adult rat diaphragm, soleus and tibialis anterior muscles. *Differentiation*, **45**, 185–191.

Edwall, D., Schalling, M., Jennische, E. and Norstedt, G. (1989) Induction of insulin-like GF-1 messenger ribonucleic acid during regeneration of rat skeletal muscle. *Endocr.*, **124**, 820–5.

Eftimie, R., Brenner, H.R. and Buonanno, A. (1991) Myogenin and MyoD join a family of skeletal muscle genes regulated by electrical activity. *Proc. Natl. Acad. Sci. USA*, **88**, 1349–53.

Enesco, M. and Puddy, D. (1964) Increase in number of nuclei and weight in skeletal muscle of rats of various ages. *Am. J. Anat.*, **114**, 235–44.

Ewton, D.Z. and Florini, J.R. (1990) Effects of insulin-like growth factors and transforming growth factor-beta on the growth and differentiation of muscle cells in culture. *Proc. Soc. Exp. Biol. Med.*, **194**, 76–80.

Ewton, D.Z., Spizz, G.F., Olson, E.N. and Florini, J.R. (1988) Decrease in transforming growth factor-ß binding and action during differentiation of muscle cells. *J. Biol. Chem.*, **263**, 4029–32.

Feit, H., Kawai, M. and Mostafapour, A.S. (1989) Increased resistance of the collagen in avian dystrophic muscle to collagenolytic attack: evidence for increased crosslinking. *Muscle Nerve*, **12**, 476–86.

Feldman, J.L. and Stockdale, F.E. (1991a) The temporal appearance of satellite cells during development. *J. Cell. Biochem.*, **15C**, 52.

Feldman, J.L. and Stockdale, F.E. (1991b) Skeletal muscle satellite cell diversity: Satellite cells from fibers of different types in cell culture. *Dev. Biol.*, **143**, 320–34.

Fisher, P.B., Miranda, A.F., Babiss, L.E. *et al.* (1983) Opposing effects of interferon produced in bacteria and of tumor promoters on myogenesis in human myoblast cultures. *Proc. Natl. Acad. Sci. USA*, **80**, 2961–5.

Flaumenhaft, R., Moscatelli, D., Saksela, O. and Rifkin, D.B. (1989) Role of extracellular matrix in the action of basic fibroblast growth factor: matrix as a source of growth factor for long term stimulation of plasminogen activator production and DNA synthesis. *J. Cell. Physiol.*, **140**, 75–81.

Florini, J.R. (1987) Hormonal control of muscle growth. *Muscle Nerve*, **10**, 577–98.

Florini, J.R., Ewton, D.Z., Falen, S.L. and Van Wyk, J.J. (1986) Biphasic concentration dependency of stimulation of myoblast differentiation by somatomedins. *Am. J. Physiol.*, **250**, C771–C778.

Florini, J.R. and Magri, K.A. (1989) Effects of growth factors on myogenic differentiation. *Am. J. Physiol.*, **256**, C701–C711.

Florini, J.R., Ewton, D.Z. and Magri, K.A. (1991a) Hormones, growth factors, and myogenic differentiation. *Ann. Rev. Physiol.*, **53**, 201–16.

Florini, J.R., Magri, K.A., Ewton, D.Z. *et al.* (1991b) 'Spontaneous' differentiation of skeletal myoblasts is dependent upon autocrine secretion of insulin-like growth factor-II. *J. Biol. Chem.*, **266**, 15917–23.

Foster, R.F., Thompson, J.M. and Kaufman, S.J. (1987) A laminin substrate promotes myogenesis in rat skeletal muscle cultures: analysis of replication and development using antidesmin and anti-BrdU monoclonal antibodies. *Dev. Biol.*, **122**, 11–20.

Franzini-Armstrong, C. (1979) Satellite and invasive cells in frog sartorius, in *Muscle Regeneration* (eds A. Mauro *et al.*), Raven Press, New York, pp. 233–8.

Fuchtbauer, E-M. and Westphal, H. (1992) MyoD and myogenin are co-expressed in regenerating skeletal muscle of the mouse. *Devel. Dynam.*, **193**, 34–9.

Funanage, V.L., Schroedl, N.A., Moses, P.A. *et al.* (1989) Hemin enhances differentiation and maturation of cultured regenerated skeletal myotubes. *J. Cell Physiol.*, **141**, 591–7.

Gatchalian, C.L., Schachner, M. and Sanes, J.R. (1989) Fibroblasts that proliferate near denervated synaptic sites in skeletal muscle synthesize the adhesive molecules tenascin (J1), N-CAM, fibronectin, and heparan sulfate proteoglycan. *J. Cell Biol.*, **108**, 1873–90.

George-Weinstein, M., Decker, C. and Horwitz, A. (1988) Combinations of monoclonal antibodies distinguish mesenchymal, myogenic, and chondrogenic precursors of the developing chick embryo. *Dev. Biol.*, **125**, 34–50.

Gibson, M.C. and Schultz, E. (1983) Age-related differences in absolute numbers of skeletal muscle satellite cells. *Muscle Nerve*, **6**, 574–80.

Goodman, S.L., Deutzmann, R. and Nurcombe, V. (1989) Locomotory competence and laminin-specific cell surface binding sites are lost during myoblast differentiation. *Development*, **106**, 795–802.

Godfrey, E.W., Siebenlist, R.E., Wallskog, P.A. *et al.* (1988) Basal lamina components are concentrated in premuscle masses and at early acetylcholine receptor clusters in chick embryo hindlimb muscles. *Dev. Biol.*, **130**, 471–86.

Gospodarowicz, D. (1990) Fibroblast growth factor: chemical structure and biologic function. *Clin. Orth. Rel. Res.*, **257**, 231–48.

Grant, A.L., Helferich, W.G., Merkel, R.A. and Bergen, W.G. (1990) Effects of phenethanolamines and propranolol on the proliferation of cultured chick breast muscle satellite cells. *J. Anim. Sci.*, **68**, 652–8.

Greene, E.A. and Allen, R.E. (1991) Growth factor regulation of bovine satellite cell growth. *J. Anim. Sci.*, **69**, 146–52.

Grounds, M.D. (1987) Phagocytosis of necrotic muscle in muscle isografts is influenced by the strain, age and sex of host mice. *J. Pathol.*, **153**, 71–82.

Grounds, M.D. (1990a) Factors controlling skeletal muscle regeneration in vivo, in *Pathogenesis and Therapy of Duchenne and Becker Muscular Dystrophy*, (eds B.A. Kakulas and F.L. Mastaglia), Raven Press, New York, pp. 177–85.

Grounds, M.D. (1990b) The proliferation and fusion of myoblasts in vivo, in *Myoblast Transfer Therapy*, (eds A.B. Eastwood, G. Karpati and R. Griggs), Plenum Press, New York, pp. 101–6.

Grounds, M.D. (1991) Towards understanding skeletal muscle regeneration. *Pathol. Res. Pract.*, **118**, 1–22.

Grounds, M.D. and McGeachie, J.K. (1989a) Myogenic cells of regenerating adult chicken muscle can fuse into myotubes after a single cell division in vivo. *Exp. Cell Res.*, **180**, 429–39.

Grounds, M.D. and McGeachie, J.K. (1989b). A comparison of muscle precursor replication in crush injured skeletal muscle of Swiss and BALBc mice. *Cell Tiss. Res.*, **255**, 385–91.

Grounds, M.D. and McGeachie, J.K. (1991) Muscle precursor replication in minced skeletal muscle isografts of Swiss and BALBc mice. *Muscle Nerve*, **123**, 305–13.

Grounds, M.D. and McGeachie, J.K. (1992) Skeletal muscle regeneration in mdx mice: an autoradiographic study. *Muscle Nerve*, **15**, 580–6.

Grounds, M.D., Garrett, K.L. and Beilharz, M.W. (1991a) The expression of MyoD1 and myogenin genes in thymic cells in vivo. *Exp. Cell Res.*, **198**, 357–61.

Grounds, M.D., Garrett, K.L., Lai, M.C. *et al.* (1991b) Identification of skeletal muscle precursor cells in vivo by use of MyoD1 and myogenin probes. *Cell Tiss. Res.*, **267**, 99–104.

Groux-Muscatelli, B., Bassaglia, Y., Barritault, D. *et al.* (1990) Proliferating satellite cells express acidic fibroblast growth factor during in vitro myogenesis. *Dev. Biol.*, **142**, 380–5.

Gutheridge, M., Wilson, M., Cowling, J. *et al.* (1992) The role of basic fibroblast growth factor in skeletal muscle regeneration. *Growth Factors*, **6**, 53–63.

Gutman, E. and Carlson, B.M. (1976) Regeneration and transplantation of muscles in old rats and between young and old rats. *Life Sci.*, **18**, 109–14.

Ham, R.G., St Clair, J.A., Webster, C. and Blau, H.M. (1988) Improved media for normal human muscle satellite cells: Serum-free clonal growth and enhanced growth with low serum. *In Vitro Cell Dev. Biol.*, **24**, 833–44.

Hartley, R.S. and Yablonka-Reuveni, Z. (1990) Long-term maintenance of primary myogenic cultures on a reconstituted basement membrane. *In Vitro Cell. Dev. Biol.*, **26**, 955–61.

Hartley, R.S.., Bandman, E. and Yablonka-Reuveni, Z. (1991) Myoblasts from fetal and adult skeletal muscle regulate myosin expression differently. *Dev. Biol.*, **148**, 249–60.

Hartley, R.S., Bandman, E. and Yablonka-Reuveni, Z. (1992) Skeletal muscle satellite cells appear during late chicken embryogenesis. *Dev. Biol.*, **153**, 206–16.

Hathaway, M.R., Hembree, J.R., Pampusch, M.S. and Dayton, W.R. (1991) Effect of transforming growth factor beta-1 on ovine satellite cell proliferation and fusion. *J. Cell Physiol.*, **146**, 435–41.

Hauschka, S.D. (1974a) Clonal analysis of vertebrate myogenesis. II. Environmental influences upon human muscle differentiation. *Dev. Biol.*, **37**, 329–44.

Hauschka, S.D. (1974b) Clonal analysis of vertebrate myogenesis. III. Developmental changes in the muscle-colony-forming cells of the human fetal limb. *Dev. Biol.*, **37**, 345–68.

Hauschka, S.D. and Konigsberg, I.R. (1966) The influence of collagen on the development of muscle clones. *Proc. Natl. Acad. Sci.*, **55**, 119–26.

Hauschka, S.D., Linkhart, T.A., Clegg, C.H. and Merrill, G.F. (1979) Clonal studies of human and mouse muscle, in *Muscle Regeneration*, (ed. A. Mauro), Raven Press, New York, pp. 311–21.

Helliwell, T.F. (1988) Lectin binding and desmin staining during bupivacaine-induced necrosis and regeneration in rat skeletal muscle. *J. Pathol.*, **155**, 317–26.

Herrera, A.A. and Banner, L.R. (1990) The use and effects of vital fluorescent dyes: Observation of motor nerve terminals and satellite cells in living muscles. *J. Neurocytology*, **19**, 67–83.

Hurko, O., McKee, L., Zuurveld, J.G.E.M. *et al.* (1986) Proliferative capacity of Duchenne and wild-type myoblasts derived from a DMD-G6PD double heterozygote, in *Molecular Biology of Muscle Development*, (eds C. Emerson, D. Fischmann, B. Nadal-Ginard and M.A.Q. Siddiqui), Alan R. Liss, New York, pp. 921–8.

Ionasescu, V. and Ionasescu, R. (1982) Increased collagen synthesis by Duchenne myogenic clones. *J. Neurol. Sci.*, **54**, 79–87.

Ishikawa, H. (1970) Satellite cells in developing muscle and tissue culture, in *Regeneration of Striated Muscle and Myogenesis*, (eds A. Mauro, S.A. Shafiq and A.T. Milhorat), Exerpta Medica, Amsterdam, 167–79.

Jennische, E. (1989) Sequential immunohistochemical expression of IGF-I and the transferrin receptor in regenerating rat muscle in vivo. *Acta Endocrinol. (Copenh.)*, **121**, 733–8.

Jennische, E. and Andersson, G.L. (1991) Expression of GH receptor mRNA in regenerating skeletal muscle of normal and hypophysectomised rats as demonstrated by a simple in situ hybridisation method. *Acta Endocr.*, **125**, 595–602.

Jennische, E. and Hansson, H.A. (1987) Regenerating skeletal muscle cells express insulin-like growth factor 1. *Acta Physiol. Scand.*, **130**, 327–32.

Jennische, E., Skottner, A. and Hansson, H-A. (1987) Satellite cells express the trophic factor IGF-I in regenerating skeletal muscle. *Acta Physiol. Scand.*, **129**, 9–15.

Jin, P., Rahm, M., Claesson-Welsh, L. *et al.* (1990) Expression of PDGF-A chain and beta-receptor genes during rat myoblast differentiation. *J. Cell Biol.*, **110**, 1665–72.

Jin, P., Sejerson, T. and Ringertz, N.R. (1991a) Recombinant PDGF-BB stimulates growth and inhibits differentiation of rat myoblasts. *J. Cell Biochem.*, **15C**, 39.

Jin, P., Sejersen, T. and Ringertz, N.R. (1991b) Recombinant platelet-derived growth factor-BB stimulates growth and inhibits differentiation of rat L6 myoblasts. *J. Biol. Chem.*, **266**, 1245–9.

Jones, G.E., Murphy, S.J. and Watt, D.J. (1990) Segregation of the myogenic cell lineage in mouse muscle development. *J. Cell. Sci.*, **97**, 285–93.

Jones, G.E., Murphy, S.J., Wise, C. and Watt, D.J. (1991) Macrophage-colony-stimulating factor (CSF-1) stimulates proliferation of myogenic cells. *J. Cell. Biochem.*, **15C**, 39.

Joseph-Silverstein, J., Consigli, S.A., Lyser, K.M. and Ver Pault, C. (1989) Basic fibroblast growth factor in the chick embryo: immunolocalization to striated muscle cells and their precursors. *J. Cell Biol.*, **108**, 2459–66.

Kahn, E.B. and Simpson, S.B. (1974) Satellite cells in mature, uninjured skeletal muscle of the lizard tail. *Dev. Biol.*, **37**, 219–23.

Kardami, E., Spector, D. and Strohman, R.C. (1988) Heparin inhibits skeletal muscle growth in vitro. *Dev. Biol.*, **126**, 19–28.

Kardami, E., Murphy, L., Liu, L. *et al.* (1990) Characterisation of two preparations of immunoglobulins to basic fibroblast growth factor exhibit distinct patterns of localisation. *Growth Factors*, **4**, 69–80.

Karpati, G. (1991) Myoblast transfer in Duchenne muscular dystrophy: A perspective, in *Muscular Dystrophy Research*, (eds C. Angelini, G.A. Danielli and D. Fontanan) Excerpta Medica, Amsterdam, New York, Oxford, pp. 101–7.

Kaufman, S.J. and Foster, R.F. (1988) Replicating myoblasts express a muscle-specific phenotype. *Proc. Natl. Acad. Sci. USA*, **865**, 9606–10.

Kaufman, S.J., George-Weinstein, M. and Foster, R.F. (1991) In vitro development of precursor cells in the myogenic lineage. *Dev. Biol.*, **146**, 228–38.

Kelly, A.M. (1978) Satellite cells and myofibre growth in the rat soleus and extensor digitorum longus muscles. *Dev. Biol.*, **65**, 1–10.

Kelly, A.M. and Zacks, S.I. (1969) The histogenesis of rat intercostal muscle. *J. Cell Biol.*, **42**, 135–53.

Kennedy, J.M., Eisenberg, B.R., Reid, S.K. *et al.* (1988) Nascent muscle fiber appearance in overloaded chicken slow-tonic muscle. *Amer. J. Anat.*, **181**, 203–15.

Kieny, M. and Mauger, A. (1984) Immunofluorescent localization of extracellular matrix components during muscle morphogenesis. I. In normal chick embryos. *J. Exp. Zool.*, **232**, 327–41.

Konigsberg, I.R. (1963) Clonal analysis of myogenesis. *Science*, **140**, 1273–84.

Krieg, T. and Heckmann, M. (1989) Regulatory mechanism of fibroblast activity. *Rec. Prog. Medicina*, **80**, 594–8.

Kuhl, U., Ocalan, M., Timpl, R. and von der Mark, K. (1986) Role of laminin and fibronectin in selecting myogenic versus fibrogenic cells from skeletal muscle cells in vitro. *Dev. Biol.*, **117**, 628–35.

Kujawa, M.J., Pechak, D.G. Fiszman, M.Y. and Caplan, A.I. (1986) Hyaluronic acid bonded to cell culture surfaces inhibits the program of myogenesis. *Dev. Biol.*, **113**, 10–16.

Lang, R., Metcalf, D., Cuthbertson, R.A. *et al.* (1987) Transgenic mice expressing a hemopoietic growth factor gene (GM-CSF) develop accumulations of macrophages, blindness, and a fatal syndrome of tissue damage. *Cell*, **51**, 675–86.

Lim, R.W. and Hauschka, S.D. (1984) A rapid decrease in epidermal growth factor-binding capacity accompanies the terminal differentiation of mouse myoblasts in vitro. *J. Cell Biol.*, **98**, 739–41.

Lipton, B.H. (1979) Skeletal muscle regeneration in muscular dystrophy, in *Muscle Regeneration*, (ed. A. Mauro), Raven Press, New York, pp. 101–14.

MacConnachie, H.F., Enesco, M. and Leblond, C.P. (1964) The mode of increase in the number of skeletal muscle nuclei in the postnatal rat. *Am. J. Anat.*, **14**, 245–53.

Manda, P. and Kakulas, B.A. (1986) The effect of the myotoxic agents iodoacetate on dystrophic mice 129/Re. *J. Neurol. Sci.*, **75**, 23–32.

Marshall, P.A., Williams, P.E. and Goldspink, G. (1989) Accumulation of collagen and altered fibre-type ratios as indicators of abnormal muscle gene expression in the mdx dystrophic mouse. *Muscle Nerve*, **12**, 528–37.

Massague, J., Cheifetz, S., Endo, T. and Nadal-Ginard, B. (1986) Type ß transforming growth factor is an inhibitor of myogenic differentiation. *Proc. Natl. Acad. Sci. USA*, **83**, 8206–10.

Massague, J. (1990). The transforming growth factor-ß family. *Annu. Rev. Cell Biol.*, **6**, 597–641.

Mastaglia, F.L., Papadimitriou, J.M. and Kakulas, B.A. (1970) Regeneration of muscle in Duchenne muscular dystrophy. An electron microscope study. *J. Neurol. Sci.*, **11**, 425–44.

Mastaglia, F.L., Dawkins, R.L. and Papadimitriou, J.M. (1975) Morphological changes in skeletal muscle after transplantation. *J. Neurol. Sci.*, **25**, 227–47.

Mauro, A. (1961) Satellite cells of skeletal muscle fibres. *J. Biophys. Biochem. Cytol.*, **9**, 493–4.

Mayne, R., Swasdison, S., Sanderson, R.D. and Irwin, M.H. (1989) Extracellular matrix, fibroblasts, and the development of skeletal muscle, in *Cellular and Molecular Biology of Muscle Development*, (eds L.H. Kedes and F.E. Stockdale), Alan R. Liss, New York, pp. 107–16.

Mazanet, R. and Franzini-Armstrong, C. (1980) The satellite cell, in *Myology*, (eds A.G. Engel and B.Q. Banker), McGraw-Hill Book Co, New York, **1**, 285–307.

McFarland, D.C., Pesall, J.E., Norberg, J.M. and Dvoracek, M.A. (1991) Proliferation of the turkey myogenic satellite cell in a serum-free medium. *Comp. Biochem. Physiol.*, **99A**, 163–7.

Miller, J.B. (1991) Myoblasts, myosin, MyoDs, and the diversification of muscle fibres. *Neuromusc. Disord.*, **1**, 7–17.

Miller, J.B. and Stockdale, F. (1989) Multiple cellular processes regulate expression of slow myosin heavy chain isoforms during avian myogenesis in vitro. *Dev. Biol.*, **136**, 393–404.

Moore, J.W., Dionne, C., Jaye, M. and Swain, J. (1991) The mRNAs encoding acidic FGF, basic FGF and FGF receptors are coordinately downregulated during myogenic differentiation. *Development*, **111**, 741–8.

Morlet, K., Grounds, M.D. and McGeachie, J.K. (1989) Muscle precursor replication after repeated regeneration of skeletal muscle. *Anat. Embryol.*, **180**, 471–8.

Moschella, M.C. and Ontell, M. (1987) Transient and chronic neonatal denervation of murine muscle: A procedure to modify the phenotypic expression of muscular dystrophy. *J. Neuroscience*, **7**, 2145–52.

Moss, F.P. and Leblond, C.P. (1971) Satellite cells as the source of nuclei in muscles of growing rats. *Anat. Rec.*, **170**, 421–36.

Mouly, V., Toutant, M. and Fiszman, M.Y. (1987) Chick and quail limb bud myoblasts, isolated at different times during muscle development, express stage-specific phenotypes when differentiated in culture. *Cell Diff.*, **20**, 17–25.

Multhauf, C. and Lough, J. (1986) Interferon-mediated inhibition of differentiation in a murine myoblast cell line. *J. Cell. Physiol.*, **126**, 211–15.

Nameroff, M. and Rhodes, L.D. (1989) Differential response among cells in the chick embryo myogenic lineage to photosensitization by Merocyanine 540. *J. Cell Physiol.*, **141**, 475–82.

Nandan, D., Clarke, E.P., Ball, E.H. and Sanwal, B.D. (1990) Ethyl-3,4-dihydroxybenzoate inhibits myoblast differentiation: Evidence for an essential role for collagen. *J. Cell Biol.*, **110**, 1673–9.

Nathan, C.F. (1987) Secretory products of macrophages. *J. Clin. Invest.*, **79**, 316–26.

Nathanson, M.A. (1979) Skeletal muscle metaplasia: formation of cartilage by differentiated skeletal muscle, in *Muscle Regeneration* (eds A. Mauro *et al.*), Raven Press, New York, pp. 83–90.

Ocalan, M., Goodman, S.L., Kuhl, U. *et al.* (1988) Laminin alters cell shape and stimulates mobility and proliferation of murine skeletal myoblasts. *Dev. Biol.*, **125**, 158–67.

Olson, E.N., Sternberg, E., Shan Hu, J. *et al.* (1986) Regulation of myogenic differentiation by Type ß transforming growth factor. *J. Cell Biol.*, **103**, 1799–805.

Olson, E.N., Brennan, T.R.J., Chakraborty, T. *et al.* (1991) Molecular control of myogenesis: antagonism between growth and differentiation. *Mol. Cell. Biochem.*, **104**, 7–13.

Olwin, B. and Hauschka, S.D. (1988) Cell surface fibroblast growth factor and epidermal growth factor receptors are permanently lost during skeletal muscle terminal differentiation in culture. *J. Cell Biol.*, **107**, 761–9.

Ontell, M. (1974) Muscle satellite cells: A validated technique for light microscopic identification and a quantitative study of changes in their population following denervation. *Anat. Rec.*, **178**, 211–28.

Ontell, M. (1979) The source of 'new' muscle fibres in neonatal muscle, in *Muscle Regeneration* (eds A. Mauro *et al.*), Raven Press, New York, 137–47.

Ontell, M. Feng, K.C., Klueber, K. *et al.* (1984) Myosatellite cells, growth, and regeneration in murine dystrophic muscle: A quantitative study. *Anat. Rec.*, **208**, 159–74.

Ontell, M., Hughes, D., Hauschka, S.D. and Ontell, M. (1991) Dystrophic satellite cell senescence is prevented by neonatal denervation. *J. Cell Biochem.*, **15C**, 62.

Ozawa, E. (1989). Transferrin as a muscle trophic factor. *Dev. Physiol. Biochem. Pharmacol.*, **113**, 89–141.

Ozawa, E. and Hagiwara, Y. (1981) Avian and mammalian transferrins are required for chick and rat myogenic cell growth in vitro, respectively. *Proc. Japan Acad. Ser. B*, **57**, 406–9.

Partridge, T.A. (1991a). Animal models of muscular dystrophy – What can they teach us? *Neuropathol. Appl. Neurobiol.*, **17**, 353–63.

Partridge, T.A. (1991b) Myoblast transplantation: possible therapy for inherited myopathies. *Muscle Nerve*, **14**, 197–212.

Podleski, T.R., Greenberg, I., Schlessinger, J. and Yamada, K.M. (1979) Fibronectin delays the fusion of L6 myoblasts. *Exp. Cell Res.*, **122**, 317–26.

Popiela, H. (1976) Muscle satellite cells in urodele amphibians: facilitated identification of satellite cells using ruthenium red staining. *J. Exp. Zool.*, **198**, 57–64.

Przyblska, J.R. (1983) Satellite cells of chicken's muscles. *Folia Morphol. (Warsaw)*, **42**, 217–27.

Quinn, L.S., Nameroff, M. and Holtzer, H. (1984) Age-dependent changes in myogenic precursor cell compartment sizes. *Exp. Cell Res.*, **154**, 65–82.

Quinn, L.S., Holtzer, H. and Nameroff, M. (1985) Generation of chick skeletal muscle cells in groups of 16 from stem cells. *Nature (London)*, **313**, 692–4.

Quinn, L.S., Ong, L.D. and Roeder, R.A. (1990) Paracrine control of myoblast proliferation and differentiation by fibroblasts. *Dev. Biol.*, **140**, 8–19.

Rieger, F., Grumet, M. and Edelman, G.M. (1985) N-CAM at the vertebrate neuromuscular junction. *J. Cell Biol.*, **101**, 285–93.

Roberts, P. and McGeachie, J.K. (1990) Endothelial cell activation during angiogenesis in freely transplanted skeletal muscles in mice and its relationship to the onset of myogenesis. *J. Anat.*, **169**, 197–207.

Roberts, P., McGeachie, J.K., Smith, E.R. and Grounds, M.D. (1989) The initiation and duration of myogenesis in transplants of intact skeletal muscle: an autoradiographic study in mice. *Anat. Rec.*, **224**, 1–6.

Robertson, T., Papadimitriou, J.M., Mitchell, C.A. and Grounds, M.D. (1990) Fusion of myogenic cells in vivo: an ultrastructural study of regenerating murine skeletal muscle. *J. Struct. Biol.*, **105**, 170–82.

Robertson, T.A., Grounds, M.D. and Papadimitriou, J.M. (1992) Elucidation of aspects of murine skeletal muscle regeneration using local and whole body irradiation. *J. Anat.*

Ross, J.J., Duxson, M.J. and Harris, A.J. (1987) Formation of primary and secondary myotubes in rat lumbrical muscles. *Development*, **100**, 395–409.

Sadeh, M. (1988) Effects of ageing on skeletal muscle regeneration. *J. Neurol. Sci.*, **87**, 67–74.

Sanes, J.R. and Cheney, J.M. (1982) Laminin, fibronectin, and collagen in synaptic and extrasynaptic portions of muscle fibre basement membrane. *J. Cell Biol.*, **93**, 442–51.

Sanes, J.R., Schachner, M. and Covault, J. (1986) Expression of several adhesive macromolecules (N-CAM, L1, J1, NILE, uvomorulin, laminin, fibronectin, and a heparan sulfate proteoglycan) in embryonic, adult, and denervated adult skeletal muscle. *J. Cell Biol.*, **102**, 420–31.

Schubert, W., Zimmerman, K., Cramer, M. and Starzinski-Powitz, A. (1989) Lymphocyte antigen Leu-19 as a molecular marker of regeneration in human skeletal muscle. *Proc. Natl. Acad. Sci.*, **86**, 307–11.

Schultz, E. (1979) Quantification of satellite cells in growing muscle using electron microscopy and fibre whole mounts, in *Muscle Regeneration* (ed. A. Mauro), Raven Press, New York, 131–5.

Schultz, E. and Lipton, B.H. (1982) Skeletal muscle satellite cells changes in proliferation potential as a function of age. *Mech. Ageing, Dev.*, **20**, 377–83.

Schultz, E., Jaryszak, B.A. and Valliere, C.R. (1985) Response of satellite cells to focal skeletal muscle injury. *Muscle Nerve*, **8**, 217–22.

Schweitzer, J.A., Dichter, M.A. and Kaufman, S.J. (1987) Fibroblasts modulate expression of Thy-1 on the surface of skeletal myoblasts. *Exp. Cell Res.*, **172**, 1–20.

Seed, J. and Hauschka, S.D. (1988) Clonal analysis of vertebrate myogenesis. VIII. Fibroblast growth factor (FGF)-dependent and FGF-independent muscle colony types during chick wing development. *Dev. Biol.*, **128**, 40–9.

Senni, M.I., Castrignano, F., Poiana, G. *et al.* (1987) Expression of adult fast pattern of acetylcholinesterase molecular forms by mouse satellite cells in culture. *Differentiation*, **36**, 194–8.

Snow, M.H. (1977) The effects of ageing on satellite cells in skeletal muscles of mice and rats. *Cell Tiss. Res.*, **185**, 399–408.

Snow, M.H. (1979) Origin of regenerating myoblasts in mammalian skeletal muscle, in *Muscle Regeneration* (eds A. Mauro *et al.*), Raven Press, New York, pp. 91–100.

Sporn, M.B., Roberts, A.B., Wakefield, L.M. and de Crombrugghe, B. (1987) Some recent advances in the chemistry and biology of transforming growth factor-beta. *J. Cell Biol.*, **105**, 1039–45.

Stedman, H.H., Sweeney, H.L., Shrager, J.B. *et al.* (1991) The mdx mouse diaphragm reproduces the degenerative changes of Duchenne muscular dystrophy. *Nature*, **352**, 536–9.

Stockdale, F.E. and Miller, J.B. (1987) The cellular basis of myosin heavy chain isoform expression during the development of avian skeletal muscles. *Dev. Biol.*, **123**, 1–9.

Studitsky, A.N. (1988) in *Transplantation of muscles in animals*, (ed. B.A. Carlson), Amerind Publ. Co. Pvt. Ltd., New Dehli (originally published in Russian in 1977) pp. 211.

Summers, P.J. and Parsons, R. (1981) An electron microscopic study of satellite cells and regeneration in dystrophic mouse muscle. *Neuropathol. Appl. Neurobiol.*, **7**, 257–68.

Sweeney, P.R. and Brown, R.G. (1981) The aetiology of muscular dystrophy in mammals – a new perspective and hypothesis. *Comp. Biochem. Physiol.*, **70B**, 27–33.

Tassin, A-M., Mege, R-M., Goudou, D. *et al.* (1991) Modulation of expression and cell surface distribution of N-CAM during myogenesis in vitro. *Neurochem*, **18**, 97–106.

Trupin, G.L., Hsu, L. and Hsieh, Y-H. (1979) Satellite cell mimics in regenerating skeletal muscle, in *Muscle Regeneration*, (eds A. Mauro *et al.*), Raven Press, New York, pp. 101–14.

Ullman, M., Ullman, A., Sommerland, H. *et al.* (1990) Effects of growth hormone on muscle regeneration and IGF-I concentration in older rats. *Acta Physiol. Scand.*, **140**, 521–5.

Valentine, B.A. and Cooper, B.J. (1991) Canine X-linked muscular dystrophy: selective involvement of muscles in neonatal dogs. *Neuromusc. Disord.*, **1**, 31–8.

Vandenburgh, H., Karlisch, P. and Farr, L. (1988) Maintenance of highly contractile tissue-cultured avian skeletal myotubes in collagen gel. *In vitro Cell Dev. Biol.*, **24**, 166–74.

Venable, J.H. and Lorenz, M.D. (1970) Trial analysis of the cytokinetics of a rapidly growing skeletal muscle, in *Regeneration of Striated Muscle and Myogenesis*, (eds A. Mauro, S.A. Shafiq and A.T. Milhorat), Exerpta Medica, Amsterdam, 271–8.

Vivarelli, E., Brown, W.E., Whalen, R.G. and Cossu, G. (1988) The expression of slow myosin during mammalian somitogenesis and limb bud differentiation. *J. Cell Biol.*, **107**, 2191–7.

Von der Mark, K. and Ocalan, M. (1989) Antagonistic effects of laminin and fibronectin on the expression of the myogenic phenotype. *Differentiation*, **40**, 150–7.

Wakshull, E., Bayne, E.K., Chiquet, M. and Fambrough, D.M. (1983) Characterization of a plasma membrane glycoprotein common to myoblasts, skeletal muscle satellite cells, and glia. *Dev. Biol.*, **100**, 464–77.

Walsh, F.S., Dickson, G., Moore, S.E. and Barton, C.H. (1989) Unmasking N-CAM. *Nature*, 339, 516.

Walton, J. (1982) The role of limited cell replicative capacity in pathological age change: a review. *Mech. Age. Develop.*, **19**, 217–44.

Webster, C., Filippi, G., Rinaldi, A. *et al.* (1986) The myoblast defect identified in Duchenne muscular dystrophy is not a primary expression of the DMD mutation. *Hum. Genet.*, **74**, 74–80.

Webster, C., Pavlath, G.K., Parks, D.R. *et al.* (1988) Isolation of human myoblasts with the fluorescence-activated cell sorter. *Exp. Cell Res.*, **174**, 252–65.

Weintraub, H., Davis, R., Tapscott, S. *et al.* (1991) The MyoD gene family: nodal point during specification of the muscle cell lineage. *Science*, **251**, 761–6.

White, N.K., Bonner, P.H., Nelson, D.R. and Hauschka, S.D. (1975) Clonal analysis of vertebrate myogenesis. IV. Medium-dependent classification of colony-forming cells. *Dev. Biol.*, **44**, 346–61.

White, T.P. and Essser, K.A. (1989) Satellite cell and growth factor involvement in skeletal muscle growth. *Med. Sci. Sports Exerc.*, **21**, S158–S163.

Wilson, S.J., McEwan, J.C., Sheard, P.W. and Harris, A.J. (1992) Early stages of myogenesis in a large mammal: formation of successive generations of myotubes in sheep tibialis cranialis muscle. *J. Musc. Res. Cell Motility.*, **13**, 534–50.

Witowski, J.A. (1986) Tissue culture studies of muscle disorders: Part 1. Techniques, cell growth, morphology, cell surface. *Muscle Nerve*, **9**, 191–207.

Wright, W.E. (1985) Myoblast senescence in muscular dystrophy. *Exp. Cell Res.*, **157**, 343–54.

Wright, W.E. and Shay, J.W. (1992) Telomere positional effects and the regulation of cellular senescence. *Trends Genet.*, **8**, 193–7.

Wright, W.E., Sassoon, D.A. and Lin, V.K. (1989) Myogenin, a factor regulating myogenesis has a domain homologous to MyoD1. *Cell*, **56**, 607–17.

Yablonka-Reuveni, Z. (1988a) Identification of satellite cell-specific antigens. *J. Cell Biochem.*, Suppl. **12C**, 334.

Yablonka-Reuveni, Z. (1988b) Discrimination of myogenic and nonmyogenic cells from embryonic skeletal muscle by 900 light scattering. *Cytometry*, **9**, 121–5.

Yablonka-Reuveni, Z. (1989) Application of density centrifugation and flow cytometry for the isolation of fibroblast-like cells from embryonic and adult skeletal muscle, in *Cellular and Molecular Biology of Muscle Development*, (eds L.H. Kedes and F.E. Stockdale), Alan R. Liss, New York, pp. 869–79.

Yablonka-Reuveni, Z. and Nameroff, M. (1987) Skeletal muscle cell populations: separation and partial characterization of fibroblast-like cells from embryonic tissue using density centrifugation. *Histochemistry*, **87**, 27–38.

Yablonka-Reuveni, Z., Quinn, L.S. and Nameroff, M. (1987) Isolation and clonal analysis of satellite cells from chicken pectoralis muscle. *Dev. Biol.*, **119**, 252–9.

Yablonka-Reuveni, Z., Anderson, S.K., Bowen-Pope, D.F. and Nameroff, M. (1988) Biochemical and morphological differences between fibroblasts and myoblasts from embryonic chicken skeletal muscle. *Cell Tissue Res.*, **252**, 339–48.

Yablonka-Reuveni, Z. and Nameroff, M. (1990) Temporal differences in desmin expression between myoblasts from embryonic and adult skeletal muscle. *Differentiation*, **45**, 21–8.

Yablonka-Reuveni, Z., Balestreri, T.M. and Bowen-Pope, D.F. (1990a) Regulation of proliferation and differentiation of myoblasts derived from adult mouse skeletal muscle by specific isoforms of PDGF. *J. Cell Biol.*, **111**, 1623–9.

Yablonka-Reuveni, Z., Bowen-Pope, D.F. and Hartley, R.S. (1990b) Proliferation and differentiation of myoblasts: The role of platelet-derived growth factor and the basement membrane, in *The Dynamic State of Muscle Fibers*, (ed. D. Pette), Walter de Gruyter, Berlin, 693–706.

Yablonka-Reuveni, Z., Bowen-Pope, D.F. and Balestreri, T.M. (1990c) Regulation of proliferation and differentiation of myoblasts derived from adult mouse skeletal muscle by specific isoforms of PDGF. *J. Cell Biol.*, 5, 32a.

Yablonka-Reuveni, Z., and Seifert, R.A. (1993) Proliferation of chicken myoblasts is regulated by specific isoforms of platelet-derived growth factor: evidence for differences between myoblasts from mid and late stages of embryogenesis. In press.

Yaffe, D. (1973) Rat skeletal muscle myoblasts, in *Tissue culture: Methods and Applications*, (eds P.F. Kruse and M.K. Patterson), Academic Press, New York, pp. 106–14.

Yaffe, D. and Saxel, O. (1977) Serial passaging and differentiation of myogenic cells isolated from dystrophic mouse muscle. *Nature*, **270**, 725–7.

Yamada, S. Buffinger, N., DiMario, J. and Strohman, R.C. (1989) Fibroblast growth factor is stored in fiber extracellular matrix and plays a role in regulating muscle hypertrophy. *Med. Sci. Sports Exerc.*, **21**, S173–S180.

Yamaguchi, A., Katagiri, T., Ikeda, T. *et al.* (1991) Recombinant human bone morphogenetic protein-2 stimulates osteoblastic maturation and inhibits myogenic differentiation in vitro. *J. Cell Biol.*, **113**, 681–7.

Yoshimura, M. (1985) Changes in hyaluronic acid synthesis during differentiation of myogenic cells and its relation to transformation of myoblasts by rous sarcoma virus. *Cell Diff.*, **16**, 175–85.

Zacharias, J.M. and Anderson, J.E. (1991) Muscle regeneration after imposed injury is better in younger than older mdx mice. *J. Neurol. Sci.*,**104**, 190–6.

Zacks, S.I. and Sheff, M.F. (1982) Age-related impeded regeneration of mouse minced anterior tibial muscle. *Muscle Nerve*, **5**, 152–61.

Zalin, R.J. (1979) The cell cycle, myoblast differentiation and prostaglandin as a developmental signal. *Dev. Biol.*, **71**, 274–88.

Zalin, R.J. (1987) The role of hormones and prostanoids in the in vitro proliferation and differentiation of human myoblasts. *Exp. Cell Res.*, **172**, 265–81.

Zerba, A., Komorowski, T.E. and Faulkner, J.A. (1990) Free radical injury to skeletal muscles of young and old rats. *Am. J. Physiol.*, **258** (Cell Physiol 27) C429–C435.

10

Molecular mechanisms of muscle damage

MALCOLM J. JACKSON

10.1 INTRODUCTION

Prior to the recognition that the protein dystrophin is missing in Duchenne dystrophy muscle and of abnormal size or reduced quantity in muscle from Becker dystrophy patients (Hoffman *et al.*, 1987a), the assumption was made that recognition of the defective gene product(s) would lead to a rapid understanding of the manner in which this defect caused the characteristic muscle degeneration and weakness seen in these conditions (e.g. see reviews by King Engel, 1977 and Rowland, 1984). It is now apparent that this thesis presupposed that the normal function of the product of the defective gene would be known or easily inferred by comparison to other known muscle proteins. Unfortunately, because of the manner in which the elucidation of the defective gene product was achieved, this has not occurred and, although it is possible to infer a certain amount about the likely function of the protein from its location in the cell and its general structure, the manner in which a lack or defect in the protein leads to muscle damage and degeneration is still not apparent.

Prior to the discovery of dystrophin a number of theories of possible pathogenic mechanisms in Duchenne dystrophy were postulated such as a fundamental abnormality in muscle microcirculation, in known muscle-specific proteins, or in nerve–muscle interactions (see Rowland, 1980, 1984 and references therein for details). These and other theories have data in support of them, but were largely abandoned because they could not be substantiated by further experiments. Rowland (1980) drew attention to the possibility of a defect in the muscle surface membrane as the primary site of the defect. This 'membrane theory' (Rowland, 1984) was stated as 'the abnormal gene product in Duchenne dystrophy is an enzyme or structural protein that is either missing or defective in function. The altered function of this protein results in abnormal composition and abnormal function of

Molecular and Cell Biology of Muscular Dystrophy
Edited by Terence Partridge
Published in 1993 by Chapman & Hall, London. ISBN 0 412 43440 7

muscle surface membranes leading, in turn, to the two fundamental disorders that characterize the disease: progressively more severe weakness and progressively more severe degeneration of muscle fibres'. Of the lines of evidence presented in support of this theory, increased serum activity of sarcoplasmic enzymes, altered morphology of surface membranes, and altered function of sarcolemmal membrane-bound enzymes, it was stated that 'the serum enzyme abnormality must be very close to the genetic defect'. The recognition that dystrophin is a component of the membrane cytoskeleton of muscle cells has, in general terms, verified the theory of Rowland.

The aim of this review is to examine this and other mechanisms of cellular damage which may be involved in muscle degeneration and necrosis in dystrophin-deficient muscle. It should be stated immediately that in the author's opinion the processes of damage and degeneration in dystrophic muscle are not fully explained by any of the current theories. In order to fully appreciate why this is so and the potential role of certain pathways it will also be necessary to review a number of studies examining the mechanisms by which damage occurs in normal muscles following various stresses.

10.2 THE MECHANISMS OF DAMAGE TO NORMAL SKELETAL MUSCLE SUBJECTED TO VARIOUS DAMAGING STRESSES

A number of words are used in a variable manner to express the processes or stages in the processes by which cells die. The most commonly used of these are damage, injury, degeneration, degradation, viability, necrosis and death. In any consideration of such processes it is important to be clear about the definitions and manner in which these are used. In the work to be described here damage and injury will be used synonymously, meaning harm to the cell which may or may not be reversible in terms of leading to cell death. Degeneration and degradation of cells will also be treated as synonyms meaning a breakdown of organized structure; these terms will be applied in both a generalized (e.g. cellular degeneration), and a specific (e.g. protein degradation) manner. It is also relevant to note that degeneration of at least some part of the cell will occur during the processes of cell damage, and will also occur as the damaged cell progresses to cell death. Death and necrosis will also be treated as synonyms, being defined as irreversible loss of all vital functions by the cell. Necrosis will primarily be used as a term for the description of dead cells seen by microscopy. A non-viable cell will be treated as one which is sufficiently damaged that it must die although at the point of time in question it may still possess some vital functions.

Damage to skeletal muscle can be assessed in a number of different ways. The most popular of these are assessment of force production by the muscle, examination of the structure of the muscle (by light microscopy or electron microscopy), examination of the efflux of cytosolic components from the muscle, examination of certain biochemical processes known to be influenced during cellular degeneration (such as muscle protein turnover), or assessment of substances normally excluded by muscle cells. Using these assessment protocols, damage to normal muscle has been examined in a number of different physiological and pathological conditions and in intact subjects or animals, in isolated muscle and isolated cell preparations; likewise the nature of the damaging stress has varied greatly dependent upon the interest of the investigator although most studies of damage to normal muscle have examined the effects of exercise, toxins or metabolic poisons. Much work has concentrated on the effects of eccentric and excessive concentric exercise on skeletal muscle with the recognition that eccentric exercise is considerably more damaging to skeletal muscle than concentric exercise (see Armstrong *et al.*, 1991 for a review). These studies have demonstrated that regulation of mechanical forces may play a very important role in maintenance of muscle cell membrane integrity, but it is apparent that the mechanisms underlying this form of damage are still obscure and the relevance of this damage to that suffered by normal subjects undergoing an excessive or unaccustomed amount of a common form of exercise such as running is also currently unclear.

The studies with which the author has been involved have attempted to simplify the problem of elucidation of mechanisms involved in muscle damage by examination of intact muscle *in vitro*. Such studies are possible only with certain rodent muscles which can be removed with little trauma to the muscles and which are essentially pennate providing ready access to tendons for mounting of the muscles. In practice rat or mouse soleus or extensor digitorum longus (EDL) muscles have been used for these studies. The original description of the apparatus was presented by Jones *et al.* (1983) although it has been modified to permit study of muscle energy metabolism by ^{31}P-nuclear magnetic resonance (NMR) techniques (West-Jordan *et al.*, 1990, 1991).

The experimental system suffers from the well-described defects of all *in vitro* muscle incubation systems, such as poor long-term viability of the muscles and incomplete oxygenation of central fibres, but provides excellent accessibility to monitor and manipulate biochemical processes. However, it must be borne in mind that such studies, although potentially useful, cannot provide a complete picture in that extra-muscle factors involved in damaging processes are not taken into account.

Initial studies with the system (Jones *et al.*, 1983) indicated that excessive, electrically stimulated isometric contractile activity of mouse soleus muscles could induce both structural damage to muscles and cytosolic enzyme efflux in a short period of time (3 h), although there was an unexplained time lag between the period of contractile activity and maximum release of cytosolic enzymes. Interestingly the time-course of this release was mimicked almost exactly by treatment of muscles with inhibitors of the glycolytic pathway (iodoacetate) or mitochondrial energy production (potassium cyanide) suggesting an association between contractile activity-induced damage to skeletal muscle and a failure of muscle energy supply. These initial studies therefore demonstrated the potential of such a model for study of the mechanisms of skeletal muscle damage and we have extensively utilized this system to examine a number of possible mechanisms by which normal muscle cells become damaged following a variety of stresses.

10.2.1 Role of calcium

An accumulation of calcium in intracellular sites has been implicated in the mechanisms of cellular damage in various cell types. In cardiac tissue there has been a large amount of interest in this area with damage due to hypoxia and re-oxygenation being shown to be associated with an increase in tissue calcium content (Nayler *et al.*, 1979) while treatment of isolated myocytes with metabolic inhibitors has been shown to lead to increases in the free intracellular calcium concentration (Allshire *et al.*, 1987). In hepatocytes loss of cell viability following incubation with various toxins was reported to be greatly reduced when the external calcium was removed from the incubation fluid (Schanne *et al.*, 1979) which prompted Schanne *et al.* to propose a general requirement for external calcium in the processes leading to cell death. This hypothesis and the experiments on which it was based have subsequently become the subject of much controversy (e.g. see Smith *et al.*, 1981; Farris *et al.*, 1985), but the general hypothesis that cell injury or death induced by various factors is mediated by a rise in the free intracellular calcium content is an attractive one because of the large difference in intracellular and extracellular calcium concentrations. Thus, a minor change in the plasma membrane could modify the permeability to calcium sufficiently to allow a relatively large amount of calcium to enter the cell down the large concentration gradient for the element.

In skeletal muscle, various studies have demonstrated the potential of calcium to influence cellular damage. High external calcium concentrations (3–10 mmol/l) have been shown to increase cytosolic creatine kinase (CK) release from isolated animal (Soybell *et al.*, 1978) and human (Anand and Emery, 1980) skeletal muscle while the work of Duncan and colleagues has demonstrated that an elevation of intracellular calcium can cause ultrastructural damage to muscle (Publicover *et al.*, 1978; Duncan *et al.*, 1979). Slow calcium-channel blocking agents have also been found to reduce CK release from human skeletal muscle (Anand and Emery, 1982).

Our studies using the *in vitro* system have indicated that the release of cytosolic enzymes following various different stresses (excess contractile activity, treatment with low dose detergents or with mitochondrial inhibitors) could be greatly reduced by removal of the external calcium during the damaging period (Jackson *et al.*, 1984; Jones *et al.*, 1984). Other workers have also described protective effects of an absence of external calcium on bupivicaine-induced muscle injury in a similar isolated muscle system (Steer *et al.*, 1986). We also observed that removal of the external calcium was effective in protection of the muscle against the histological and ultrastructural damage induced by excess contractile activity (Jones *et al.*, 1984), although later work suggests that structural damage induced by other stresses was not reduced by removal of external calcium (Duncan and Jackson, 1987). Other experiments have demonstrated that excess contractile activity or treatment with mitochondrial inhibitors is associated with a dramatic increase in total muscle calcium (Claremont *et al.*, 1984) occurring via a rapid influx of calcium from the external medium (McArdle *et al.*, 1992).

These studies therefore suggested that the processes underlying cytosolic enzyme efflux from damaged muscle, and those responsible for ultrastructural changes in certain circumstances, are mediated by a net influx of extracellular calcium down the large extracellular–intracellular concentration gradient for this element and furthermore that this increased intracellular calcium content then mediates degenerative pathways in the muscle cell.

10.2.2 Calcium-activated degenerative processes in muscle

The possible manner in which a failure of intracellular calcium homoeostasis leads to damage to skeletal muscle has been the subject of a large number of experimental studies. In 1976 Wrogeman and Pena proposed that increased intracellular calcium leads to mitochondrial overload compromising oxidative phosphorylation and hence cellular energy supplies. They suggested that this was a general mechanism for cell necrosis in muscle diseases. It is undoubtedly true that cellular calcium overload causes damage to mitochondria because this can be seen by electron microscopic examination following treatment of muscle with the calcium ionophore (e.g. see Duncan and Jackson, 1987), however this does not appear to play a role in the relatively rapid processes by which muscle cells become non-viable following calcium overload. It is apparent from inhibitor studies that accumulation of calcium by mitochondria is not a prerequisite for the damage leading to cytosolic enzyme efflux from muscles (Jackson *et al.*, 1984) and, in studies of isolated muscles by ^{31}P-NMR techniques, West-Jordan *et al.* (1990) were able to demonstrate a stimulation of creatine kinase efflux from skeletal muscle by treatment with the calcium ionophore, although no short-term effects on muscle ATP and phosphocreatine content were seen.

Another process which has been extensively quoted as a mechanism by which a raised intracellular calcium can cause skeletal muscle damage is by activation of calcium-dependent proteases in the cytoplasm (Ebashi and Sugita, 1979). Calcium-dependent proteases are thought to cause certain characteristic patterns of damage in skeletal muscle including selective loss of Z line material (Busch *et al.*, 1972) and this form of damage is apparent in a number of experimental models of muscle damage (Jones *et al.*, 1984; Duncan *et al.*, 1979). However inhibitors of these proteases, such as leupeptin, are incapable of preventing various features of calcium-induced damage to muscle cells (Jackson *et al.*, 1984; Duncan *et al.*, 1979) suggesting that this process is not the only mechanism by which calcium caused cellular damage in these systems.

We have been particularly interested in the possibility that increased intracellular calcium results in cytosolic enzyme release from skeletal muscle by influencing membrane lipid metabolism. Particular attention has been paid to the possible role of phospholipase A_2 enzymes in this respect. It was initially observed that several non-specific inhibitors of phospholipase A enzymes prevented or greatly reduced cytosolic enzyme efflux from muscles treated with mitochondrial inhibitors (Jackson *et al.*, 1984) and this work was subsequently extended by the demonstration that prostaglandins E_2 and $F_{2\alpha}$ were released from muscle during calcium overload (Jackson *et al.*, 1987a). Since the rate-limiting step for prostaglandin production by tissues is release of the precursor fatty acid (arachidonic acid) from phospholipids by phospholipase A_2 or C activity, this provided confirmation that calcium accumulation by muscle cells activated phospholipases. Further inhibitor studies in this area suggested that the fatty acids released by phospholipase activity might be acted on by lipoxygenase enzymes to produce a product or products which influenced membrane permeability (Jackson *et al.*, 1987a), but our inability to detect lipoxygenase activity in skeletal muscle and the multiple actions of putative inhibitors of lipoxygenase (Jackson, 1989) suggest that this is unlikely to be true. It therefore appears that calcium overload activates phospholipase enzymes (probably phospholipase A_2) forming lysophospholipids and free fatty acids from membrane phospholipids. Following physiological activation of this enzyme, rapid reacylation of the lysophospholipid will occur, but with a pathological rise in intracellular calcium, reacylation may not be able to compensate for continued activation of the enzyme. In cardiac tissue it has been suggested that the processes of reacylation are also inhibited during the damaging process making net breakdown of membrane phospholipids more likely (Chien *et al.*, 1984; Das *et al.*, 1985) but this possibility does not appear to have been examined in skeletal muscle. Since lysophospholipids are disruptive of membrane organization and free fatty acids can have detergent effects on membranes, accumulation of these substances in cells is inevitably damaging.

Studies of the effects of putative inhibitors of phospholipase enzymes on the ultrastructural features of calcium-induced damage have not demonstrated any protective effects even where evidence for inhibition of cytosolic enzyme efflux was presented (Duncan and Jackson, 1987) indicating that such processes cannot explain all of the features of calcium-induced damage.

It is apparent from the preceding discussion that none of the proposed mechanisms provides an explanation for all of the varied effects of calcium overload on the muscle cell. A number of other processes activated by calcium are likely to be involved, such as damage to the cytoskeleton (Needleman *et al.*, 1976) or activation of lysosomal enzymes (Gilbert *et al.*, 1975), but again it is inherently unlikely that any will be of greater importance than those discussed in detail. Rather it is much more likely that an accumulation of intracellular calcium initiates a series of processes all of which are damaging to various parts of the muscle cell. Thus those proposed to be particularly important by various investigators are more likely to be dependent upon the experimental model system chosen and method of assessment of damage rather than any primary role in the processes of damage. Under more physiological conditions it is possible that there will be an order of activation of such processes dependent upon the concentration of calcium required to activate the particular process, but in the type of experimental conditions under which damaging processes are usually studied (e.g. treatment with metabolic inhibitors or calcium ionophore) intracellular calcium will rapidly rise to approximately extracellular concentrations and many processes would be expected to be activated more or less simultaneously.

10.2.3 Role of free radicals

An increase in free radical mediated reactions leading to muscle damage during exercise has been proposed by a number of workers (Dillard *et al.*, 1978; Brady *et al.*, 1979; Gee and Tappel, 1981; Davies *et al.*, 1982), but the role of these reactive materials in damage to skeletal muscle is still not clear. The reason for this is that most studies of free radical activity in tissues rely on non-specific, indirect indicators of free radical activity (Jackson, 1987). In cardiac tissue it has been proposed that increased free radical activity can lead to a failure of calcium homoeostasis with consequent damage to myocytes (Ferrari *et al.*, 1986), but our initial studies suggested that, in skeletal muscle, calcium overload could lead to an activation of free radical-mediated processes. In early studies we examined the effects of variation in the muscle vitamin E content on contractile activity-induced damage to skeletal muscle and demonstrated an apparent protective effect of this vitamin (Jackson *et al.*, 1983). Since the major role of vitamin E appears to be as a lipid-soluble antioxidant, inhibiting free radical-mediated lipid peroxidation, this suggested that free radicals may be involved in the processes of damage in this system. Studies using election spin resonance (e.s.r.) spectrometry initially supported the involvement of free radicals as primary damaging factors in contractile activity-induced damage (Jackson *et al.*, 1985a), but later studies suggested all of the e.s.r. changes could be induced by intracellular calcium overload, possibly by interference with normal mitochondrial processes (Johnson *et al.*, 1988). Recent findings, examining the effects of extracellular α-tocopherol against skeletal muscle damage, confirm the original observations, but they also suggest that it may be acting by a non-antioxidant mechanism (Phoenix *et al.*, 1989). Nevertheless a considerable number of studies reporting protective effects of antioxidants against various forms of skeletal muscle damage have been presented (e.g. see Zerba *et al.*, 1990; Packer and Viguie, 1990) and it seems likely that free radicals may play some role in damage to muscle induced by exercise although the relationship and importance of this compared to a failure of intracellular calcium homoeostasis is unclear.

10.2.4 Mechanisms of muscle damage induced by excessive contractile activity

In order to facilitate later discussion concerning the possible mechanisms of damage in dystrophin-deficient muscle it is relevant to discuss our current understanding of the manner in which excessive isometric or concentric contractile activity induces efflux of intracellular creatine kinase enzymes from normal skeletal muscle. An outline of the processes involved is presented in Figure 10.1 while the evidence that these processes occur is shown in brief in Figure 10.2. In essence it appears that excessive contractile activity causes a fall in muscle ATP content which does not recover in some fibres or parts of some fibres (West-Jordan *et al.*, 1991). This leads to a failure of calcium homoeostasis with a rapid influx of external calcium into the cell (McArdle *et al.*, 1992) and consequent cellular calcium overload. As previously stated this can initiate a series of degenerative pathways of which activation of phospholipase enzymes appears to be important in the processes leading to release of cytosolic enzymes. As shown in Figure 10.2 there is a lag between the accumulation of intracellular calcium and the activation of phospholipase enzymes (as monitored indirectly by release of prostaglandin E_2). The reason for this time lag is not clear, but release of the PGE_2 occurs with a very similar time-course to release of intracellular creatine kinase (McArdle *et al.*, 1991a).

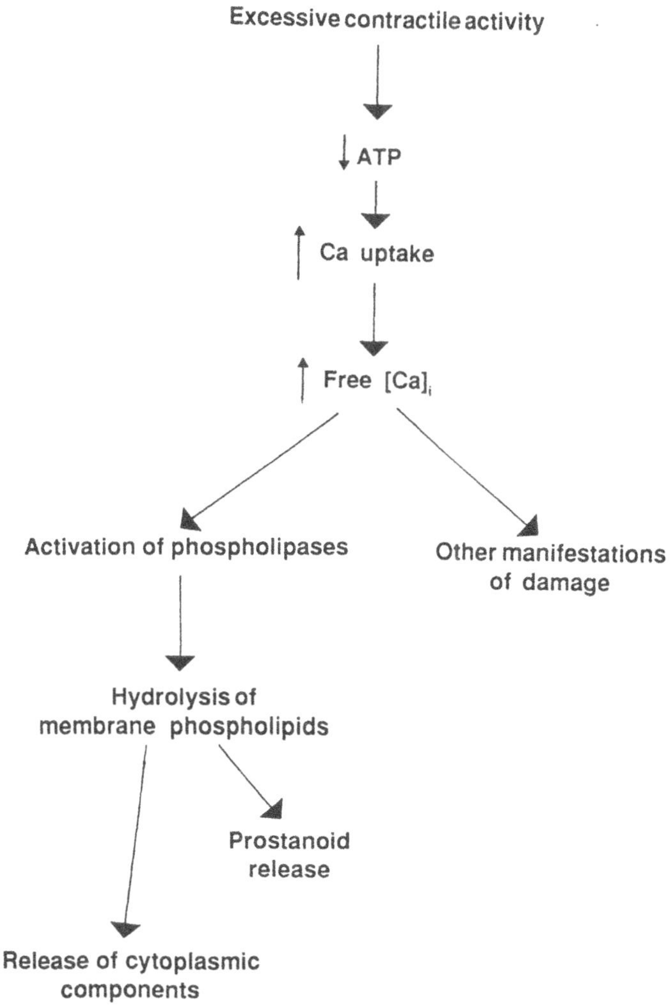

Figure 10.1 Schematic representation of the mechanisms by which excessive contractile activity may induce cytosolic enzyme release from isolated muscles.

There are a number of important aspects of the above process which are unclear. Perhaps the most important of these is the process by which cytosolic enzymes are released. It is commonly assumed that release of large cytoplasmic proteins reflects loss of cell viability, but there are a number of pieces of evidence which argue against this for skeletal muscle. There are a number of clinical studies where patients are found to have elevated circulating activities of muscle-derived enzymes in the absence of other

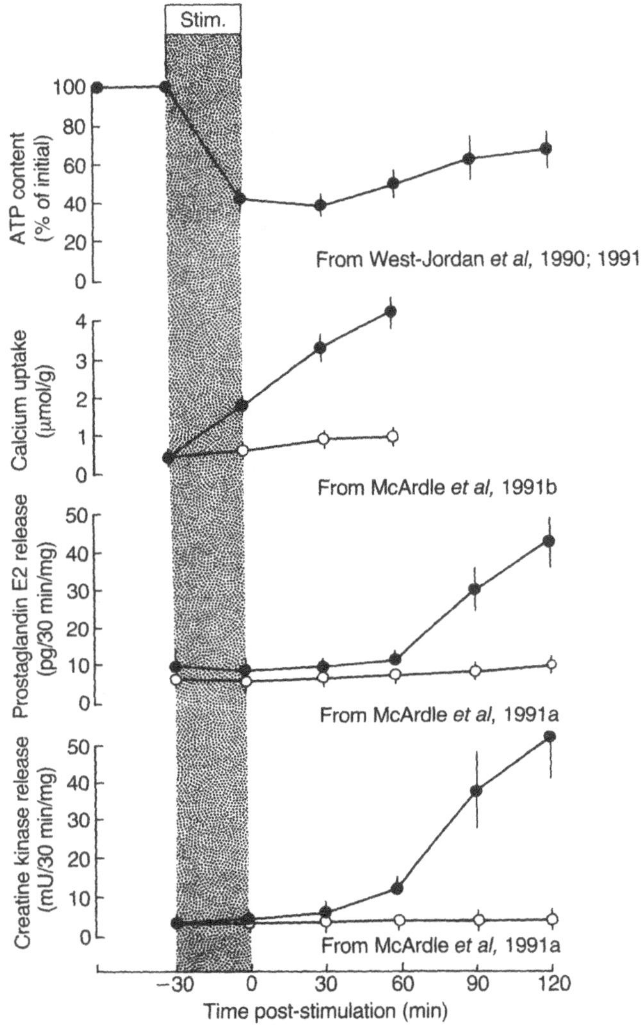

Figure 10.2 Effect of repetitive electrically-stimulated tetani (for 0.5 s every 2 s) on ATP content, calcium uptake, prostaglandin E_2 release and creatine kinase release from isolated rodent muscles. Stimulated muscles (●), non-stimulated control muscles (○). Data derived from West-Jordan *et al.*, 1990, 1991; McArdle *et al.*, 1991a, 1991b, 1992.

evidence of muscle damage (Rowland, 1984). In addition, West-Jordan *et al.* (1990) have reported release of creatine kinase from muscles with normal ATP and phosphocreatine content (see Table 10.1), while we have

Table 10.1 PCr/ATP ratio of isolated soleus muscles treated with agents inducing cytosolic enzyme release

	Time after beginning of treatment (min)		
	0	2	30
Control, untreated	3.5 ± 0.3	3.3 ± 0.2	3.3 ± 0.3
2.4 dinitrophenol (200 μM)	3.5 ± 0.3	0.7 ± 0.4	0
Calcium ionophore A23187 (20 μM)	4.0 ± 0.3	3.7 ± 0.4	3.5 ± 0.2

Data presented as mean ± SEM, from West-Jordan *et al.* (1990)

shown that release of cytosolic enzymes from muscle subjected to various different stresses is not size dependent and appears to represent only a specific portion of cytosolic proteins (Jackson *et al.*, 1991a). It therefore appears that the process of cytosolic enzyme efflux may be more complicated than merely a leakage from necrotic cells, representing perhaps calcium-induced release of a specific group of proteins for unknown reasons.

10.3 MECHANISMS OF DAMAGE IN DYSTROPHIN-DEFICIENT MUSCLE

Following the discovery of dystrophin as the protein product of the Duchenne muscular dystrophy gene (Hoffman *et al.*, 1987a) there was a rapid publication of a number of papers concerned with the cellular location of the protein and its putative function. The current position concerning the cellular location of dystrophin will be discussed fully in other chapters of this book and hence will not be repeated here. However it is relevant to briefly mention some studies in that they explain how some workers arrived at certain conclusions concerning putative functions of dystrophin. In particular initial subcellular fractionation studies suggested that dystrophin was associated with the triads (membrane structures formed by the convergence of the T-tubules and pairs of terminal cisternae of the sarcoplasmic reticulum – Hoffman *et al.*, 1987b; Knudson *et al.*, 1988), while immunocytochemical studies localized dystrophin to the sarcolemmal membrane of the muscle cell (Zubrzycka-Gaarn *et al.*, 1988; Watkins *et al.*, 1988). These studies prompted Hoffman and co-workers (1987b) to suggest a role for triad-localized dystrophin in regulation of muscle calcium metabolism and Karpati and Carpenter (1988) to propose a role for sarcolemmal-localized dystrophin to provide mechanical stability to the sarcolemma of the muscle cell. One of the immediate advances from the discovery of dystrophin as the missing protein in Duchenne dystrophy has been the recognition of certain animal models as being dystrophin-deficient (i.e. the same biochemical defect as patients with Duchenne). Details of the *mdx* mouse and *XMD* dog are provided elsewhere in this book, but from the point of view of the mechanism of myofibre degeneration induced by dystrophin deficiency, the *mdx* mouse provides a unique model in which to test hypotheses and examine potential pathways involved.

10.3.1 Role in providing mechanical stability to the muscle sarcolemma

Following the recognition that dystrophin is localized on the interior of the muscle sarcolemma (Watkins *et al.*, 1988; Cullen *et al.*, 1990) Karpati and Carpenter (1988) proposed that the protein might provide stability to the membrane during normal contractile activity, a protective effect missing in dystrophin-deficient muscle. This theory appears to be largely based on the observation of Edwards *et al.* (1984) that muscles regularly used in an eccentric manner (i.e. postural muscles) suffer the most rapid decline in force in patients with Duchenne and other dystrophies. They proposed that this was due to the initiation of damage by mechanical factors in these muscles and that this could explain the characteristic pattern of weakness in such patients. Karpati and Carpenter (1988) modified this 'mechanical damage hypothesis' to take account of the location of dystrophin and its putative role as a cytoskeletal protein to propose that the protein prevented disruption of the plasmalemma during eccentric contractions by anchoring it to the cytoskeleton (see Figure 10.3 for a schematic representation). In dystrophin-deficient muscle it was proposed that this protective effect was missing, leading to enhanced sensitivity to contraction (particularly eccen-

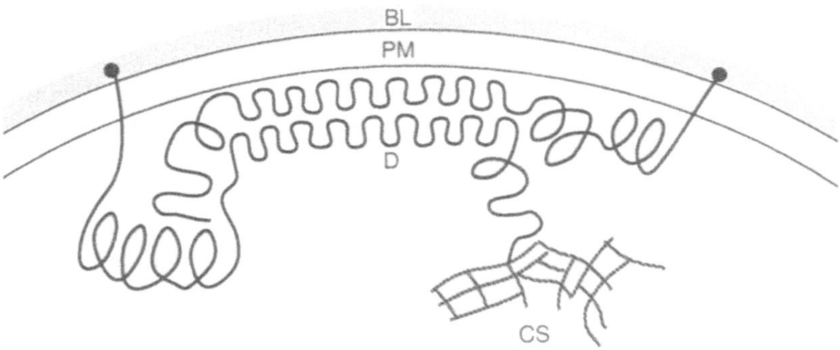

Figure 10.3 Schematic representation of the hypothetical manner in which dystrophin (as a putative antiparallel homodimer) associates with the sarcolemma and the cytoplasmic cytoskeleton to provide mechanical stability to the muscle plasma membrane against the stresses of contractile activity (after Karpati and Carpenter, 1988). D, dystrophin; BL, basal lamina; PM, plasma membrane; CS, cytoskeleton.

tric contraction) induced damage in dystrophin-deficient muscle. Support for such a mechanism comes from studies in the dystrophin-deficient *XMD* dog which appears to be very sensitive to exercise-induced muscle damage (Valentine *et al.*, 1988), but similar effects have not been seen in exercising patients with Duchenne dystrophy (Jackson *et al.*, 1987b).

In an attempt to clarify whether dystrophin-deficient muscle is more susceptible to contraction-induced damage we have examined the response of dystrophin-deficient *mdx* and control mouse extensor digitorum longus muscles to excessive repetitive contractile activity *in vitro* with and without concurrent stretching of the muscle to 130% of its resting length (McArdle *et al.*, 1991a). No evidence for enhanced damage to the *mdx* muscle was observed with either form of activity and indeed *mdx* muscle appeared to be relatively resistant to this form of damage. In related studies Saccho *et al.* (1992) have studied the effect of eccentric contractions on force loss from muscles of *mdx* and control mice *in vivo* and could find no evidence for an enhanced loss of force in the dystrophin-deficient muscle. Taken together these data indicate that there does not appear to be any enhancement of the susceptibility of dystrophin-deficient *mdx* mouse muscle fibres to contractile activity-induced damage. It should be noted that it is entirely reasonable to expect that muscles which show substantial degeneration with replacement of muscle tissue will be more susceptible to stretch or contraction-induced damage because of the enhanced mechanical forces on the remaining intact fibres (this was a key point in the original 'mechanical damage' hypothesis proposed by Edwards *et al.*, 1984) and may explain the differences between the findings in the severely affected *XMD* dog and those in the *mdx* mouse. However, if this were correct, the effects in the *XMD* dog would be secondary to the dystrophic process rather than a primary effect of a lack of dystrophin.

In an alternative approach to this problem Menke and Jockusch (1991) have reported a decreased stability of dystrophin-deficient *mdx* muscle fibres to hypo-osmotic shock. These authors have utilized individual muscle fibres isolated using collagenase for these studies which may have influenced the response of the muscle fibres. Pericellular collagen and the surrounding muscle fibres may provide substantial support for muscle fibres *in vivo* and hence isolation of muscle fibres may have exposed a non-physiological difference between *mdx* and control muscle fibres. Thus, although these data may be important in our understanding of the effects of a lack of dystrophin it will be important to ascertain if there are any physiological correlates to hypo-osmotic shock (Jackson *et al.*, 1991b). In recent studies Hutter and colleagues (Hutter, 1992) have been unable to confirm the findings of Menke and Jockusch (1991) and have undertaken alternative studies to ascertain whether the dystrophin-deficient *mdx* muscle membrane shows reduced stability to stress induced by suction *in vitro*. These have not indicated any differences from control tissue (Hutter *et al.*, 1991; Bovell *et al.*, 1990; Franco and Lansman, 1990).

10.3.2 Role in regulation of muscle calcium metabolism

A role for calcium in the damage to and degeneration of muscle seen in Duchenne muscular dystrophy has been proposed for a considerable time. In 1977 King Engel proposed that a putative sarcolemmal defect in Duchenne muscle would allow a pathological influx of calcium from the extracellular fluid to occur with consequent degeneration and death of the cell. These workers described excess calcium within muscle (and other) cells as the 'ultimate molecular assassin' pushing damaged muscle fibres past their point of no return. A similar mechanism was envisaged by Rowland, (1980, 1984) and Wrogemann and Pena (1976) who envisaged that a defective membrane would allow external calcium to enter the cell initiating degenerative pathways. It should be noted that calcium is playing a secondary role in these processes; their validity was supported by a number of studies examining the calcium content of dystrophic muscle by a histochemical (Bodensteiner and Engel, 1978), electron microscopic (Maunder-Sewry *et al.*, 1980) and biochemical (Jackson *et al.*, 1985b) techniques. All studies demonstrated an increase in calcium content of Duchenne muscle. Examples of the benefits of manipulation of muscle calcium content were also presented by Bertorini and co-workers (Palmieri *et al.*, 1981) examining the (so called) dystrophic hamster, which is now known to have no defect in dystrophin expression.

The initial location of dystrophin to the triads of skeletal muscle appeared to rationalize a role for calcium. Hoffman and co-workers (1987b) proposed that the location of the dystrophin in these structures (which are thought to be involved in regulation of calcium release by sarcoplasmic reticulum) suggested a role for the protein in regulation of muscle intracellular calcium kinetics. However, the location of dystrophin to the triads has not been supported by subsequent studies (Zubrzycka-Gaarn *et al.*, 1988; Watkins *et al.*, 1988; Salviati *et al.*, 1989). Alternative theories put forward to explain how plasma membrane-located dystrophin might modulate cellular calcium content are that it is involved in regulation of stretch-activated (Duncan, 1989) or inactivated (Franco and Lansman, 1990) ion channels of the muscle plasma membrane. The former of these theories does not appear to have any data in support of it and the previously described experiments involving subjecting *mdx* mouse muscles to eccentric contractions (Saccho *et al.*, 1992) and stretching of *mdx* muscle *in vitro* (McArdle *et al.*, 1991a) appear to argue against such a role. Franco and Lansman (1990) provided evidence for novel stretch-inactivated ion channels in the form of 'patch clamp' recordings of ion channel activity from *mdx* muscle surface membranes. Their work suggests an increased chance that ion channels capable of allowing calcium entry to the muscle cell will be open at any point of time in *mdx* surface membranes. Data suggesting an increased activity of an alternative form of channel, a 'calcium-leak' channel, in *mdx* surface membrane has been presented by Steinhardt's group (Fong *et al.*, 1990). This group have also reported that the intracellular free calcium concentration in *mdx* muscle fibres is elevated (Turner *et al.*, 1988) and that this elevation is dependent upon a normal extracellular calcium content. Furthermore these workers were also able to present data indicating that this elevated free intracellular calcium concentration mediated an elevated rate of protein degradation in *mdx* muscles *in vitro*. This latter finding was confirmed by MacLennan *et al.* (1991) who also reported that the elevated rate of muscle protein degradation in *mdx* mice *in vitro* was compensated by a further increase in muscle protein synthesis rates (MacLennan and Edwards, 1990).

In an attempt to define the effect of a lack of dystrophin on *mdx* muscle calcium uptake, we have examined ^{45}calcium uptake by intact *mdx* and control EDL muscles *in vitro*. No evidence for increased ^{45}calcium uptake by *mdx* muscles was observed during incubation alone or during excessive contractile activity (McArdle *et al.*, 1991b, 1992), indeed in common with the studies of muscle damage (as assessed by creatine kinase release) reported earlier (McArdle *et al.*, 1991a) there appeared to be less ^{45}calcium uptake by the *mdx* muscles in response to contractile activity.

It is therefore apparent that the situation concerning a role of dystrophin in modulating muscle calcium accumulation is currently unclear. Although there is evidence in support of defective ion channels in *mdx* muscle membranes leading to an increased inward movement of calcium ions, these data are contradictory, and this influx is not of a sufficient magnitude to be measured by ^{45}calcium uptake. It is arguable that, if the net inward movement of calcium by defective ion channels is small, this cannot be responsible for the gross evidence of ongoing muscle damage seen in patients with Duchenne dystrophy and that, since muscle has a very sophisticated system for regulation of the free intracellular calcium content, it may well be able to compensate for a small net increase in inward flux. However this argument cannot account for the elevated free intracellular calcium content of *mdx* fibres described by Turner *et al.* (1988) which is perhaps the best evidence presented so far in support of a crucial role for calcium in the muscle damage initiated by a lack of dystrophin.

There is however, one further point which must be borne in mind in interpretation of studies utilizing the *mdx* mouse. Adult *mdx* mice consistently display large numbers of muscle fibres which are in the process of regeneration, a factor which should be excluded as a cause of claimed differences between *mdx* and control tissue, but which has been ignored by many authors.

10.3.3 Myofibre 'leakage' as a primary effect of a lack of dystrophin

In discussion of the 'membrane theory' of Duchenne dystrophy, Rowland (1980, 1984) stated that 'the serum enzyme abnormality must be close to the genetic defect'. He did not specifically attribute this abnormality to 'leakage' from the muscle fibre, but concluded that the enzyme efflux might be due to an abnormality in the process by which enzyme efflux occurs from normal muscle. In a recent review Hoffmann and Gorospe (1992) stated that 'the "membrane hypothesis" of Rowland (1980) is verified by the recognition of dystrophin-deficient animals, where plasma membrane leakage is the primary cellular defect' and expanded this idea to suggest that the 'leakage does not necessarily lead to myofibre necrosis'. They also proposed that this 'leakage' manifested itself in the loss of muscle cytoplasmic constituents (such as creatine kinase) to the blood and in an increased permeability to calcium, allowing increased entry from the extracellular fluid.

This simple but attractive hypothesis has also been examined by our group in recent experiments. Studies of CK efflux from isolated strips of Duchenne biceps muscle (Jackson *et al.*, 1991c) indicate that although Duchenne muscle has a substantially increased release of CK compared to control at the commencement of incubation *in vitro*. This release rapidly 'normalizes' such that by 120 mins of incubation the release is not significantly higher than from control. In related work McArdle *et al.* (1991a) have also looked at CK efflux from *mdx* EDL muscles *in vitro* and found that it was generally not significantly elevated compared to that from control muscles. It was also observed that some individual *mdx* muscles released increased amounts of CK during the initial part of the incubation, but this rapidly normalized during the incubation.

It is therefore apparent that despite obvious and gross elevations of circulating CK activities *in vivo* in Duchenne patients and *mdx* mice (which presumably reflect increased release of CK from muscle) there is no evidence of increased release from isolated strips of Duchenne muscles or *mdx* muscles *in vitro* except in the very early stages of incubation. It is apparent that this reduction of CK efflux during incubation is not simply due to depletion of a small mobilizable pool (Jackson *et al.*, 1991c), but may reflect a true reduction in this manifestation of the dystrophic process during short-term incubation of dystrophic muscle *in vitro*.

The other aspect of the 'myofibre leakage' hypothesis is that there is an increased 'leakage' of extracellular calcium into the cell. However measurements of *mdx* muscle calcium uptake using ^{45}calcium do not support an elevated influx of calcium into *mdx* muscles at rest *in vitro* (McArdle *et al.*, 1992). Overall we therefore find no evidence in support of the concept of myofibre 'leakage' as a key result of a lack of dystrophin. However this conclusion must be judged with care since all of the negative findings are derived from *in vitro* studies. It is clear that increased release of CK from Duchenne and *mdx* is a feature seen *in vivo*, but the lack of this finding *in vitro* appears to be worthy of further consideration in terms of providing clues to the mechanisms underlying calcium accumulation and CK efflux from dystrophic muscle (this topic will be considered further in Section 10.4).

10.3.4 'Anchoring' or stabilization of key membrane components as a functional role for dystrophin

Campbell's group have reported that dystrophin is associated in a complex with membrane glycoproteins, at least part of which is missing in the plasma membrane of dystrophin-deficient *mdx* mice (Ervasti *et al.*, 1990; Ervasti and Campbell 1991). Although the cellular function of this glycoprotein complex is not yet clear it appears possible that the function of dystrophin in normal muscle is to 'anchor' these glycoproteins in the plasma membrane. It appears that part of the glycoprotein complex binds laminin (Ibraghimov-Beskrovnaya *et al.*, 1992) and further elucidation of the nature of these glycoproteins may provide clues as the manner in which dystrophin deficiency leads to cellular degeneration. A further effect of dystrophin-deficiency on membrane glycoproteins was described by Evans *et al.* (1990) who reported that the lateral mobility of membrane glycoproteins was abnormally high in *mdx* muscle. This may reflect a generalized change in membrane conformation or fluidity since Barany and Venkasubramian (1988) reported elevated phospholipid mobility in muscle plasma membranes of patients with Duchenne muscular dystrophy. Such changes in the mobility of membrane components may well have significant effects on membrane metabolism.

Membrane phospholipid metabolism, particularly prostaglandin metabolism, appears to be very susceptible to changes in membrane structure in that, for instance, phospholipase A_2 of intracellular origin generally possesses only low activity against bilayer phospholipids, but perturbation of membrane structure can dramatically influence activity (Quinn, 1989). We have shown that prostaglandin E_2 release is abnormally elevated from Duchenne (Jackson *et al.*, 1991c) and *mdx* (McArdle *et al.*, 1991a) muscle, incubated *in vitro* in response to contractile activity or to a deliberate elevation of intracellular calcium with the calcium ionophore. This elevated release does not appear to be due to the presence of infiltrating cells, such as macrophages, in the Duchenne muscle (Jackson *et al.*, 1991c). Since the rate limiting step in the production of prostaglandins by tissues is phospholipid hydrolysis these results imply that phospholipase A_2 (or possibly phospholipase C) activity in dystrophin-deficient muscle is elevated in response to an equivalent stimulus for activation of the enzyme.

Skeletal muscle tissue can produce several prostaglandins (e.g. PGI_2, thromboxane B_2, PGE_2 and $PGF_2\alpha$ – Tian and Baracos, 1989) and it is apparent that stimulation of muscle will lead to increased local production of a range of prostaglandins in dystrophin-deficient muscle compared to control. The effects of prostanoids on muscle are unclear. Previous claims for dramatic regulatory effects of prostaglandins on muscle protein turnover (Reeds and Palmer 1986; Rodemann *et al.*, 1982) have not been substantiated (Barnett and Ellis, 1987; Sugden and Fuller, 1991), however experiments with inhibitors of prostaglandin synthesis suggest that prostanoids may play an important role in muscle development (McLennan, 1985, 1988), in some forms of muscle wasting (Tian and Baracos, 1989) and in certain effects of insulin on muscle (Leighton *et al.*, 1990).

Control of phospholipase activity in muscle *in vivo* is dependent upon activation of the enzyme by calcium, the nature of the substrate and possibly by endogenous inhibitory proteins such as 'lipocortins'. Lipocortins are glycoproteins which act to inhibit phospholipase A_2 activity *in vitro* and possibly *in vivo* (Flower, 1990), although the nature of their action and their specificity is the subject of considerable controversy (Sebaldt *et al.*, 1990; Conricode and Ochs, 1989). They are now recognized to be part of the same 'family' of proteins as calpactins (Davidson *et al.*, 1987). It may be relevant that lipocortin is induced by steroids and that it has now been demonstrated that corticosteroid therapy is of benefit in patients with Duchenne dystrophy (Mendell *et al.*, 1989).

It is therefore apparent that prostaglandin metabolism is abnormal in both human Duchenne and *mdx* mouse muscle and that this may reflect a change in membrane structure or mobility due to a lack of dystrophin. However the exact nature of the abnormality in prostaglandin metabolism or the relevance of this to the pathogenesis of Duchenne or Becker muscular dystrophy is not clear.

10.4 CONCLUSIONS

From the preceding section it will be apparent that although there are a number of theories concerning the manner in which a lack of dystrophin causes muscle damage and degeneration no firm mechanism has yet been established. In the opinion of the author it is clear that calcium overload of the muscle cells is the final pathway leading to the degeneration of the dystrophic muscle cell (Turner *et al.*, 1988; Jackson *et al.*, 1991c), but the mechanism by which the increase in calcium occurs is by no means clear.

There is some evidence that certain ion channels may be abnormal in *mdx* muscle (Franco and Lansman, 1990; Fong *et al.*, 1990) but these have been assessed in *mdx* myofibres *in vitro* and our data indicates that CK efflux is suppressed during incubation of Duchenne and *mdx* muscle *in vitro*. Also it is not possible to detect a significant increase in ^{45}calcium uptake by *mdx* muscle *in vitro* (McArdle *et al.*, 1991b). In recent experiments to try and clarify this position we have studied technetium 99m (^{99m}Tc) pyrophosphate uptake by *mdx* muscles *in vivo* and *in vitro* (Evriviades *et al.*, 1991). ^{99}Tc prophosphate is commonly used to image degenerating muscle in clinical studies (see Jones *et al.*, 1986) and was found to accumulate to a greater extent in *mdx* than control muscles following intravenous injection *in vivo* but not during *in vitro* incubation, thus supporting the idea of a suppression of the rate of degeneration of dystrophic muscle once the muscles are *ex vivo*.

The finding of Campbell's group (Ervasti *et al.*, 1990; Ervasti and Campbell, 1991; Ibraghimov-Beskrovnaya *et al.*, 1992) that dystrophin is associated with membrane glycoproteins and that part of an oligomeric glycoprotein complex is missing in *mdx* muscle appears to be very important. This glycoprotein complex may itself be of crucial importance to membrane function or conversely may be relatively inconsequential in this respect, but provide an indicator of other abnormalities of membrane function. It appears that if a defect in the glycoprotein complex is important in the mechanisms by which cell damage occurs then the most likely effect would be to allow external calcium to enter the cell. One such possibility would be that the glycoprotein complex was (part of) the regulatory apparatus for an ion channel or (part of) the ion channel itself, although recent data from Campbell's group argues against this (Ibraghimov-Beskrovnaya *et al.*, 1992); another would be that the glycoprotein complex interacts with some form of receptor on the cell surface acting to stabilize or control the activity of the receptor. The major prerequisite for such a putative receptor would be that its action in response to agonist was to elevate intracellular calcium by allowing calcium to enter the cell from the external fluid. It could be envisaged that in the absence of dystrophin the receptor became destabilized or uncontrolled allowing excessive calcium to enter the cell in response to agonist. Such a theory could also explain the apparent suppression of CK efflux from *mdx* and Duchenne muscles once they are incubated *in vitro*, since this might remove the muscles from the influence of any putative agonist of the receptor normally stabilized by dystrophin. A schematic representation of how such an interaction might occur is shown in Figure 10.4. This tentative theory provides an explanation for a number of apparently divergent results (i.e. the defective glycoprotein complex in dystrophic muscle, the external calcium dependence of the damage, the abnormal ion channel activity and the suppression of degener-

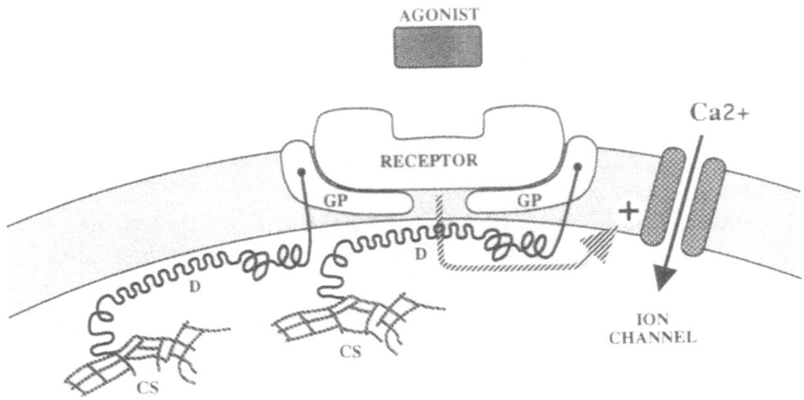

Figure 10.4 Hypothetical role of dystrophin as a stabilizer of a putative receptor in muscle surface membranes. In the presence of agonist this receptor could normally act to allow a transient rise of intracellular calcium by entry of extracellular calcium via ion-channels, whereas in the absence of dystrophin this function could be disturbed by loss of the stabilizing glycoprotein complex allowing excessive entry of extracellular calcium and consequent degeneration of the muscle fibres. D, dystrophin molecules; CS, cytoskeletal proteins; GP, stabilizing glycoproteins.

ation during incubation of muscle *in vitro*), but has no direct data in support of it. It is presented in the hope of stimulating further consideration of possible alternative roles for dystrophin and the search for increased understanding of the mechanism of cellular degeneration in these debilitating diseases.

ACKNOWLEDGEMENTS

The author would like to thank all of his past and present collaborators in this work for their help and co-operation and Professor R.H.T. Edwards for his continued advice and helpful discussions throughout the period during which this work was undertaken. Financial support from the Muscular Dystrophy Group of Great Britain and Northern Ireland, F. Hoffmann La Roche and Co. Ltd and the NATO scientific programme is also gratefully acknowledged.

REFERENCES

Allshire, A., Piper, H.M., Cuthbertson, K.S.R. and Cobbold, P.H. (1987) Cytosolic free Ca^{2+} in single heart cells during anoxia and reoxygenation. *Biochem. J.*, **244**, 381–5.

Anand, R. and Emery, A.E.H. (1980) Calcium stimulated enzyme efflux from human skeletal muscle. *Res. Comm. Chem. Path. Pharmacol.*, **28**, 541–50.

Anand, R. and Emery, A.E.H. (1982) Verapamil and calcium-stimulated enzyme efflux from skeletal muscle. *Clin. Chem.*, **28**, 1482–4.

Armstrong, R.B., Warren, G.L. and Warren, J.A. (1991) Mechanisms of exercise-induced muscle fibre injury. *Sports Medicine*, in press.

Barany, M. and Venkasubramanian, P.N. (1988) Estimation of tissue phospholipids by natural abundance ^{13}C-NMR. *Biochim. Biophys. Acta*, **923**, 339–46.

Barnett, J.G. and Ellis, S. (1987) Prostaglandin E_2 and the regulation of protein degradation in skeletal muscle. *Muscle Nerve*, **10**, 556–9.

Bodensteiner, J.B. and Engel, A.G. (1978) Intracellular calcium accumulation in Duchenne dystrophy and other myopathies: a study of 567 000 muscle fibres in 114 biopsies. *Neurology*, **28**, 439–46.

Bovell, D.L., Burton, F.L. and Hutter, O.F. (1990) Mechanical properties of sarcolemma of murine and human skeletal muscle. *J. Physiol.*, **429**, 4P.

Brady, P.S., Brady, L.J. and Ullrey, D.E. (1979) Selenium, vitamin E and the response to swimming stress in the rat. *J. Nutr.*, **109**, 1103–9.

Busch, W.A., Stromer, M.H., Goll, D.E. and Suzuki, A. (1972) Ca^{2+}-specific removal of Z lines from rabbit skeletal muscle. *J. Cell Biol.*, **52**, 367–81.

Chien, K.R., Han, A., Sen, A. *et al.* (1984) Accumulation of unesterified arachidonic acid in ischaemic canine myocardium. *Circ. Res.*, **54**, 313–22.

Claremont, D., Jackson, M.J. and Jones, D.A. (1984) Accumulation of calcium in experimentally damaged mouse muscles. *J. Physiol.*, **353**, 57P.

Conricode, K.M. and Ochs, R.S. (1989) Mechanism for the inhibitory and stimulatory actions of proteins on the activity of phospholipase A_2. *Biochim. Biophys. Acta*, **1003**, 36–43.

Cullen, M.J., Walsh, J., Nicholson, L.V.B. and Harris, J.B. (1990) Ultrastructural localisation of dystrophin in human muscle using gold immunolabelling. *Proc. Roy. Soc. Lond.*, **B240**, 197–210.

Das, D.K., Engelman, R.M., Ronson, J.A. *et al.* (1985) Role of membrane phospholipids in myocardial injury induced by ischaemia and reperfusion. *Am. J. Physiol.*, **251**, H71–H79.

Davidson, F.F., Dennis, E.A., Powell, M. and Glenney, J.R. Jr. (1987) Inhibition of phospholipase A_2 by 'lipocortins' and calpactins. *J. Biol. Chem.*, **262**, 1698–705.

Davies, K.J.A., Quintanilha, A.H., Brooks, G.A. and Packer, L. (1982) Free radicals and tissue damage produced by exercise. *Biochem. Biophys. Res. Comm.*, **107**, 1198–205.

Dillard, C.J., Litov, R.E., Savin, W.M. *et al.* (1978) Effects of exercise, vitamin E and ozone on pulmonary function and lipid peroxidation. *J. App. Physiol.*, **45**, 927–32.

Duncan, C.J. (1989) Dystrophin and the integrity of the sarcolemma in Duchenne muscular dystrophy. *Experientia*, **45**, 175–7.

Duncan, C.J., Smith, J.L. and Greenway, H.C. (1979) Failure to protect frog skeletal muscle from ionophore-induced damage by the use of the protease inhibitor leupeptin. *Comp. Biochem. Physiol.*, **63C**, 205–7.

Duncan, C.J. and Jackson, M.J. (1987) Different mechanisms mediate structural changes and intracellular enzyme efflux following damage to skeletal muscle. *J. Cell. Science*, **87**, 183–8.

Ebashi, S. and Sugita, H. (1979) The role of calcium in physiological and pathological processes of skeletal muscle, in *Current Topics in Nerve and Muscle Research*, (eds A.J. Aguayo and G. Karpati), Excerpta Medica, Amsterdam, pp. 73–84.

Edwards, R.H.T., Jones, D.A., Newham, D.H. and Chapman, S.J. (1984) Role of mechanical damage in pathogenesis of proximal myopathy in man. *Lancet*, 548–51.

Ervasti, J.M. and Campbell, K.P. (1991) Membrane organisation of the dystrophin-glycoprotein complex. *Cell*, **66**, 1121–31.

Ervasti, J.M., Ohlendieck, K., Kahl, S.D. *et al.*, (1990) Deficiency of a glycoprotein component of the dystrophin complex in dystrophic muscle. *Nature*, **345**, 315–19.

Evans, C.A., Gordon, J.F., Hutter, O.F. and Kusel, J.R. (1990) Lateral mobility of membrane glycoproteins in chick myotubes and in skeletal muscle fibres isolated from normal and mdx mice. *J. Physiol.*, **429**, 54P.

Evriviades, D., Swift, A., McArdle, A. *et al.* (1991) Reduction in the rate of degeneration of dystrophic muscle during incubation *in vitro*. *Clin. Sci.*, **81**, 32P.

Farris, M.W., Pascoe, G.A. and Reed, D.J. (1985) Vitamin E reversal of the effect of extracellular calcium on chemically-induced toxicity in hepatocytes. *Science*, **227**, 751–4.

Ferrari, R., Ceconi, C., Curello, S. *et al.* (1986) Intracellular effects of myocardial ischaemia and reperfusion: role of calcium and oxygen. *Europ. Heart J.*, **7** (Suppl. A.), 3–12.

Flower, R.D., (1990) Lipocortin. *Prog. Clin. Biol. Res.*, **349**, 11–25.

Fong, P.Y., Turner, F.R., Denerclaw, W.F. and Steinhardt, R.A. (1990) Increased activity of calcium leak channels in myotubes of Duchenne human and *mdx* mouse origin. *Science*, **250**, 673–6.

Franco, A. and Lansman, J.B. (1990) Calcium entry through stretch-inactivated ion channels in *mdx* myotubes. *Nature*, **344**, 670–3.

Gee, D.L. and Tappel, A.L. (1981) The effect of exhaustive exercise on expired pentane as a measure of *in vivo* lipid peroxidation in the rat. *Life Sciences*, **28**, 2425–9.

Gilbert, D.S., Newby, B.J. and Anderton, B.H. (1975) Neurofilament disguise, destruction and discipline. *Nature*, **256**, 586–9.

Hoffman, E.P., Brown, R.H. and Kunkel, L.M. (1987a) Dystrophin: The protein product of the Duchenne muscular dystrophy locus. *Cell*, **51**, 919–28.

Hoffman, E.P., Knudson, C.M., Campbell, K.P. and Kunkel, L.M. (1987b) Subcellular fractionation of dystrophin to the triads of skeletal muscle. *Nature*, **330**, 754–8.

Hoffman, E.P. and Gorospe, J.R.M. (1992) The animal models of Duchenne muscular dystrophy: windows on the pathophysiological consequences of dystrophin deficiency, in *Current Topics in Membrane Research*, Vol 38, (eds

M. Mooseker and J. Morrow), Academic Press, New York. pp. 113–54.

Hutter, O.F. (1992) The membrane hypothesis of Duchene muscular dystrophy; quest for functional evidence. *J. Inherit. Metab. Dis.*, in press.

Hutter, O.F., Burton, F.L. and Bovells, D.L. (1991) Mechanical properties of normal and *mdx* mouse sarcolemma: bearing on function of dystrophin. *J. Musc. Res. Cell Motil.*, **12**, 585–9.

Ibraghimov-Beskrovnaya, O., Ervasti, J.M., Leveille, C.J. *et al.* (1992) Primary structure of dystrophin-associated glycoproteins linking dystrophin to the extracellular matrix. *Nature*, **355**, 696–702.

Jackson, M.J., Jones, D.A. and Edwards, R.H.T. (1983) Vitamin E and skeletal muscle, in *Biology of Vitamin E*, Ciba Foundation Symposium No. 101, Pitman, London, pp. 224–39.

Jackson, M.J., Jones, D.A. and Edwards, R.H.T. (1984) Experimental skeletal muscle damage: The nature of the calcium activated degenerative processes. *Europ. J. Clin. Invest.*, **14**, 369–74.

Jackson, M.J., Edwards, R.H.T. and Symons, M.C.R. (1985a) Electron spin resonance studies of intact mammalian skeletal muscle. *Biochim. Biophys. Acta*, **847**, 185–90.

Jackson, M.J., Jones, D.A. and Edwards, R.H.T. (1985b) Measurements of calcium and other elements in needle biopsy samples of muscle from patients with neuromuscular disorders. *Clin. Chim. Acta*, **147**, 215–21.

Jackson, M.J. (1987) Muscle damage during exercise – possible role of free radicals and protective effect of vitamin E. *Proc. Nutr. Soc.*, **46**, 77–80.

Jackson, M.J. (1988) Use of inhibitors in studies of the processes of cytosolic enzyme release from skeletal muscle. *Biochem. J.*, **257**, 621.

Jackson, M.J., Wagenmakers, A.J.M. and Edwards, R.H.T. (1987a) The effect of inhibitors of arachidonic acid metabolism on efflux of intracellular enzymes from skeletal muscle during experimental damage. *Biochem. J.*, **241**, 403–7.

Jackson, M.J., Round, J.M., Newham, D.J. and Edwards, R.H.T. (1987b) An examination of some factors influencing creatine kinase activities in the blood of patients with muscular dystrophy. *Muscle Nerve*, **10**, 15–21.

Jackson, M.J., Page, S. and Edwards, R.H.T. (1991a) The nature of the proteins lost from skeletal muscle during experimental damage. *Clin. Chim. Acta*, **197**, 1–8.

Jackson, M.J., McArdle, A., Edwards, R.H.T. and Jones, D.A. (1991b) Muscle damage in *mdx* mice. *Nature*, **350**, 664.

Jackson, M.J., Brooke, M.H., Kaiser, K. and Edwards, R.H.T. (1991c) Creatine kinase and prostaglandin E_2 release from isolated Duchenne muscle. *Neurology*, **41**, 101–4.

Johnson, K., Sutcliffe, L., Edwards, R.H.T. and Jackson, M.J. (1988) Calcium ionophore enhances the electron spin resonance signal from isolated skeletal muscle. *Biochim. Biophys. Acta*, **964**, 285–8.

Jones, D.A., Jackson, M.J. and Edwards, R.H.T. (1983) The release of intracellular enzymes from an isolated mammalian skeletal muscle preparation. *Clin. Sci.*, **65**, 193–201.

Jones, D.A., Jackson, M.J., McPhail, G. and Edwards, R.H.T. (1984) Experimental muscle damage: The importance of external calcium. *Clin. Sci.*, **66**, 317–22.

Jones, D.A., Newham, D.J., Round, J.M. and Tolfree, S.E. (1986) Experimental human muscle damage: Morphological changes in relation to other indices of damage. *J. Physiol.*, **375**, 435–48.

Karpati, G. and Carpenter, S.C. (1988) The deficiency of a sarcolemmal cytoskeletal protein (dystrophin) leads to necrosis of skeletal muscle fibres in Duchenne-Becker dystrophy, in *Neuromuscular Junction*, (eds L.S. Sellin, R. Libellius and S. Thesloff), Elsevier, Amsterdam, pp. 429–36.

King Engel, W. (1977) Integrative histochemical approach to the defect of Duchenne muscular dystrophy, in *Pathogenesis of Human Muscular Dystrophies*, (ed L.P. Rowland), Excerpta Medica, pp. 277–309.

Knudson, C.M., Hoffman, E.P., Kahl, S.D. *et al.* (1988) Evidence for the association of dystrophin with the transverse tubular system in skeletal muscle. *J. Biol. Chem.*, **263**, 8480–4.

Leighton, B., Challiss, R.A. and Newsholme, E.A. (1990) The role of prostaglandins as modulators of insulin-stimulated glucose metabolism in skeletal muscle. *Horm. Metab. Res. Suppl.*, **22**, 89–95.

McArdle, A., Edwards, R.H.T. and Jackson, M.J. (1991a) Effects of contractile activity on muscle damage in the dystrophin-deficient *mdx* mouse. *Clin. Sci.*, **80**, 367–71.

McArdle, A., Edwards, R.H.T. and Jackson, M.J. (1991b) ^{45}Calcium accumulation by isolated muscles from dystrophin-deficient *mdx* mice. *J. Physiol.*, **434**, 62P.

McArdle, A., Edwards, R.H.T. and Jackson, M.J. (1992) Accumulation of calcium by normal and dystrophin-deficient muscle during contractile activity *in vitro*. *Clin. Sci.*, **82**, 455–9.

McLennan, I.S., (1985) Inhibition of prostaglandin synthesis produces a muscular dystrophy-like myopathy. *Exptl. Neurol.*, **89**, 616–21.

McLennan, I.S. (1988) Characterization of a myopathy caused by prostaglandin dysfunction. *Aust. Paed. J.*, **24** (Suppl. 1), 21–3.

MacLennan, P. and Edwards, R.H.T. (1990) Protein turnover is elevated in muscles of *mdx* mice. *Biochem. J.*, **268**, 795–7.

MacLennan, P., McArdle, A. and Edwards, R.H.T. (1991) Effects of Ca^{2+} on the protein turnover of incubated muscles from *mdx* mice. *Am. J. Physiol.*, **260**, E594–E598.

Maunder-Sewry, C.A., Gorodetsky, R., Yaron, R. and Dubowitz, V. (1980) Elemental analysis of skeletal muscle in Duchenne muscular dystrophy. *Muscle Nerve*, **3**, 502–8.

Mendell, J.R., Moxley, R.T., Griggs, R.C. *et al.* (1989) Randomised double-blind six month trial of prednisolone in Duchenne's muscular dystrophy. *New Eng. J. Med.*, **320**, 1592–7.

Menke, A. and Jockusch, H. (1991) Decreased osmotic stability of dystrophin-less muscle cells from the *mdx* mouse. *Nature*, **349**, 69–71.

Nayler, W.G., Poole-Wilson, P.A. and Williams, A. (1979) Hypoxia and calcium. *J. Mol. Cell Cardiol.*, **11**, 683–706.

Needleman, P., Moncada, S., Bunting, S. *et al.* (1976) Identification of an enzyme in platelet microsome which generates thromboxane A_2 from prostaglandin endoperoxides. *Nature*, **261**, 558–60.

Packer, L. and Viguie, C. (1989) Human exercise: oxidative stress and antioxidant therapy, in *Advances in Myochemistry: 2*, (ed G. Benzi), John Libbey, Eurotext, pp. 1–17.

Palmieri, G.M.A., Nutting, D.F., Bhattacharya, S.K. *et al.* (1981) Parathyroid ablation in dystrophic hampsters. *J. Clin. Invest.*, **68**, 646–54.

Phoenix, J., Edwards, R.H.T. and Jackson, M.J. (1989) Inhibition of calcium-induced cytosolic enzyme efflux from skeletal muscle by vitamin E and related compounds. *Biochem. J.*, **287**, 207–13.

Publicover, S.J., Duncan, C.J., Smith, J.L. (1978) The use of A23187 to demonstrate the role of intracellular calcium in causing ultrastructural damage in mammalian muscle. *J. Neuropath. Exptl. Neurol.*, **37**, 554–7.

Quinn, P.J. (1989) Membrane lipid phase behaviour and lipid–protein interactions. *Biochem. Soc. Trans.*, **18**, 133–6.

Reeds, P.J. and Palmer, R.M. (1986) The role of prostaglandins in the control of muscle protein turnover, in *Control and Manipulation of Animal Growth*, (eds P.J. Buttery, N.B. Hayes and D.B. Lindsey), Butterworths, London, pp. 161–5.

Rodemann, H.P., Waxman, L. and Goldberg, A.L. (1982) The stimulation of protein degradation in muscle by Ca^{2+} is mediated by prostaglandin E_2 and does not require the calcium-activated protease. *J. Biol. Chem.*, **257**, 8716–23.

Rowland, L.P. (1980) Biochemistry of muscle membranes in Duchenne muscular dystrophy. *Muscle Nerve*, **3**, 3–20.

Rowland, L.P. (1984) The membrane theory of Duchenne dystrophy: Where is it? *Ital. J. Neurol. Sci.*, Suppl. 3, 13–28.

Saccho, P., Jones, D.A., Dick, J. and Vrbova, G. (1991) Contractile properties and susceptibility to exercise induced damage of normal and *mdx* mouse tibialis anterior muscle. *Clin. Sci.*, **82**, 227–36.

Salviati, G., Betto, R., Ceoldo, S. *et al.* (1989) Cell fractionation studies indicate that dystrophin is a protein of surface membranes of skeletal muscle. *Biochem. J.*, **258**, 837–41.

Schanne, F.X., Kane, A.B., Young, A.B. and Forber, J.L. (1979) Calcium dependence of toxic cell death: a final common pathway. *Science*, **206**, 700–2.

Sebaldt, R.J., Sheller, J.R., Oates, J.A. *et al.* (1990) Inhibition of eicosanoid biosynthesis by glucocorticoids in humans. *Proc. Natl. Acad. Sci. U.S.A.*, **87**, 6974–8.

Smith, M.T., Thor, H. and Orrenius, S. (1981) Toxic injury to isolated hepatocytes is not dependent on extracellular calcium. *Science*, **213**, 1257–9.

Soybell, D., Morgan, J. and Cohen, L. (1978) Calcium augmentation of enzyme leakage from mouse skeletal muscle and its possible site of action. *Res. Comm. Chem. Path. and Pharmacol.*, **20**, 317–29.

Steer, J.H., Mastaglia, F.L., Papadimitriou, J.M. and Von Bruggen, I. (1986) Bupivicaine-induced muscle injury: The role of extracellular calcium. *J. Neurol. Sci.*, **73**, 205–17.

Sugden, P.H. and Fuller, S.J. (1991) Regulation of protein turnover in skeletal and cardiac muscle. *Biochem. J.*, **273**, 21–37.

Tian, S. and Baracos, V.E. (1989) Prostaglandin-dependent muscle wasting during infection in the broiler chick (*Callces domesticus*) and the laboratory rat (*Rattus norvegicus*). *Biochem J.*, **263**, 485–90.

Turner, P.R., Westwood, T., Regan, C.M. and Steinhardt, R.A. (1988) Increased protein degradation results from elevated free calcium levels found in muscle from *mdx* mice. *Nature*, **335**, 735–8.

Valentine, B.A., Cooper, B.J., De La Hunta, A. *et al.* (1988) Canine X-linked muscular dystrophy: An animal model of Duchenne muscular dystrophy: Clinical studies. *J. Neurol. Sci.*, **88**, 89–91.

Watkins, S.C., Hoffman, E.P., Slayter, H.S. and Kunkel, L.M. (1988) Immuno electron microscopic location of dystrophin myofibres. *Nature*, **333**, 863–6.

Weller, B., Karpati, G. and Carpenter, S. (1990) Dystrophin-deficient *mdx* muscle fibres are preferentially vulnerable to necrosis induced by experimental lengthening contractions. *J. Neurol. Sci.*, **100**, 9–13.

West-Jordan, J.A., Martin, P.A., Abraham, R.J. *et al.* (1990) Energy dependence of cytosolic enzyme efflux from rat skeletal muscle. *Clin. Chim. Acta*, **189**, 163–72.

West-Jordan, J.A., Martin, P.A., Abraham, R.J. *et al.* (1991) Energy metabolism during damaging contractile activity in isolated skeletal muscle: A ^{31}P-NMR study. *Clin. Chim. Acta*, **203**, 119–34.

Wrogemann, K. and Pena, S.D.J. (1976) Mitochondrial calcium overload: a general mechanism for cell necrosis in muscle diseases. *Lancet*, **1**, 672–4.

Zerba, E., Komorowski, T.E. and Faulkner, J. (1990) Free radical injury to skeletal muscles of young, adult and old mice. *Am. J. Physiol.*, **258**, c429–c435.

Zubrzycka-Gaarn, E.E., Bulman, D.E., Karpati, G. *et al.* (1988) The Duchenne muscular dystrophy gene product is localised to the sarcolemma of human skeletal muscle fibres. *Nature*, **333**, 466–9.

11

Human dystrophin gene transfer: genetic correction of dystrophin deficiency

GEORGE DICKSON and MATTHEW DUNCKLEY

11.1 SOMATIC GENE THERAPY: AN INTRODUCTION

Somatic gene therapy, as generally conceived, involves reconstituting a biological function by adding a normal gene to somatic, i.e. non-germline, cells which are genetically deficient in that gene product (Friedmann, 1989). Single gene or Mendelian disorders which are recessive (autosomal or X-linked) are particularly attractive candidates for such a therapeutic approach since successful introduction of one normal gene copy would be expected to prevent the pathological phenotype. Moreover, as has been noted for many inborn errors of metabolism, even relatively low levels of residual function (5–10% of normal) may prevent the major clinical pathology associated with complete deficiency (Ledley, 1990). Thus, full restoration of physiological levels of deficient gene product, while desirable, may not be necessary for the clinical efficacy of any particular gene therapeutic strategy.

At the level of the whole organism somatic gene replacement can, in principle, be achieved by transplantation strategies in which a genetically-deficient organ is replaced or repopulated by donor tissue or cells possessing a normal genetic constitution (Hugh-Jones *et al.*, 1984; Partridge *et al.*, 1989). This type of approach in relation to Duchenne muscular dystrophy (DMD) is currently the subject of considerable clinical and experimental interest and is reviewed in detail elsewhere in this volume (see Chapter 12 by Morgan and Watt). In addition, recombinant DNA and molecular cloning technologies have introduced the possibility of selectively replacing or augmenting abnormal genetic make-up by direct somatic gene transfer procedures (Miller, 1990; Verma, 1990). This latter approach represents a truly therapeutic option in which a fundamental biochemical abnormality

Molecular and Cell Biology of Muscular Dystrophy
Edited by Terence Partridge
Published in 1993 by Chapman & Hall, London. ISBN 0 412 43440 7

would be corrected in a patient's own cells (Ledley, 1987a).

In recent years a number of clear-cut stages have been defined in the theoretical and experimental development of gene transfer approaches to disease treatment. The first is the suitability of targeted diseases: ideally recessive, single-gene disorders with significant morbidity for which no adequate therapy currently exists, and for which the cellular site(s) of phenotypic deficiency is accessible to available gene transfer techniques (Ledley, 1987b). Duchenne muscular dystrophy clearly satisfies most of these criteria; however, as discussed below, dystrophin (the protein which is mutated or absent in DMD) is normally expressed in all striated and smooth muscle cells, and also in the brain, (Chelly *et al.*, 1988; Hoffman *et al.*, 1988; Lidov *et al.*, 1990) so accessing all of these widely distributed sites of biochemical deficiency presents enormous technological hurdles. The second requirement in progressing towards gene therapeutic options is for appropriate cloned genes to be available. These should be contained in vectors or delivery vehicles capable of achieving efficient gene transfer and adequate expression of the active gene product in target cells.

Finally, the recombinant gene products must have been demonstrated experimentally to clearly reverse or significantly complement the pathological phenotype in appropriate human cell culture and animal models of genetic deficiency.

Only once these various experimental criteria have been met and evaluated critically can clinical considerations of (i) therapeutic gene transfer protocols, (ii) long-term safety of adopted procedures and (iii) related risk–benefit analysis in individual cases begin to be fully assessed.

11.2 THERAPEUTIC TARGETS IN DUCHENNE MUSCULAR DYSTROPHY

Dystrophin, the normal product of the DMD gene locus, is a large, sarcolemma-associated polypeptide, thought to be a major component of the cortical cytoskeleton in skeletal myofibres (Zubrzycka-Gaarn *et al.*, 1988; Campbell and Kahl, 1989; Ohlendieck and Campbell, 1991a). In addition to skeletal muscle, dystrophin or closely related isoforms are also present in cardiac and smooth muscle cells and in CNS neuronal populations (Hoffman *et al.*, 1988).

While smooth muscle and neural defects have been variably observed in DMD patients, the major debilitating pathophysiology resulting from dystrophin deficiency is the progressive degeneration and loss of skeletal myofibres which becomes life-threatening as respiratory muscle function is compromised. The development of cardiomyopathic features in later stages of the disease is also a major cause of mortality (Rowland, 1985; Wessel, 1990). Thus, dystrophin gene transfer into selected skeletal and (potentially) cardiac muscles would perhaps be considered a minimum therapeutic target in designing gene delivery systems of clinical utility.

In the case of skeletal muscle, a variety of experimental gene transfer approaches have been successfully applied, including DNA-mediated transfection and retroviral-mediated transduction both *in vivo* and *in vitro* (Wolff et al., 1990; Thomason and Booth, 1990; Dickson *et al.*, 1991; Dunckley *et al.*, 1992). With heart cells, however, the only successful approach in terms of long-term gene transfer has been by simple direct injection of DNA into ventricular regions *in vivo* (Acsadi *et al.*, 1991; Kitsis *et al.*, 1991). Unfortunately, the efficiency of these various procedures so far has been, in general, too low to offer any clinical potential. Furthermore, with respect to the smooth muscle and neural features of dystrophin deficiency, procedures allowing generalized and targeted stable gene transfer to these regions have yet to be described.

11.3 RECOMBINANT DYSTROPHIN GENES: THEIR NATURE AND AVAILABILITY

The DMD gene locus on the X chromosome is extremely large, spanning about 2.5 Mb and consisting of over 70 exons (Hoffman *et al.*, 1987; Koenig *et al.*, 1988; Brown and Hoffman, 1988). In addition, evidence has accumulated that it forms a complex transcriptional unit containing specific neural and striated muscle promoter structures (Nudel *et al.*, 1989; Barnea *et al.*, 1990) and that it exhibits alternative splicing pathways, especially in relation to exons encoding the C-terminal region (Domain D) of the dystrophin molecule (Feener *et al.*, 1989; Walsh *et al.*, 1989).

While the extreme size of the native gene structure of course precludes direct cloning within conventional plasmid, phage or eukaryotic viral vectors, it is possible to incorporate the entire gene within a yeast artificial chromosome (YAC) construct (Coffey *et al.*, 1992; Monaco *et al.*, 1992). Given that microinjection or lipid-mediated fusion of YAC constructs into eukaryotic cells has been shown to represent a feasible route of gene transfer (Gnirke and Huxley, 1991; Gnirke *et al.*, 1991; Strauss and Jaenisch, 1992) it may thus in future be possible to entertain the prospect of manipulating and transferring the entire DMD locus isolated from normal tissue into dystrophin deficient cells. However, at present this remains only a theoretical possibility and most effort to date has been invested in constructing and expressing recombinant dystrophin genes based on cDNA clones.

Full-length recombinant cDNAs have now been described encoding authentic human (Dickson *et al.*, 1991) and mouse (Lee *et al.*, 1991) skeletal muscle dystrophin. In the case of the human cDNA the starting point was a panel of five overlapping partial cDNA clones which were pieced together by a sub-cloning strategy to yield a final 12 kb product spanning the entire open reading frame of the human mRNA (~ 14 kb) as it is expressed normally in adult human skeletal muscle (Figure 11.1). In the case of the mouse cDNA, specific reverse transcription primers based on the sequence of mouse dystrophin mRNA were used to produce directly two large overlapping cDNA segments corresponding to 5′ and 3′ halves of the mRNA which were then joined centrally. Both the cDNAs were initially cloned into constitutive, high-level eukaryotic expression plasmids and, following transfection (see below) into 3T3 fibroblasts and COS cells, were indeed shown to encode recombinant dystrophin polypeptides of appropriate size (427 kD) corresponding to the major form of dystrophin observed in adult skeletal muscle.

11.4 PHYSICAL GENE TRANSFER TECHNIQUES

Mammalian cells can be induced to directly take up and express functional recombinant genes by a range of physical procedures involving exposure to DNA co-precipitated with calcium phosphate (Shih *et al.*, 1979), complexed

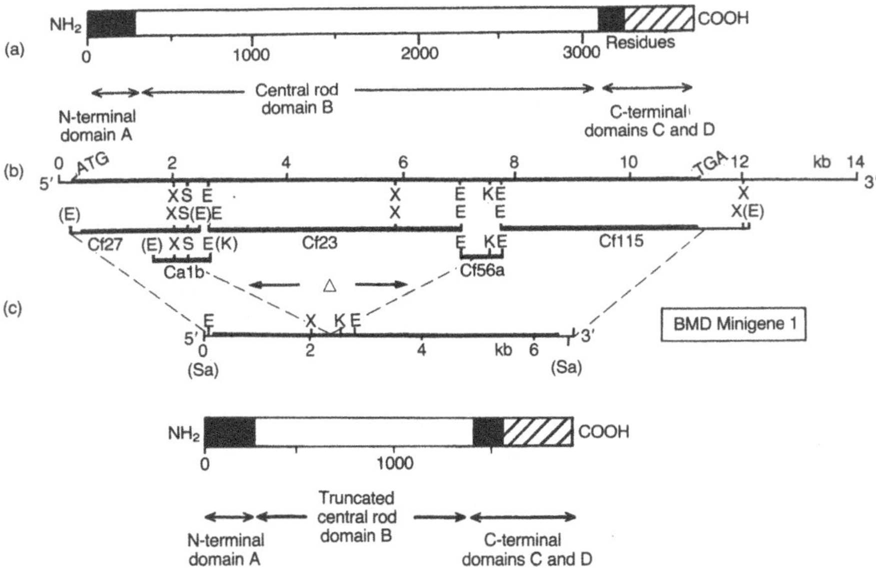

Figure 11.1 Schematic diagram of human dystrophin cDNAs and corresponding proteins. (A) Linear representation of human dystrophin indicating predicted polypeptide domains A–D (Koenig *et al.*, 1988). (B) The intact full-length human cDNA was constructed from an existing panel of cloned cDNA fragments (Dickson *et al.*, 1991). (C) Cloning of a smaller (6.3 kb) dystrophin cDNA based on a mild BMD phenotype. The central in-frame deletion (5.1 kb) is indicated (△) which results in production of a truncated 229 kD dystrophin lacking a large portion of domain B (England *et al.*, 1990; Acsadi *et al.*, 1991).

with polycations and polycationic lipids (Rose *et al.*, 1991), encapsidated into fusogenic liposomes (Mannino and Gould-Fogerite, 1988), or in erythrocyte ghosts (Wiberg *et al.*, 1983). In addition, exposure of cells to rapid pulses of high-voltage current (electroporation) to induce transient membrane permeability (Potter *et al.*, 1984), direct introduction of DNA into cells by microinjection (Graessmann and Graessmann, 1983) or carriage on high-velocity microprojectiles (Johnston *et al.*, 1991) also results in gene transfer and expression in target cells. The efficiency of physical transfection can exceed 1% in suitable recipient cells *in vitro* and the direct transfer of genes into whole animals *in vivo* has also been demonstrated.

With regard to dystrophin cDNA-based gene constructs, direct DNA-mediated gene transfer has been accomplished by various approaches. In cell culture both transient and stable gene transfer has been achieved into fibroblast and myoblast cell lines resulting in expression of appropriately-sized recombinant dystrophin with apparent cell-surface localization (Figure 11.2) (Dickson *et al.*, 1991). Furthermore, transient transfection of dystrophin-deficient mouse (*mdx*) and human (DMD) primary myoblast cultures results clearly in the correct location of transgene product to the plasma membrane of differentiated myotubes following fusion. The most convincing demonstration of the correct targeting of recombinant human dystrophin to the skeletal muscle sarcolemma has been following direct *in vivo* gene transfer via intramuscular injection of DNA constructs into *mdx* mice (Figure 11.3) (Acsadi *et al.*, 1991). This *in vivo* approach was based on earlier observations that injection of purified plasmid containing the *Escherichia coli* β-galactosidase or firefly luciferase genes into rodent skeletal muscle resulted in relatively long-term gene expression in individual myofibres in the area of the injection (Wolff *et al.*, 1990, 1991). The cellular and molecular basis of this gene transfer phenomenon remains poorly understood at present. However, injected DNA appears to persist in skeletal muscle, at least in an unintegrated extra-chromosomal state, and expression at relatively constant levels for over 12 months in the case of the luciferase gene has been obtained (Dr J. Wolff, personal communication). When dystrophin gene constructs were injected into *mdx* skeletal muscle, recombinant gene product was observed at the sarcolemma of myofibres in clusters around the injection site for up to 14 days (the latest time-point examined).

At present this procedure for *in vivo* somatic gene transfer into skeletal muscle is of too low efficiency to be clinically useful: only 1–2% of myofibres were genetically corrected to the dystrophin-positive phenotype following a single injection of DNA into the rectus femoris muscle in the *mdx* mouse. There is, however, considerable scope for technological improvement of this type of approach; for instance, the developmental and physiological status of targeted skeletal muscle tissue may profoundly affect

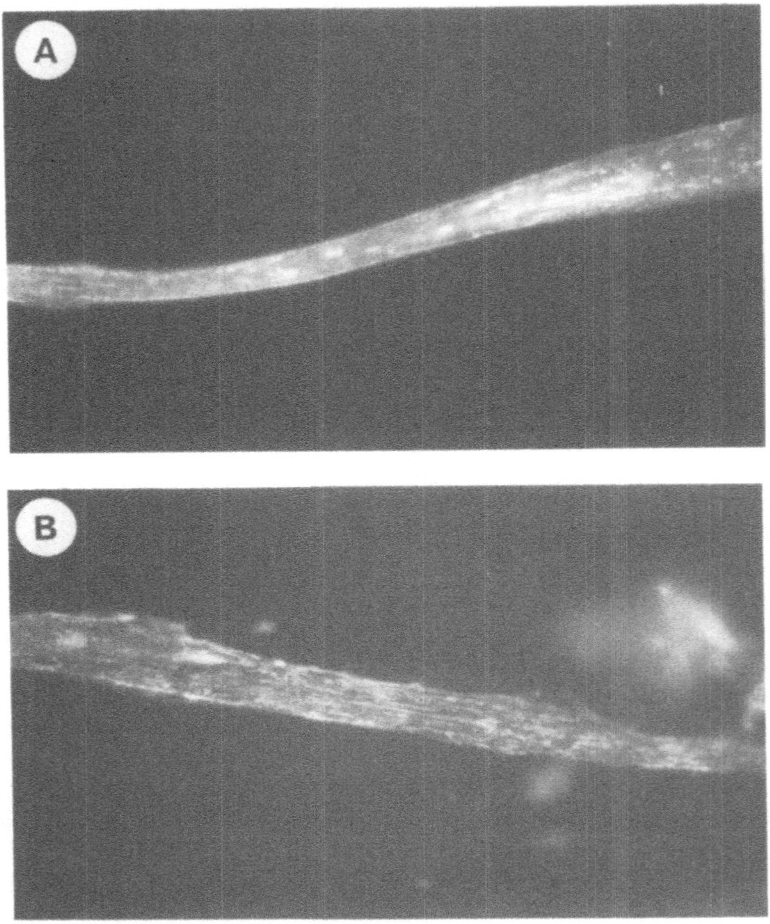

Figure 11.2 Recombinant human dystrophin expression in transfected *mdx* mouse myotubes *in vitro* demonstrating its localization to the sarcolemma in a veined, reticular distribution.

gene transfer efficiency (Dr D. Wells, personal communication). In addition, the use of fusogenic liposomes or tungsten microprojectiles as DNA carriers may enhance *in vivo* DNA delivery (Williams *et al.*, 1991).

Taken at the simplest experimental level the combined *in vitro* and *in vivo* gene transfer studies reported to date have verified the functional integrity of the dystrophin cDNA gene constructs, confirming the nature and sarcolemmal-localization of the recombinant products. Furthermore, in *mdx* myofibres transfected *in vivo* with dystrophin gene constructs, evidence for reduced myofibre regeneration compared with control

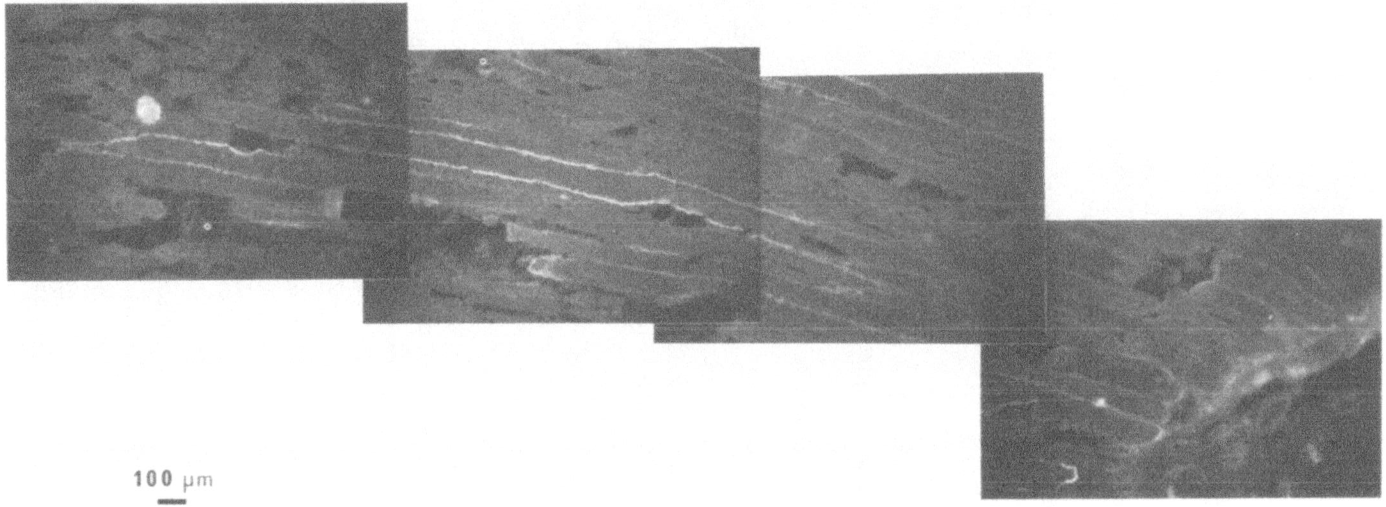

Figure 11.3 Longitudinal section of *mdx* mouse quadriceps following direct *in vivo* injection of a plasmid vector (pDMD1; Dickson *et al*., 1991) containing the full-length human dystrophin cDNA. A number of dystrophin-positive fibres are seen with expression extending up to 1 mm along myofibres.

dystrophin-negative myofibres has been obtained (Acsadi *et al.*, 1991), suggesting that the recombinant products are indeed biologically active.

11.5 VIRAL-MEDIATED GENE TRANSFER STRATEGIES

While direct DNA-mediated gene transfer into mammalian cells using the physical procedures or agents described in the previous section have been enormously valuable experimentally, these processes are currently below the level of efficacy required for any clinical application. However, in recent years the biological properties of various viral systems have been adapted to develop potential high-efficiency somatic gene transfer vectors, including adenoviral- (Berkner, 1988; Rosenfeld *et al.*, 1991) herpes simplex- (Geller *et al.*, 1990), vaccinia- (Fuerst *et al.*, 1986) and retroviral-based approaches (Eglitis *et al.*, 1985). With the exception of the retroviral systems, these vectors have been generally designed and utilized for transient, yet high efficiency, gene transfer studies although it has recently been suggested in the case of adenoviral vectors that a degree of truly long-term transgene expression may be feasible (Rosenfeld *et al.*, 1992). Nevertheless, it remains unclear whether natural immune responses to adenoviral-transduced cells could be circumvented.

The biology of the retroviral life-cycle offers outstandingly attractive features for the development of vectors which lead to highly efficient and stable somatic gene transfer (Miller and Rosman, 1989). Thus replication-defective, yet infectious recombinant retroviral particles can be prepared (Figure 11.4) which successfully infect a wide range of cell-types from diverse species, including humans. Following infection, the RNA within the viral genome is reverse transcribed to DNA and permanently integrated into the host cells' chromosomes from where it can be expressed at high efficiency.

The main disadvantage associated with retroviral and other, e.g. adenoviral, vector systems is their limited ability to accept foreign DNA for gene transfer. The packaging capacity of the retroviral capsid is maximally in the region of ~ 11 kb (Morgenstern and Land, 1991), about 3 kb of which must be accounted for in essential viral promoter, integration and packaging signals. Thus the amount of foreign DNA which can be successfully incorporated into a functional recombinant proviral structure is < 8 kb.

In the case of dystrophin gene transfer, this limitation of retroviral packaging effectively excludes the use of retroviral vectors to transfer the full-length cDNA sequence. One approach to circumvent this problem has, however, been offered by the observation of several cases of Becker muscular dystrophy which involve large in-frame deletion of extensive regions of internal coding sequence corresponding to the central rod domain of the native dystrophin molecule yet nevertheless manifest as an extremely

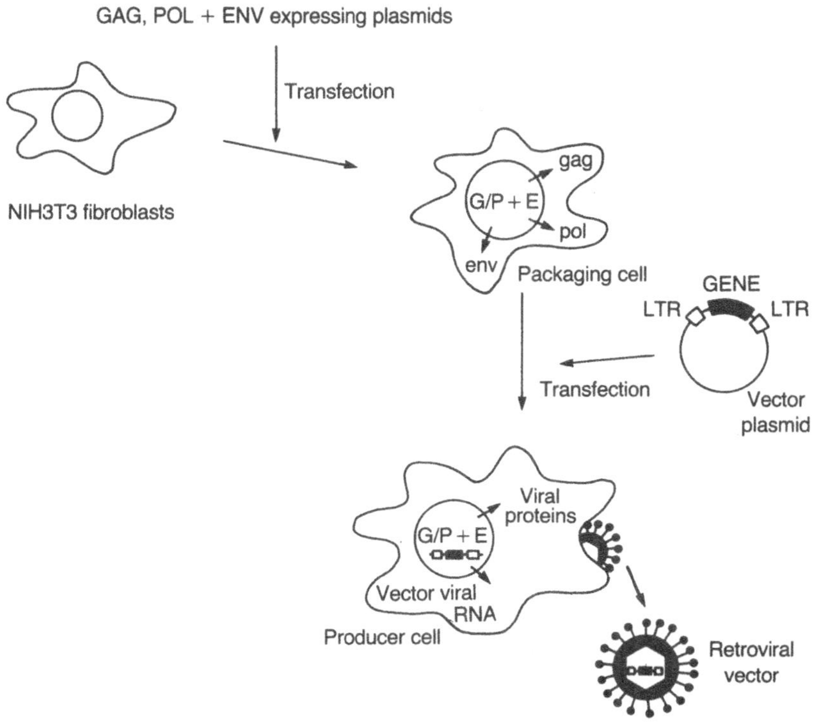

Figure 11.4 Production of recombinant retrovirus using packaging cell lines. The retroviral vector plasmid is transfected into packaging cells possessing appropriate *gag*, *pol*, and *env* genes, where the provirus (between the LTR sequences) is incorporated into an infectious particle or virion. (After Eglitis and Anderson, 1988.)

mild phenotype (England *et al.*, 1990). The protein products of these mutant genes represent naturally-occurring dystrophin molecules of much reduced polypeptide, and hence cDNA, chain length which still appear to be highly biologically active.

One such Becker cDNA has in fact been cloned (Figure 11.1), shown to encode a sarcolemma-associated product of appropriately reduced size (229 kd) and found to be compatible with retroviral-mediated gene transfer (Dunckley *et al.*, 1992). This cDNA of ~ 6.3 kb was cloned into a murine leukaemia virus-based retroviral plasmid and successfully packaged into infectious virions at a primary titre of ~ 10^5 colony-forming units (cfu) per ml. When *mdx* myoblasts in culture were infected with Becker-retrovirus significant transduction of some 10% of myotubes was found when cultures were allowed to differentiate and membrane association of the product was observed. Furthermore, our recent studies have confirmed that direct intramuscular injection of retrovirus in the *mdx* mouse *in vivo* results in the conversion of a proportion of myofibres to dystrophin positivity (Figure 11.5). Thus, the feasibility of retroviral-mediated dystrophin gene transfer has been clearly established.

Since retroviral integration requires active cell division (Miller *et al.*, 1990), it is likely that the *in vivo* transduction of skeletal myofibres is predominantly occurring via infection of proliferating satellite cells (myoblast stem cells) in actively regenerating regions of *mdx* muscle. Indeed, studies with retroviral reporter gene constructs indicated that experimentally-induced muscle regeneration can greatly enhance retroviral-mediated gene transfer (Thomason and Booth, 1990). The on-going pathophysiology of myofibre degeneration and regeneration occurring in young *mdx* mice and during the early myopathic phase in DMD may thus favour incorporation of retroviral vectors into the most severely affected sites of tissue damage. However, for a target tissue so widely distributed as skeletal muscle, intramuscular injection as a clinical approach to administration of gene transfer vectors is feasible only in the context of selected muscle groups. It will therefore be important to examine alternative systemic

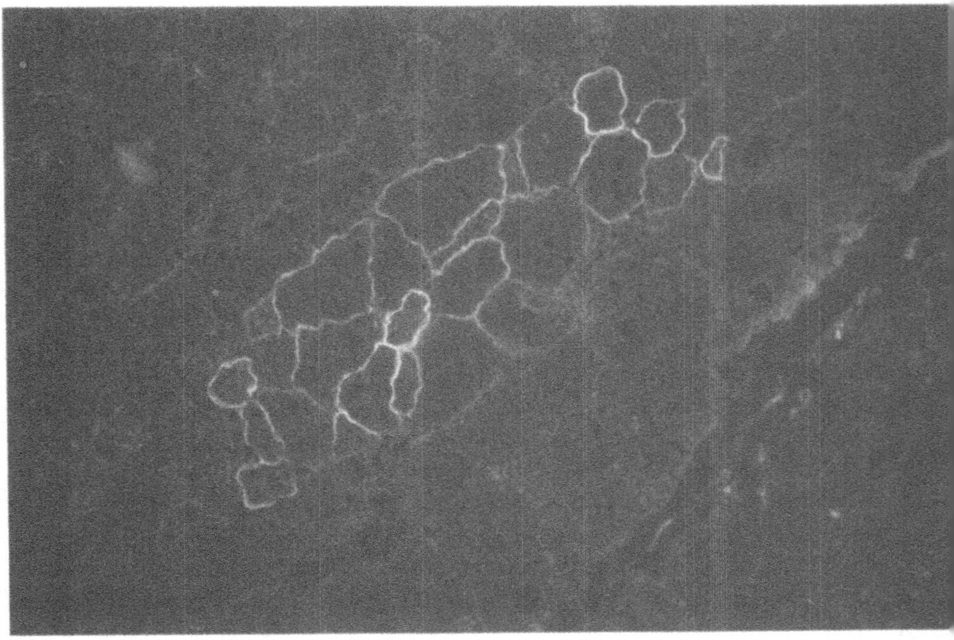

Figure 11.5 Myofibres from an *mdx* mouse showing expression of the BMD dystrophin minigene two weeks after an *in vivo* injection of recombinant murine retrovirus into the quadriceps muscle.

delivery routes for retroviral particles, such as by intravenous or intraperitoneal administration, as described for hepatic gene transfer (Hatzoglou *et al.*, 1990; Dichek *et al.*, 1991). In addition, it might be feasible to implant retroviral producer cells in some form of 'caged' environment or as part of the myofibre syncytium to serve as a chronic endogenous source of recombinant infectious retrovirus. It is, however, likely that prolonged exposure to retroviral particles or repeated dosage would evoke an immunological response in the host organism, and in these circumstances experimental or clinical manipulation of immune surveillance might prove necessary.

The potential systemic dissemination of recombinant retrovirus also raises the question of targeting infectivity and/or gene expression to appropriate cell-types, e.g. skeletal muscle. Native envelope proteins of standard retroviral vectors confer widespread cell-type infectivity, generally of either ecotropic (rodent) or amphotropic (most mammalian species) host ranges (Danos and Mulligan, 1988). However, it remains theoretically possible to produce modified *env* protein specificities (Watanabe *et al.*, 1991; Battini *et al.*, 1992), perhaps by fusion to specific cell-binding ligands, such as immunoglobulins and cell adhesion molecules. Furthermore, the core promoter/enhancer elements of the dystrophin gene itself, which confer muscle- or brain-specific expression, have been isolated (Nudel *et al.*, 1989; Klamut *et al.*, 1990) and appear to be sufficiently small in size to permit inclusion into retroviral constructs (Miller and Rosman, 1989; Dunckley *et al.*, unpublished observation). Thus, it may be possible to design retroviral vectors which possess both cell-type specific infectivity and gene expression.

11.6 STUDIES IN TRANSGENIC ANIMALS: GERM LINE GENE TRANSFER

The various gene transfer procedures and strategies covered so far in this chapter are designed towards the uptake and expression of exogenous DNA in normal diploid somatic cells in a fashion which will not result in transmission of a new genotype to progeny. An alternative gene transfer strategy involves the stable transfer of genetic material into the germ line of recipient organisms to produce transgenic animals with a heritably-altered genotype (Hogan *et al.*, 1986). For practical and ethical reasons this approach is not applicable to humans, but in the *mdx* mouse we have an experimental animal model of DMD (Stedman *et al.*, 1991; Partridge, 1991) and germ line transfer of putative therapeutic gene constructs is required to fully demonstrate the biological activity of corresponding recombinant products.

The BMD-based minigene which, as described above, has considerable promise as a therapeutic gene transfer reagent in conjunction with viral vector systems, has been examined by production of transgenic *mdx* mice, (Wells *et al.*, 1992). Although full physiological levels of recombinant gene expression were not obtained in the animals studied, transgene expression was nevertheless correlated with a marked reduction in the skeletal muscle necrosis and regeneration which is the main pathological feature of dystrophin-deficiency in the *mdx* mouse (Figure 11.6). Thus, the Becker gene indeed expresses a highly functionally-active polypeptide despite its reduced size (Figure 11.1). In addition, the transgenic study indicates that restoration of full physiological levels of dystrophin expression may not be necessary to effect a significant functional correction of the dystrophic phenotype.

Similar germ line transfer studies with the full-length mouse and human cDNAs are ongoing at present (Lee *et al.*, 1990; Dickson *et al.*, unpublished), but it is apparent from many studies that repeated sub-cloning of, in particular, large segments of DNA in bacteria can result in the introduction of inadvertent deletions and rearrangements (Yanisch-Perron *et al.*, 1985). The quality control of novel gene constructs incorporating improved gene structure and expression characteristics, in terms of restriction mapping, direct sequencing and functional complementation analysis in transgenic mice will remain vital checks of construct integrity, authenticity and safety. Furthermore, the efficacy and suitable regulation of selected promoter elements can be examined (Klamut *et al.*, 1990) and questions addressed as to the potential deleterious effects of inappropriate or excessive dystrophin production.

11.7 DYSTROPHIN: STRUCTURE AND FUNCTION

Based on primary sequence homologies and predicted secondary structure analysis, the dystrophin molecule is thought to consist of four clearly defined sub-domains (Hoffman *et al.*, 1987; Koenig *et al.*, 1988): an N-terminal putative actin-binding domain (domain A) (Levine *et al.*, 1990); an extended central rod of coiled coil structure (domain B); a cysteine-rich region with a probably-degenerate Ca^{++}-binding EF-hand motif; and a C-terminal domain which undergoes extensive alternative splicing (Feener *et al.*, 1989). In brain tissue, use of an alternative promoter and first exon produces an altered N-terminus (Nudel *et al.*, 1989) and more recently evidence of other promoter structures deep within the gene have emerged (Bar *et al.*, 1990). On the basis of sub-cellular fractionation, immunolocalization, and biochemical properties, skeletal muscle dystrophin appears to be a major component (~ 5%) of the sub-sarcolemmal cytoskeleton in this tissue (Ohlendieck and Campbell, 1991a) and its absence in the *mdx* mouse or human DMD muscle cells has been suggested to result in reduced membrane stability (Menke and Jockusch, 1990), and elevated intracellu-

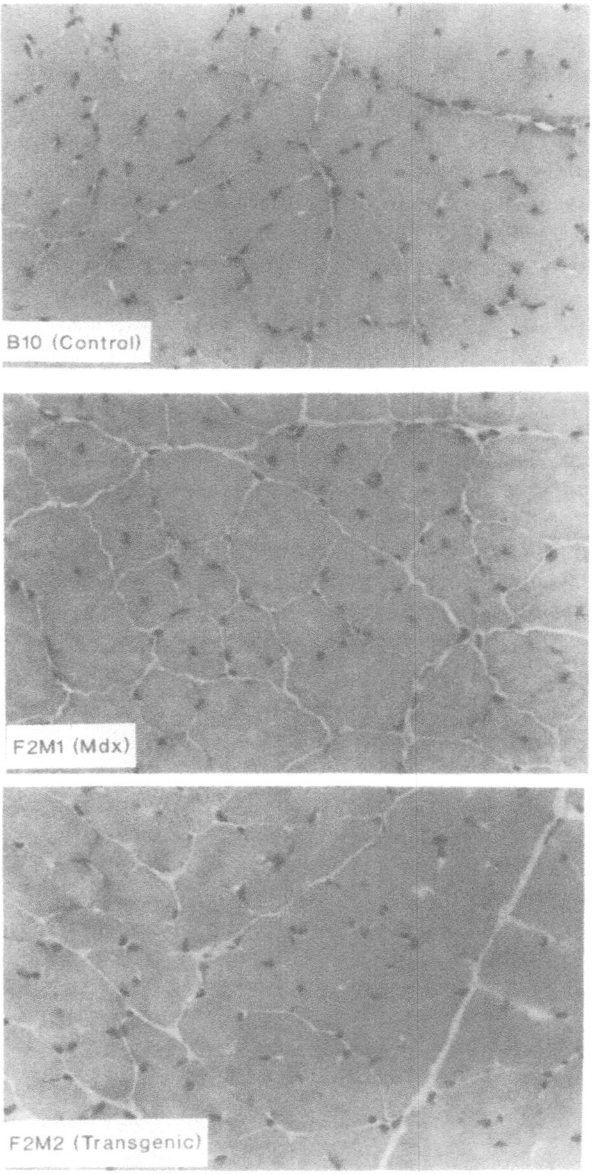

Figure 11.6 Haematoxylin and eosin stains of quadriceps muscle from (a) normal C57/B10, (b) 4-week-old *mdx* (F2M1), and (c) 4-week-old *mdx* transgenic (F2M2) mice demonstrating a significant reduction in the number of centrally nucleated myofibres (characteristic of regenerated muscle) to near-normal levels in mice expressing recombinant human dystrophin.

lar Ca^{++} levels (Franco and Lansman, 1990; Fong *et al.*, 1990; Turner *et al.*, 1991).

Dystrophin appears to be associated at the molecular level with a group of sarcolemma-associated proteins (Ervasti and Campbell, 1991) which are severely reduced in dystrophin-deficient muscle (Ohlendieck and Campbell, 1991b). In addition, recent evidence suggests that it may be linked across the lipid bilayer with laminin in the myofibre basal lamina (Ibraghimov-Beskrovnaya *et al.*, 1992; Dickson *et al.*, 1992). It still remains unclear, however, whether the increased membrane susceptibility to osmotic shock and the observed elevations in intracellular Ca^{++} in *mdx* myofibres *in vitro* result directly from dystrophin deficiency or are secondarily attributable to the regenerative state of the fibres.

At the molecular level the domain(s) of the dystrophin molecule which interacts with dystrophin-associated glycoproteins remains to be defined and while the C-terminal regions of dystrophin have been suggested to mediate this membrane association, recent studies have indicated that some patients deleted for this region still retain dystrophin immunoreactivity at the sarcolemma (Hoffman *et al.*, 1991; Helliwell *et al.*, 1992). It is likely that gene transfer studies to complement dystrophin deficiency in conjunction with analysis of membrane stability and intracellular Ca^{++} levels, e.g. in cultured myofibres, will result in a clear definition of the primary cellular lesions resulting from dystrophin deficiency. Also, site-directed deletion mutation analyses or domain swapping experiments (Critchley *et al.*, 1991), coupled with gene transfer *in vitro* and *in vivo*, will allow the regions of the dystrophin molecule which mediate interaction with other sarcolemmal and cyoskeletal components to be defined. In this way, gene transfer approaches will advance our understanding of the structural biochemistry and cellular function of dystrophin in skeletal muscle. Such experiments will also allow properties of different dystrophin isoforms which result from alternative promoter use to be compared, and for the biochemical and functional consequences of RNA splicing patterns to be examined.

11.8 GENE THERAPY FOR DUCHENNE MUSCULAR DYSTROPHY: FUTURE PROSPECTS

Much progress has been made in recent years in developing appropriate gene constructs and gene transfer techniques suitable for correction of dystrophin deficiency in skeletal muscle fibres. In terms of clinically useful gene therapeutic strategies towards the treatment of DMD, however, difficulties remain in relation to both qualitative and quantitative aspects. We have in hand the appropriate reagents to be transferred. Now the major obstacles are devising vector or delivery systems to provide sufficiently widespread and high efficiency gene transfer. Improvements in physical delivery systems involving, for example, specific fusogenic liposomes (Mannino and Gould-Fogerite, 1988) prepared from the lipid constituents of fusion-competent myoblasts, or perhaps DNA-coated tungsten microprojectiles (Williams *et al.*, 1991), the so-called 'gene-gun' method, are critical requirements for developing dystrophin gene therapy for specific muscle groupings.

For retroviral-mediated techniques, improved retroviral titres are required, such that maximal widespread infectivity can be achieved with a single dose of retrovirus, avoiding potential immune responses to the recombinant virions. In this respect, it is also conceivable that an immune response might be evoked by the *de novo* expression of dystrophin itself. However, in approximately 40% of human DMD cases and in the *mdx* mouse, so-called 'revertant' myofibres expressing dystrophin immunoreactivity exist in low numbers (Gold *et al.*, 1990; Hoffman *et al.*, 1990). There is no evidence of specific immune responses to these cells and they may in themselves result in the development of appropriate immune tolerance. It is apparent that the effects and consequences of dystrophin deficiency in cardiac, smooth muscle and neural tissues remain unanswered by the present advances. Nevertheless, successful treatment of even limited skeletal muscle groupings, e.g. diaphragm, intercostal, etc., would represent a major step forward in alleviating some of the major debilitating and life-threatening features of DMD.

REFERENCES

Acsadi, G., Dickson, G., Love, D.R. *et al.* (1991) Human dystrophin expression in *mdx* mice after intramuscular injection of DNA constructs. *Nature*, **352**, 815–8.

Bar, S., Barnea, E., Levy, Z. *et al.* (1990) A novel product of the Duchenne muscular dystrophy gene which greatly differs from the known isoforms in its structure and tissue distribution. *Biochem. J.*, **272**, 557–60.

Barnea, E., Zuk, D., Simantov, R. *et al.* (1990) Specificity of expression of the muscle and brain dystrophin gene promoters in muscle and brain cells. *Neuron*, **5**, 881–8.

Battini, J-L., Heard, J.M. and Danos, O. (1992) Receptor choice determinants in the envelope glycoproteins of amphotropic, xenotropic, and polytropic murine leukemia viruses. *J. Virol.*, **66**, 1468–75.

Berkner, K.L. (1988) Development of adenovirus vectors for the expression of heterologous genes. *BioTechniques*, **6**, 616–29.

Brown, R.H. Jr. and Hoffman, E.P. (1988) Molecular biology of Duchenne muscular dystrophy. *TINS*, **11**, 480–4.

Campbell, K.P. and Kahl, S.D. (1989) Association of dystrophin and an integral membrane glycoprotein. *Nature*, **338**, 259–62.

Chelly, J. Kaplan, J-C., Maire, P. *et al.* (1988) Transcription of the dystrophin gene in human muscle and non-muscle tissues. *Nature*, **333**, 858–60.

Coffey, A.J., Roberts, R.G., Green, E.D. *et al.* (1992) Construction of a 2.6-Mb contig in yeast artificial chromosomes spanning the human dystrophin gene using an STS-based approach. *Genomics*, **12**, 474–84.

Critchley, D.R., Gilmore, A., Hemmings, L. *et al.* (1991) Cytoskeletal proteins in adherens-type cell-matrix junctions. *Biochem. Soc. Trans.*, **19**, 1028–33.

Danos, O. and Mulligan, R.C. (1988) Safe and efficient generation of recombinant retroviruses with amphotropic and ecotropic host ranges. *Proc. Natl. Acad. Sci. USA*, **85**, 6460–4.

Dichek, D.A., Bratthauer, G.L., Beg, Z.H. *et al.* (1991) Retroviral vector-mediated *in vivo* expression of low-density-lipoprotein receptors in the Watanabe heritable hyperlipidemic rabbit. *Som. Cell Mol. Genet.*, **17**, 287–301.

Dickson, G., Love, D.R., Davies, K.E. *et al.* (1991) Human dystrophin gene transfer: production and expression of a functional recombinant DNA-based gene. *Hum. Genet.*, **88**, 53–8.

Dickson, G., Azad, A., Morris, G.E. and Walsh, F.S. (1992) Co-localisation and association of dystrophin with laminin at the surface of mouse and human myotubes. *J. Cell Sci.*, **103**, 1223–33.

Dunckley, M.G., Love, D.R., Davies, K.E. *et al.* (1992) Retroviral-mediated transfer of a dystrophin minigene into *mdx* myoblasts *in vitro*. *FEBS Lett.*, **296**, 128–34.

Eglitis, M.A., Kantoff, P., Gilboa, E. and Anderson, W.F. (1985) Gene expression in mice after high efficiency retroviral-mediated gene transfer. *Science*, **230**, 1395–8.

Eglitis, M.A. and Anderson, W.F. (1988) Retroviral vectors for the introduction of genes into mammalian cells. *BioTechniques*, **6**, 608–14.

England, S.B., Nicholson, L.V.B., Johnson, M.A. *et al.* (1990) Very mild muscular dystrophy associated with deletion of 46% of dystrophin. *Nature*, **343**, 180–2.

Ervasti, J.M. and Campbell, K.P. (1991) Membrane-organisation of the dystrophin-glycoprotein complex. *Cell*, **66**, 1121–31.

Feener, C.A., Koenig, M. and Kunkel, L.M. (1989) Alternative splicing of human dystrophin mRNA generates isoforms at the carboxy-terminus. *Nature*, **338**, 509–11.

Fong, P., Turner, P.R., Denetclaw, W.F. and Steinhardt, R.A. (1990) Increased activity of calcium leak channels in myotubes of Duchenne human and *mdx* mouse origin. *Science*, **250**, 673–6.

Franco, A. Jr. and Lansman, J.B. (1990) Calcium entry through stretch-inactivated ion channels in *mdx* myotubes. *Nature*, **344**, 670–3.

Friedmann, T. (1989) Progress toward human gene therapy. *Science*, **244**, 1275–81.

Fuerst, T.R., Niles, E.G., Studier, F.W. and Moss, B. (1986) Eukaryotic transient-expression system based on recombinant vaccinia virus that synthesises bacteriophage T7 RNA polymerase. *Proc. Natl. Acad. Sci. USA*, **83**, 8122–6.

Geller, A.I., Keyomarsi, K., Bryan, J. and Pardee, A.B. (1990) An efficient deletion mutant packaging system for defective herpes simplex virus vectors: Potential applications to human gene therapy and neuronal physiology. *Proc. Natl. Acad. Sci. USA*, **87**, 8950–4.

Gnirke, A. and Huxley, C. (1991) Transfer of the human HPRT and GART genes from yeast to mammalian cells by microinjection of YAC DNA. *Som. Cell Mol. Genet.*, 17, 573–80.

Gnirke, A., Barnes, T.S., Patterson, D. *et al.* (1991) Cloning and *in vivo* expression of the human GART gene using yeast artificial chromosomes. *EMBO J.*, **10**, 1629–34.

Gold, R., Meurers, B., Reichmann, H. *et al.* (1990) Duchenne muscular dystrophy: evidence for somatic reversion of the mutation in man. *J. Neurol.*, **237**, 494–8.

Graessmann, M. and Graessmann, A. (1983) Microinjection of tissue culture cells. *Methods Enzymol.*, **101** 482–92, Academic Press, New York.

Hatzoglou, M., Lamers, W., Bosch, F. *et al.* (1990) Hepatic gene transfer in animals using retroviruses containing the promoter from the gene for phosphoenolpyruvate carboxykinase. *J. Biol. Chem.*, **265**, 17285–93.

Helliwell, T.R., Ellis, J.M., Mountford, R.C. *et al.* (1992) A truncated dystrophin lacking the C-terminal domains is localised at the muscle membrane. *Am. J. Hum. Genet.*, in press.

Hoffman, E.P., Brown, R.H. Jr. and Kunkel, L.M. (1987) Dystrophin: the protein product of the Duchenne muscular dystrophy locus. *Cell*, **51**, 919–28.

Hoffman, E.P., Hudecki, M.S., Rosenberg, P.A. *et al.* (1988) Cell and fibre-type distribution of dystrophin. *Neuron*, **1**, 411–20.

Hoffman, E.P., Morgan, J.E., Watkins, S.C. and Partridge, T.A. (1990) Somatic reversion/suppression of the mouse *mdx* phenotype *in vivo*. *J. Neurol. Sci.*, **99**, 9–25.

Hoffman, E.P., Garcia, C.A., Chamberlain, J.S. *et al.* (1991) Is the carboxyl-terminus of dystrophin required for membrane association? A novel, severe case of Duchenne muscular dystrophy. *Ann. Neurol.*, 30, 605–10.

Hogan, B., Constantini, F. and Lacy, E. (1986) *Manipulating the Mouse Embryo*, Cold Spring Harbor Laboratory, New York.

Hugh-Jones, K., Hobbs, J.R., Chambers, D. *et al.* (1984) Bone marrow transplantation in mucopolysaccharidoses, in *Molecular Basis of Lysosomal Storage Disorders*, (eds J.A. Barranger and R.O. Brady), Academic Press, New York, pp. 411–28.

Ibraghimov-Beskrovnaya, O., Ervasti, J.M., Leveille, C.J. *et al.* (1992) Primary structure of dystrophin-associated glycoproteins linking dystrophin to the extracellular matrix. *Nature*, 355, 696–702.

Johnston, S.A., Riedy, M., De Vit, M.J. *et al.* (1991) Biolistic transformation of animal tissue. *In Vitro Cell. Dev. Biol.*, **27P**, 11–14.

Kitsis, R.N., Buttrick, P.M., McNally, E.M. *et al.* (1991) Hormonal modulation of a gene injected into rat heart *in vivo*. *Proc. Natl. Acad. Sci. USA*, **88**, 4138–42.

Klamut, H.J., Gangopadhyay, S.B., Worton, R.G. and Ray, P.N. (1990) Molecular and functional analysis of the muscle-specific promoter region of the Duchenne muscular dystrophy gene. *Mol. Cell Biol.*, **10**, 193–205.

Koenig, M., Monaco, A.P. and Kunkel, L.M. (1988) The complete sequence of dystrophin predicts a rod-shaped cytoskeletal protein. *Cell*, **53**, 219–28.

Ledley, F.D. (1987a) Somatic gene therapy for human disease: background and prospects. Part I. *J. Pediatrics*, **110**, 1–8.

Ledley, F.D. (1987b) Somatic gene therapy for human disease: background and prospects. Part II. *J. Pediatrics*, **110**, 167–74.

Ledley, F.D. (1990) Clinical application of somatic gene therapy in inborn errors of metabolism. *J. Inher. Metab. Dis.*, **13**, 597–616.

Lee, C.C., Pearlman, J.A., Chamberlain, J.S. *et al.* (1990) Cloning and expression of a mouse dystrophin cDNA. *J. Neurol. Sci.*, **98** (suppl.), 130.

Lee, C.C., Pearlman, J.A., Chamberlain, J.S. and Caskey, C.T. (1991) Expression of recombinant dystrophin and its localisation to the cell membrane. *Nature*, **349**, 334–6.

Levine, B.A., Moir, A.J.G., Patchell, V.B. and Perry, S.V. (1990) The interaction of actin with dystrophin. *FEBS Lett.*, **263**, 159–62.

Lidov, H.G.W., Byers, T.J., Watkins, S.C. and Kunkel, L.M. (1990) Localisation of dystrophin to postsynaptic regions of central nervous system cortical neurons. *Nature*, **348**, 725–8.

Mannino, R.J. and Gould-Fogerite, S. (1988) Liposome-mediated gene transfer. *BioTechniques*, **6**, 682–90.

Menke, A. and Jockusch, H. (1990) Decreased osmotic stability of dystrophin-less muscle cells from the *mdx* mouse. *Nature*, **349**, 69–71.

Miller, A.D. and Rosman, G.J. (1989) Improved retroviral vectors for gene transfer and expression. *BioTechniques*, **7**, 980–90.

Miller, A.D. (1990) Progress toward human gene therapy. *Blood*, **76**, 271–8.

Miller, D.G., Adam, M.A. and Miller, A.D. (1990) Gene transfer by retrovirus vectors occurs only in cells that are actively replicating at the time of infection. *Mol. Cell Biol.*, **10**, 4239–42.

Monaco, A.P., Walker, A.P., Millwood, I. *et al.* (1992) A yeast artificial chromosome contig containing the complete Duchenne muscular dystrophy gene. *Genomics*, **12**, 465–73.

Morgenstern, J.P. and Land, H. (1991) Choice and manipulation of retroviral vectors, in *Gene Transfer and Expression Protocols*, (ed. E.J. Murray), *Methods Mol. Biol.*, **7**, Humana Press, Clifton, New Jersey, pp. 181–205.

Nudel, U., Zuk, D., Einat, P. *et al.* (1989) Duchenne muscular dystrophy gene product is not identical in muscle and brain. *Nature*, **337**, 76–78.

Ohlendieck, K. and Campbell, K.P. (1991a) Dystrophin constitutes 5% of membrane cytoskeleton in skeletal muscle. *FEBS Lett.*, **283**, 230–4.

Ohlendieck, K. and Campbell, K.P. (1991b) Dystrophin-associated proteins are greatly reduced in skeletal muscle from *mdx* mice. *J. Cell Biol.*, **115**, 1685–94.

Partridge, T.A., Morgan, J.E., Coulton, G.R. *et al.* (1989) Conversion of *mdx* myofibres from dystrophin-negative to -positive by injection of normal myoblasts. *Nature*, **337**, 176–9.

Partridge, T. (1991) Animal models of muscular dystrophy – what can they teach us? *Neuropath. Appl. Neurobiol.*, **17**, 353–63.

Potter, H., Weir, L. and Leder, P. (1984) Enhancer-dependent expression of human kappa immunoglobulin genes introduced into mouse pre-B lymphocytes by electroporation. *Proc. Natl. Acad. Sci. USA*, **81**, 7161–5.

Rose, J.K., Buonocore, L. and Whitt, M.A. (1991) A new cationic liposome reagent mediating nearly quantitative transfection of animal cells. *BioTechniques*, **10**, 520–5.

Rosenfeld, M.A., Siegfried, W., Yoshimura, K. *et al.* (1991) Adenovirus-mediated transfer of a recombinant alpha-1-antitrypsin gene to the lung epithelium *in vivo*. *Science*, **252**, 431–4.

Rosenfeld, M.A., Yoshimura, K., Trapnell, B.C. *et al.* (1992) *In vivo* transfer of the human cystic fibrosis transmembrane conductance regulator gene to the airway epithelium. *Cell*, **68**, 143–55.

Rowland, L.P. (1985) Clinical perspective: Phenotypic expression in muscular dystrophy. in *Gene Expression in Muscle*, (eds R.C. Strohman and S. Wolf), *Adv. Exp. Med. Biol.*, **182**, 3–14, Plenum Press, New York.

Shih, C., Shilo, B.Z., Goldfarb, M.P. *et al.* (1979) Passage of phenotypes of chemically transformed cells via transfection of DNA and chromatin. *Proc. Natl. Acad. Sci. USA*, **76**, 5714–8.

Stedman, H.H., Sweeney, H.L., Shrager, J.B. *et al.* (1991) The *mdx* mouse diaphragm reproduces the degenerative changes of Duchenne muscular dystrophy. *Nature*, **352**, 536–9.

Strauss, W.M. and Jaenisch, R. (1992) Molecular complementation of a collagen mutation in mammalian cells using yeast artificial chromosomes. *EMBO J.*, **11**, 417–21.

Thomason, D.B. and Booth, F.W. (1990) Stable incorporation of a bacterial gene into adult rat skeletal muscle *in vivo*. *Am. J. Physiol.*, **258**, C578–C581.

Turner, P.R., Fong, P., Denetclaw, W.F. and Steinhardt, R.A. (1991) Increased calcium influx in dystrophic muscle. *J. Cell Biol.*, **115**, 1701–12.

Verma, I.M. (1990) Gene therapy. *Scientific American*, **263**, 34–41.

Walsh, F.S., Pizzey, J.A. and Dickson, G. (1989) Tissue-specific isoforms of dystrophin. *TINS*, **12**, 235–8.

Watanabe, N., Nishi, M., Ikawa, Y. and Amanuma, H. (1991) Conversion of a Friend mink cell focus-forming virus to Friend spleen focus-forming virus by modification of the 3′ half of the *env* gene. *J. Virol.*, **65**, 132–7.

Wells, D.J., Wells, K.E., Walsh, F.S. *et al.* (1992) Human dystrophin expression corrects the myopathic phenotype in transgenic *mdx* mice. *Hum. Mol. Genet.*, **1**, 35–40.

Wessel, H.B. (1990) Dystrophin: a clinical perspective. *Ped. Neurol.*, **6**, 3–12.

Wiberg, F.C., Sunnerhagen, P., Kaltoft, K. *et al.* (1983) Replication and expression in mammalian cells of transfected DNA: description of an improved erythrocyte ghost fusion technique. *Nucl. Acids Res.*, **11**, 7287–302.

Williams, R.S., Johnston, S.A., Riedy, M. *et al.* (1991) Introduction of foreign genes into tissues of living mice by DNA-coated microprojectiles. *Proc. Natl. Acad. Sci. USA*, **88**, 2726–30.

Wolff, J.A., Malone, R.W., Williams, P. *et al.* (1990) Direct gene transfer into mouse muscle *in vivo*. *Science*, **247**, 1465–8.

Wolff, J.A., Williams, P., Acsadi, G. *et al.* (1991) Conditions affecting direct gene transfer into rodent muscle *in vivo*. *BioTechniques*, **11**, 474–85.

Yanisch-Perron, C., Vieira, J. and Messing, J. (1985) Improved M13 phage cloning vectors and host strains: Nucleotide sequences of the M13mp18 and pUC19 vectors. *Gene*, **33**, 103–19.

Zubrkzycka-Gaarn, E.E., Bulman, D.E., Karpati, G. *et al.* (1988) The Duchenne muscular dystrophy gene product is localised in sarcolemma of human skeletal muscle. *Nature*, 333, 466–9.

12

Myoblast transplantation in inherited myopathies

JENNIFER E. MORGAN and DIANA J. WATT

12.1 INTRODUCTION

Amongst the different forms of disease of human skeletal muscle are those where the lesion and subsequent pathological state is manifest within the muscle tissue itself: these are termed primary myopathies. Within this group are the muscular dystrophies, in which there is progressive degenerative wasting of striated musculature. The most debilitating of the muscular dystrophies is the X-linked Duchenne muscular dystrophy, which has an incidence of 1 in 3000 males at birth. In this disease, atrophy and loss of fibres from both locomotory and respiratory muscles results in their replacement by fibrous connective tissue and fat. There is regeneration of skeletal muscle, but this is insufficient to compensate for the degeneration. Affected boys become progressively weaker, are usually wheelchair-bound by their early teens and die in their late teens or early twenties (Walton and Gardner-Medwin, 1974).

Due to the severity of this myopathy, there has been much effort to find a cure, or, at least, a method to alleviate the condition and improve the quality of life of those suffering from this disease. In recent years attention has focused on the possibility of introducing missing genes or of replacing those which are defective within the skeletal muscle fibre of the affected individual. This sort of approach would employ the implantation of cells, containing the normal genome, into the myopathic muscle. Therapy involving the replacement of a missing gene product was first contemplated in diseases caused by defects in lysosomal storage, such as Hunter's, Sanfillipo A and Hurler's syndromes. In such cases, however, the disease affects so many different tissues of the body, that attempted treatment by subcutaneous implantation of normal fibroblasts had little effect (Dean *et al.*, 1979,

Molecular and Cell Biology of Muscular Dystrophy
Edited by Terence Partridge
Published in 1993 by Chapman & Hall, London. ISBN 0 412 43440 7

1980, 1981; Gibbs *et al.*, 1980), although bone marrow transplantation was more successful (O'Reilly, 1983).

Skeletal muscle obviously differs from bone marrow in that total replacement of its constituent cells is not feasible. However, the way in which this tissue grows and repairs itself, does render it susceptible to incorporation of its precursor cells into the mature tissue. For, the mode of development of skeletal muscle is by fusion of mononuclear precursor cells to form immature multinucleate cells called 'myotubes' (Stockdale and Holtzer, 1961) (Figure 12.1). These mature by the continued acquisition of precursor cells during postnatal life to form the mature, functional skeletal muscle cell (Cardasis and Cooper, 1975). More importantly, in terms of treatment of primary myopathic disease by implantation of normal precursor cells, skeletal muscle regenerates after injury in a way similar to that in which it forms, i.e. by fusion of mononuclear precursor cells (Carlson, 1973). As stated earlier, the muscle fibres of DMD patients undergo extensive degeneration, and although an attempted regeneration occurs, it is insufficient over the long term to compensate for the extensive muscle loss, and fibre atrophy ensues. Thus, it is feasible to contemplate the incorporation of normal donor muscle precursor cells into the fibres of patients suffering from primary inherited diseases of muscle.

As a consequence of implanting normal muscle precursor cells (mpc) into myopathic muscles, donor mpc should become incorporated into the host fibres. These fibres would then be 'mosaic', i.e. myonuclei originating from both the implanted cells and the host fibre would be present within the same common cytoplasm and able to express their own gene products within this fibre. Normal gene products derived from the normal myonuclei should therefore compensate for products which are deficient or abnormal within the myopathic host muscle fibre. If this sort of approach is to be successful, it is first necessary to establish whether injected mpc can be incorporated into a muscle fibre *in vivo*. Early work on mouse muscle has shown that this is indeed the case. It was found that mpc were capable of fusing with regenerating (Watt *et al.*, 1982) and growing (Watt *et al.*, 1984a) muscle in mice. When donor mpc of one allotype of glucose-6-phosphate isomerase (GPI), *Gpi* $1s^a$, were implanted into host muscle of *Gpi* $1s^b$ allotype, a heterodimeric isoform of GPI was detected. This heterodimer, formed by the association of a host subunit of GPI with a donor subunit of GPI (Gearhart and Mintz, 1972) indicated the formation of 'mosaic' host/donor muscle fibres in which donor genes were being expressed.

The above experiments indicate that injected mpc form muscle *in vivo*, an essential prerequisite for cell implantation therapy. It is also imperative that the implanted cells survive and continue to function in the host fibre, producing their muscle-specific proteins. Further evidence that implanted myogenic cells survive the injection procedure and form muscle *in vivo* has

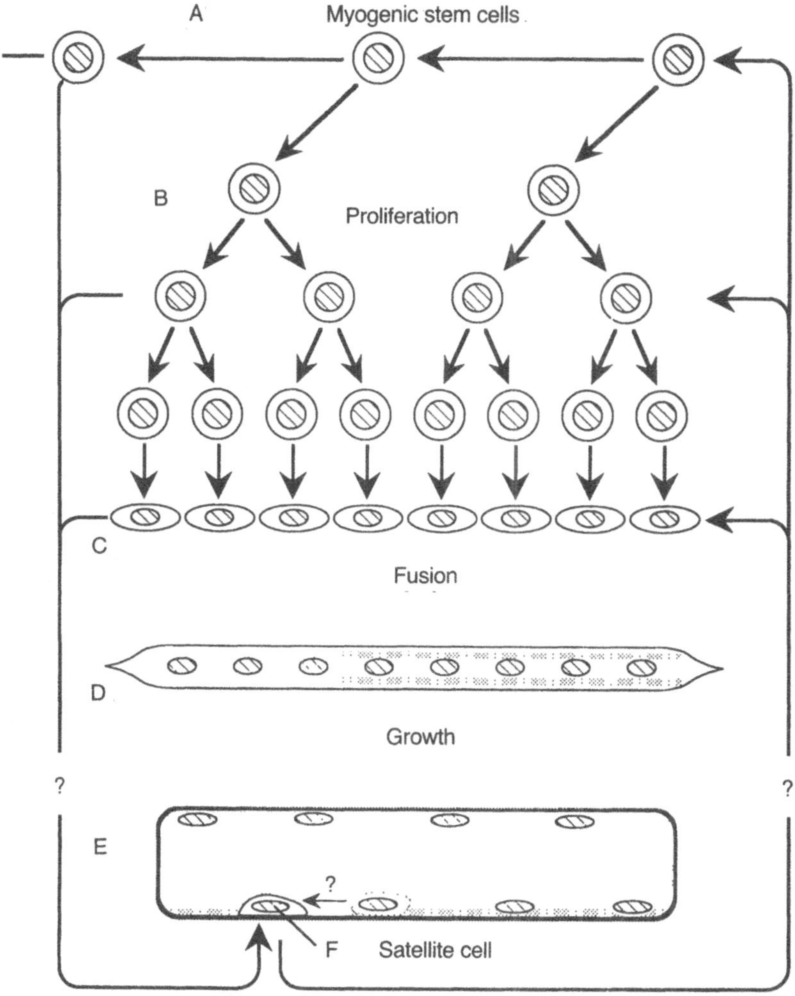

Figure 12.1 Diagrammatic representation of the formation of skeletal muscle. Myogenic stem cells (A) proliferate. Some of these daughter cells remain stem cells, while others (B) give rise to mpc committed to differentiate into skeletal muscle (C). These committed myogenic cells then fuse into multinucleated myotubes (D) which in turn mature into the fully functional striated muscle cell (E). During muscle formation, some myogenic cells remain outside the multinucleated muscle cell, but under its basal lamina (F). These satellite cells are normally quiescent. During muscle growth, however, they divide and some of their daughter cells fuse with the elongating muscle fibre. When muscle is damaged, the satellite cells again divide and fuse to make new muscle fibres. The origin of the satellite cell is uncertain, nor is it known where the progeny of the activated satellite cell fits into the myogenic lineage.

been found in other species such as the quail and the rat (Lipton and Schultz, 1979). Tritiated thymidine-labelled mpc injected into adult muscle retained their myogenic phenotype and became incorporated into host muscle fibres. When autologous mpc were introduced into either uninjured (Jones, 1979) or minced (Ghins *et al.*, 1986) adult rat muscle, implanted cells were capable of forming muscle within their new location. This was also found to be the case when such cells were introduced into whole muscle grafts where endogenous host mpc had been destroyed by X-irradiation (Alameddine *et al.*, 1989).

The effect of a normal myonucleus within a myopathic muscle fibre would be greatest if the normal gene product were disseminated widely throughout the target fibre. Certainly, GPI has the ability to become so distributed: in mosaic muscle fibres containing myonuclei of both the *GPI 1-s*^{a} and *1-s*^{b} allotypes, the isoenzymes were found to be homogeneously mixed in the cytoplasm (Frair and Peterson, 1983). This indicates that the incorporation of only a few normal nuclei should change the GPI phenotype of the entire fibre. A similar result was also obtained when normal mpc were allowed to fuse with those obtained from the muscular dysgenic mouse. In mixed cultures of normal and dysgenic precursor cells, only a very few myonuclei of normal origin were needed to give an apparently normal phenotype in terms of contraction (Peterson and Pena, 1984). In contrast, however, other gene products appear not to have the ability to disseminate so widely throughout the fibre into which they are introduced. In experiments using heterokaryons derived from the fusion of mouse mpc with either human mpc or with human non-muscle cells, Pavlath *et al.* (1989) have shown that several protein products of the human myonuclei are found to be located in close proximity to the nuclei which produced them. This was confirmed by Ralston and Hall (1989), who found that single nuclei have local domains of influence within cultured mosaic myotubes which contained unlabelled nuclei and nuclei derived from C2 cells transfected with a nuclear reporter protein.

The most significant advantage of injecting intact mpc into myopathic muscle is that a full complement of normal genes is introduced into the diseased muscle fibre. It therefore follows that this sort of therapy is applicable to all forms of muscle disease where a gene product is either missing or defective, even though the gene in question, or the biochemical defect, may not yet have been identified. In the case of Duchenne muscular dystrophy, the defective gene has been isolated and cloned (Koenig *et al.*, 1987). This very large gene, in excess of 2.3 kilobases, located on the short arm of the X chromosome, codes for a protein, dystrophin, (Hoffman *et al.*, 1987) which is normally found on the inner face of the plasmalemma of the muscle fibre. The function of dystrophin has not as yet been elucidated, although some evidence suggests that it may have a vital function in maintaining the surface resilience of the muscle fibre (Menke and Jockush, 1991). Other studies suggest that not only is dystrophin missing in the muscles of these patients, but certain glycoproteins normally located on or in the sarcolemma are also deficient (Ervasti *et al.*, 1990).

Dystrophin deficiency is not confined to man, but is also found to occur in various other species. In particular, the discovery of the first genetic and biochemical homologue of DMD (Hoffman *et al.*, 1987), the X-linked muscular dystrophic or *mdx* mouse, meant that the ability of mpc implantation to treat a human myopathy could be critically tested.

12.2 MYOBLAST TRANSPLANTATION IN ANIMAL MODELS OF INHERITED MYOPATHIES

Until the discovery of the *mdx* mouse, other animal models, particularly murine models, had been useful in establishing the general principle of transplantation of normal muscle precursor cells into diseased muscle.

The first widely studied mouse models were the dystrophic variants of the 129/ReJ and C57Bl/6J mice, designated the 129/ReJ *dy*/*dy* and C57Bl/6J dy^{2J}/dy^{2J} dystrophic mutants respectively. These animals suffer from a progressive deterioration of striated muscle, myofibres being replaced by connective and/or fatty tissue (Meier and Southard, 1970; Bourke and Ontell, 1986). The main difference between these two mutants is the less rapid onset and time course of the disease in the dy^{2J} mouse. The dy^{2J} mouse has been extensively used for implantation of precursor cells (Law and Yap, 1979; Law *et al.*, 1982, 1988a, 1990a).

As a model of a primary myopathy, the *dy* mutant is not ideal, for there is demyelination of the dorsal nerve roots (Bradley and Jenkinson, 1973), indicating a neurogenic element to this disease. Further, there is no correlation of muscle phenotype with genotype in chimaeric C57Bl/6J dy^{2J}/dy^{2J} normal mice which developed from mixed dy^{2J} and normal blastocysts (Peterson, 1974) and in addition, the gene defect and biochemical abnormality are unknown. In initial work with this mutant, Law and Yap (1979) grafted whole soleus muscles from newborn normal littermates into C57Bl/6J dy^{2J} neonates. Greater cross sectional area, twitch and tetanic tensions, and an increase in the number of fibres with high resting potentials were achieved in treated muscles. In a further refinement of this technique, Law (1982) injected mesenchyme from normal 12-day-old C57Bl/6J embryos into the solei of either normal C57Bl/6J or dystrophic C57Bl/6J dy^{2J} 20-day-old recipients, again effecting an improvement in structure and function of the injected muscle when compared to those which had not been treated. When myoblasts from the G8 cell line, a transformed cell line derived from Swiss Webster mice (Christian and Nelson, 1977), were injected into soleus muscles of C57Bl/6J $dy^{2J}dy^{2J}$ mice, three of the 14 treated mice showed behaviour patterns indistinguishable from normal mice, two were slightly improved, while the remaining nine showed no behavioural improvement whatsoever, when examined two months after injection (Law *et al.*, 1988a). Analysis of the GPI isoenzyme types indicated that although injected cells had survived, fusion with host muscle fibres to form mosaic host/donor fibres had not occurred; nevertheless, there was histological improvement in the treated dystrophic muscles, probably due to new muscle fibres of donor origin. In a further series of experiments, myogenic cells from 14-day-old embryonic normal C57Bl/6J mice were grown *in vitro* for 3 days prior to implantation into solei of the dystrophic dy^{2J} mutant. Six months after injection, half of the injected solei exhibited the heterodimeric form of GPI and an improvement in both the histology and function of the treated muscles was reported. In control experiments, where fibroblastic muscle cells were implanted, only host and donor but no heterodimeric, GPI were found and the muscle histology was not improved (Law *et al.*, 1988b, 1990a).

These results are, however, difficult to interpret in light of the findings of a series of experiments performed by Ontell and co-workers, indicating that 129/ReJ *dy/dy* myotubes must be innervated at a critical period in order to express the dystrophic phenotype. Although 129/ReJ *dy/dy* muscles which had regenerated following injection of bupivicaine, a myotoxic agent which kills the myofibres (Hall-Craggs, 1974), exhibited the histological abnormalities seen in control *dy* muscles (Martin and Ontell, 1988), whole extensor digitorum longus muscles which had regenerated after complete removal and autografting, failed to show the myofibre necrosis, connective tissue infiltration and myofibre diameter variations observed in age-matched, untraumatized *dy* control muscles examined in this experiment (Ontell, 1986; Bourke and Ontell, 1986). These findings, taken together with the fact that both chronic (Moschella and Ontell, 1987) and transient (Hermanson *et al.*, 1988) denervation of 129/ReJ *dy/dy* muscles also resulted in improved muscle function, suggests that damage to the nerve supply of *dy* muscles may prevent the manifestation of the dystrophic phenotype. Such factors may play some role in the finding of improved muscle function reported by Law and co-workers.

Critical assessment of the results of the above experiments in which normal mpc were implanted into dy^{2J} muscles is difficult, first because the *dy* mutation causes a disease which cannot, like many of the human dystrophies, be classified as a primary myopathy, and second, because the biochemical defect has not been identified.

In contrast, the ICR/IAn strain of mouse provided the first model of an X-linked, primary myopathy, in which the biochemical defect is known. Although the histology of this mouse appears almost normal, the phenotypic expression of the disease is a lack of the enzyme phosphorylase kinase

(PhK) within its skeletal muscle fibres (Rahim *et al.*, 1980). It is therefore possible to assess any increase in the missing gene product PhK, following implantation of normal mpc. Limited success has been achieved using this animal model (Morgan *et al.*, 1988a). Mosaic host/donor muscle fibres were detected in most cases (88%) where normal mpc were implanted into whole muscle autografts in the ICR/IAn host. However, the number of cells implanted was important. In grafts where between 3–5 $\times$ 10^5 cells were implanted, no PhK activity was detected, whereas when 10^6 or 3 $\times$ 10^6 cells were introduced, PhK activity, ranging from 2–12% of normal, was observed. Results were even less encouraging when normal cells were introduced into the growing muscle of infant ICR/IAn hosts, where mosaic fibres were found in only 17% of injected muscles and PhK activity did not differ significantly from control levels. These experiments showed that regenerating PhK-deficient muscle incorporated donor precursor cells more efficiently than did growing muscle. However, even though high proportions of mosaic fibres were formed in the regenerated autografts, the incorporation did not result in substantial expression of the missing gene product. The most likely explanation of this is the fact that PhK activity is decreased even in normal muscle which has degenerated and regenerated following grafting (Morgan, 1988b).

Greater success in alleviating an inherited biochemical deficiency has been achieved in the *mdx* mouse. This animal, which arose as a spontaneous mutant in the C57Bl/10ScSn strain (Bulfield *et al.*, 1984), suffers widespread degeneration of its skeletal muscles (Carnwath and Shotton, 1987; Coulton *et al.*, 1988a). However, the regenerative capacity of *mdx* skeletal muscle is high, and in older animals, the muscles are bigger and stronger than those of their normal counterparts (Anderson *et al.*, 1987; Coulton *et al.*, 1988b). As previously stated, the *mdx* mouse has been shown to be genetically homologous to Duchenne muscular dystrophy (Sicinsky *et al.*, 1989) and its muscles, like those of DMD sufferers, lack dystrophin (Hoffman *et al.*, 1987). Although dystrophin is deficient in both the muscles and nerves of the *mdx* mouse, cross grafting experiments provided evidence that the disease is, in fact, a primary myopathy (Morgan *et al.*, 1989). When intact EDL muscles taken from normal mice were grafted into *mdx* hosts and allowed to regenerate, the *mdx* phenotype was not expressed in these muscles. However, in the reverse situation where *mdx* muscles regenerated in normal hosts, the *mdx* phenotype was retained.

The *mdx* mouse is considered by many to be an adequate biochemical model of DMD and, as such, has been used to test the feasibility of myoblast transfer. Several groups have been successful in partially correcting the *mdx* myopathy by the introduction of normal precursor cells. Karpati and colleagues injected tritiated-thymidine labelled myoblasts from the C2 myogenic cell line, into the quadriceps of young *mdx* mice. They found mosaic muscle fibres, containing labelled and non-labelled nuclei, and did not observe any evidence of necrosis in these fibres (Karpati *et al.*, 1989a). A major problem with C2 cells is that they are transformed cells and behave abnormally *in vivo*, eventually forming tumours (Partridge *et al.*, 1988). In further experiments, Karpati and colleagues implanted cloned human myoblasts labelled with tritiated-thymidine into *mdx* quadriceps muscles. Up to 5% of fibres in the injected muscles were converted to dystrophin positive fibres. These results were interpreted by the authors to mean that, as none of the fibres which had converted were undergoing necrosis at the time of their examination, they had been 'rescued' by the inserted cells (Karpati *et al.*, 1989b), although any fibres which had necrosed would have lost their labelled nuclei. The low number of dystrophin-positive fibres in these muscles may be explained by the fact that host mice had not been made tolerant to the donor tissues and rejection of many of the implanted cells may have occurred.

Alleviation of the biochemical deficiency in the *mdx* mouse has been achieved by Partridge and colleagues. In these experiments, donor mpc were prepared by enzymatically disaggregating neonatal mouse muscle (Watt *et al.*, 1982). The advantage of these experiments was two-fold: first, the injected cells, unlike the C2 cell line used by Karpati, are not derived from a transformed cell line; and second, immune rejection of the implanted cells was avoided either by injection of the neonatal host *mdx* mouse with donor strain spleen cells to elicit immune tolerance (Watt *et al.*, 1984a), or by breeding the *mdx* mutation onto a nude mouse background and using these animals as hosts (Partridge *et al.*, 1989). The athymic nude mouse possesses a deficient immune system, and will accept tissue from many other sources. Precursor cells were implanted into either growing (5–7-day-old) *mdx* muscle, or into *mdx* muscle at the time of the onset of the myopathy. As with previous work, using phosphorylase kinase deficient hosts, the injected cells became incorporated far more frequently into regenerating, than into growing muscle. Fusion of precursor cells into host fibres, as assessed by the presence of donor and heterodimeric GPI, was significantly greater in the *mdx* nude host than in the neonatally tolerant animal. Muscles containing high proportions of donor and heterodimeric GPI, had 30–40% dystrophin-positive fibres, the dystrophin being located at its expected subsarcolemmal position. Further, immunoblotting showed the dystrophin to be of the correct size, and present at up to 40% of normal levels (Partridge *et al.*, 1989). The greater levels of dystrophin in treated muscles, as compared to the work of Karpati *et al.* (1989a,b) may well be explained by the immunological status of the host.

The main drawback of the *mdx* mouse as a model for DMD is that its muscles retain their ability to regenerate effectively and are not subject to the atrophy and replacement of fibres by fibrofatty connective tissue so characteristic of Duchenne. In view of the criticism levelled at the *mdx* mouse as a model for DMD, attempts have been made to change its histopathology to one more closely resembling the human disease. This has involved X-irradiating its muscles, thus effectively removing the endogenous precursor cells necessary for the formation of new fibres. Under such circumstances, regeneration of muscle fibres was inhibited, resulting in a muscle whose histopathological features more closely resembled the fibre atrophy and connective tissue replacement seen in DMD (Wakeford *et al.*, 1991). Irradiated *mdx* muscle therefore provides a good model in which to test whether the incorporation of normal mpc into an *mdx* myofibre prevents that fibre from undergoing further necrosis.

Further work with the irradiated *mdx* model (Morgan *et al.*, 1990), has involved the implantation of mpc at 19–21 days of age, at which time the host muscle is beginning to undergo massive necrosis (Coulton *et al.*, 1988a). GPI isoenzyme analysis and dystrophin immunostaining (Figure 12.2) of the treated muscles indicated widespread takeover of the host

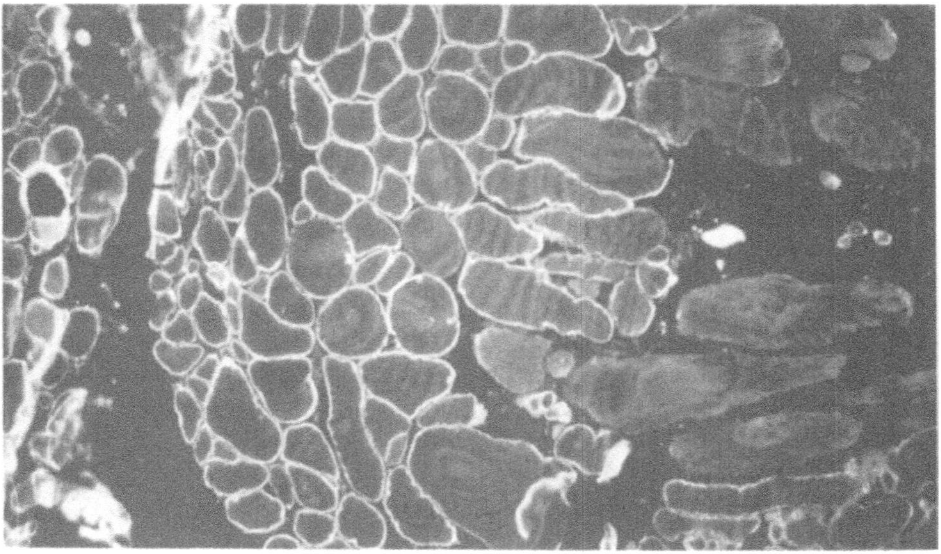

Figure 12.2 Dystrophin immunostain of a tibialis anterior muscle of an *mdx nu/nu* mouse. This muscle has been X-irradiated with 18 Gray at 16 days of age and injected with normal mpc from newborn C57Bl/10 mice 3 days later. 49 days after injection, the muscle contained donor and heterodimeric GPI isoenzymes and contained 72% dystrophin-positive fibres.

muscle by the implanted donor cells. The newly formed muscle fibres were initially purely of donor origin, reflecting the fact that the host cannot contribute to regeneration, their mpc being incapacitated by the irradiation. However at later stages, mosaic fibres were observed. The fact that mosaic fibres were detected in host muscle which is incapable of regenerating, indicates, for the first time, that implanted cells are capable of repairing degenerating muscle and that host regeneration is not necessary for the incorporation of implanted mpc. This result is very encouraging for the possibility of treatment of DMD patients, whose muscles have very little regenerative capacity. Further, the successfully injected muscles exhibited a near normal histological appearance with some preservation of muscle bulk compared with irradiated muscles which had been sham injected.

A further interesting result generated by the work of Morgan *et al.* (1990) was the lack of atrophy in X-irradiated *mdx* muscles, injected with *mdx*, rather than normal, mpc. In these muscles, no obvious muscle atrophy was observed histologically, and the contribution of the donor cells to the formation of new muscle fibres, as measured by GPI analysis, increased with time. This indicated that although the host could not contribute to muscle formation, this was being compensated by the donor cells, despite the fact that they were of the dystrophin-deficient *mdx* phenotype. The histopathological features of this muscle, however, were not improved and features characteristic of *mdx* muscle, such as a high level of central-nucleation, and foci of necrosis and regeneration, were evident. This is an exceedingly important finding, for it suggests that injection of *mdx* mpc, which possess a high regenerative capacity, prevents atrophy of the irradiated *mdx* fibres. Perhaps, if DMD muscle could maintain the capacity to regenerate itself as well as *mdx*, the lack of dystrophin might not lead to such catastrophic consequences.

Isolated, dystrophin-positive fibres have been occasionally found in irradiated *mdx* muscles which had been repopulated with *mdx* mpc. Small numbers of dystrophin-positive fibres are also found in untreated *mdx* muscles and are thought to arise as the result of a somatic change which permits expression of the defective dystrophin gene (Hoffman *et al.*, 1990). The presence of such dystrophin-positive fibres in man too (Nicholson *et al.*, 1990) emphasizes the need for good controls when analysing the effects of implanting normal donor mpc.

Besides the *mdx* mouse, a model of DMD has also arisen within a colony of golden retriever dogs (Cooper *et al.*, 1988; Kornegay *et al.*, 1988). This myopathic *xmd* dog will be invaluable in evaluating the efficacy of myoblast transfer. The condition in the dog is characterized by high creatine kinase levels (Kornegay *et al.*, 1988; Valentine *et al.*, 1989). Within the first few weeks of life, there is degeneration of fibres and a build-up of connective tissue within the skeletal muscles of these animals, resulting, as in the human situation, in a gross deficiency of actual fibres. The condition is exceedingly rapid in onset, and takes only a few weeks to develop into a devastating myopathy. The defect is in the homologous gene to that in the *mdx* mouse and DMD patients and, as in the mouse and man, there is an absence of dystrophin (Cooper *et al.*, 1988). The *xmd* dog thus lends itself well to studies of the effect of implanting precursor cells into dystrophic muscle in an animal which is closer in size to man and has histopathological features resembling human DMD patients (Cooper, 1990). Although a DMD homologue has been discovered in the cat (Carpenter *et al.*, 1989), it has not been studied in any great detail due to a lack of numbers of animals which have been identified with this abnormality. The histopathology of those which have been studied seems to resemble the features found in the mouse rather than those seen in DMD patients (Hoffman and Gorospe, 1991).

12.3 PROBLEMS ASSOCIATED WITH MYOBLAST TRANSFER

For myoblast transfer therapy to become a viable proposition in the treatment of human primary myopathies, it is essential that certain questions should first be answered.

12.3.1 How important is it to ensure that the cells being implanted are myogenic?

The need to implant myogenic, as opposed to non-myogenic cells, is evident for several reasons. The whole rationale behind the treatment is to supply the patient with muscle-specific gene products which are lacking in his own muscles. Thus, it would be reasonable to suggest that the cells to be implanted into the diseased muscle should be capable of synthesizing the proteins of the skeletal muscle fibre which the patient lacks. Hence, the cell of choice for implantation would naturally be the myogenic precursor cell (mpc), presumably obtained from muscle biopsies or from aborted human fetuses. Such tissue, however, contains a high proportion of non-myogenic cells, so it will be necessary to purify any source of donor material to produce an enriched myogenic population, capable of synthesizing the necessary muscle proteins following their implantation into the target muscle. Another important reason to purify myogenic cells prior to myoblast implantation, is to remove immunogenic antigen-presenting cells, e.g. macrophages and endothelial cells, which are likely to provoke an immune response against the implanted cells or the resultant myofibres.

Various cell separation techniques have been used experimentally to produce high yields of myogenic cells from donor muscle. Discontinuous

Percoll gradients were employed to separate out the myogenic cells from mouse (Morgan, 1988b) and chicken (Yablonka-Reuveni and Nameroff, 1987; Yablonka-Reuveni *et al.*, 1987) muscle. In muscle cells derived from embryonic chicken muscle, clonal analysis and the use of muscle-specific antibodies indicated that one cell fraction contained 70% myogenic cells (Yablonka-Reuveni and Nameroff, 1987), whereas when adult chicken muscle was used as the source of cells, 99% of the resultant clones were myogenic (Yablonka-Reuveni *et al.*, 1987). Interestingly, they showed that the embryonic myogenic cells were heterogeneous, giving rise to small and large clones, whereas cells from the adult chicken only formed large clones. In the mouse experiments, myogenicity was assessed *in vitro* by measuring the percentage of nuclei in myotubes in mass cultures and *in vivo* by the presence of donor and heterodimer GPI in injected muscle grafts. By these criteria, Percoll separation gave rise to populations which were enriched for myogenic cells, but which were no more myogenic *in vivo* than were control cultures (Morgan, 1988b).

Cloning of myogenic cells is an obvious, although labour-intensive, way of enriching for myogenic cells (Blau and Webster, 1981; Webster and Blau, 1990). In experiments already undertaken, where mpc have been implanted into a single muscle of young boys suffering from DMD, the myogenic cells for implantation were obtained by cloning from biopsied donor muscle (Law *et al.*, 1990b; Tremblay *et al.*, 1991). This cloning method yielded, from 1 g of biopsy sample, 8×10^6 cells (Law *et al.*, 1990b) or 5×10^7 cells from an unspecified weight of biopsy (Tremblay *et al.*, 1991). The disadvantage of cloning is that the cells have already undergone very many divisions prior to their implantation. Purification of myogenic cells by fluorescent activated cell sorting (FACS) using a monoclonal antibody against the neural cell adhesion molecule (N-CAM), which is present on muscle cells but not on muscle fibroblasts (Walsh and Ritter, 1981; Walsh, 1990) has been performed. This method gave rise to a cell suspension containing 99.9% myogenic cells (Miller *et al.*, 1991). The disadvantage of this technique, however, is the loss of exceedingly high proportions of viable cells from the biopsy sample while being passed through the FACS machine (Webster *et al.*, 1988). In contrast to the FACS method of separating cells, Jones *et al.* (1990) have used a modification of the 'panning method' (Wysocki *et al.*, 1978) to produce an enriched population of mouse myogenic cells for implantation into the muscle of the *mdx* strain of mouse. This method, based on the affinity of myogenic cells for Mab H28, an antibody to mouse N-CAM, resulted in the recovery of 90–95% viable cells from the original sample of muscle and produced a highly enriched population of myogenic cells. The myogenic nature of such cells was verified by cell morphology and by the presence of the muscle specific intermediate filament, desmin.

In contrast to this line of thought, there is some evidence to indicate that cells within the donor muscle which are not myogenic may play a role in the formation of new muscle fibres within the target muscle. In whole mouse muscle grafts, where endogenous tissue was killed by freezing the tissue with liquid nitrogen followed by implantation of Percoll-separated mpc into these grafts, larger amounts of muscle were formed when unseparated mpc, rather than just the myogenic fraction, was added to the graft. In addition, more new muscle, entirely of host origin, was formed when cells from the fibroblast fraction were injected, rather than when no cells at all were added (Morgan and Partridge, 1988b). This suggests that the presence of fibroblasts might augment muscle formation; certainly *in vitro*, formation of basal lamina around quail myotubes was found to be dependent on the presence of fibroblasts in the culture (Sanderson *et al.*, 1986). There is also evidence that the fusion of fibroblasts with muscular dysgenic myotubes *in vitro* (Chaudhari *et al.*, 1989) results in rescue of up to 28% of fibres by incorporation of 'non-myogenic' fibroblasts into the host muscle fibre; rescued fibres, unlike their muscular dysgenic counterparts, were able to contract. In the case of DMD, implantation of non-muscle cells which do not produce dystrophin would not be beneficial, unless their presence were, in some way, to augment myogenesis. However, it is true to say that as dystrophin is absent from the smooth muscles of DMD patients, the presence of normal smooth muscle cells in the crude suspension of donor muscle cells may also be of benefit.

12.3.2 How important in terms of myoblast implantation is the stage of myogenic differentiation of the implanted cell?

It is likely that the type of myogenic cell, rather than the absolute number, is the important factor for consideration in the treatment of primary myopathies. Although many of the workers in the field of myoblast transfer therapy have shown that injected mpc give rise to new muscle within the myopathic host (Morgan *et al.*, 1988a, 1990; Partridge *et al.*, 1989; Law *et al.*, 1988a,b; Karpati *et al.*, 1989a, b), it seems imperative for long-term benefit, that some of the injected cells should remain as stem cells. Stem cells have the ability to retain their capacity to undergo division and become incorporated into myopathic host muscle during any further bouts of necrosis or injury that occur in the dystrophic muscle during the course of the disease.

The most direct line of evidence that some implanted cells do indeed remain as stem cells after their introduction into the myopathic host muscle has been witnessed by the ability, albeit rare, to extract myogenic cells from *mdx* muscles which had first been irradiated to remove the endogenous population of host cells and then been injected with normal donor mpc. (For the rationale behind this experiment, see Figure 12.3.) Clones of myogenic cells subsequently isolated from these *mdx* muscles were all found not only to contain the donor type GPI isoenzyme, but also to form dystrophin-positive myotubes when subsequently grown *in vitro* (Morgan *et al.*, 1991).

Two further lines of evidence also indirectly indicate that normal donor cells implanted into myopathic mouse muscle do contain some cells which are capable of retaining their proliferative capacity. In a series of animals originally set up as a control group, donor mpc derived from newborn *mdx* muscle were implanted into the muscles of the same strain of host. Hence both host and donor strains were myopathic, but had been bred to carry the different isoenzyme types of GPI, which could be used to determine their relative contributions to the muscle. In such cases, the proportion of donor type GPI was found to be greater in injected muscles examined at later

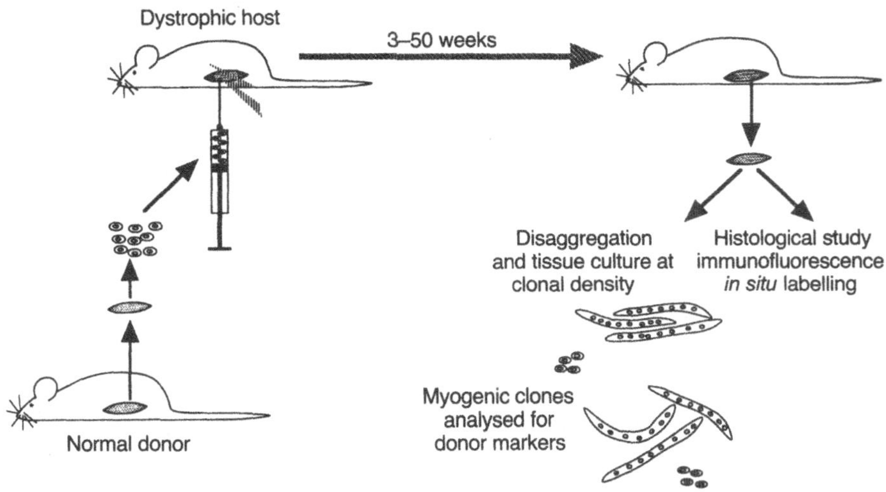

Figure 12.3 Schematic representation of the experimental protocol devised to ascertain if injected mpc give rise to muscle stem cells. Mpc derived from a donor mouse of *Gpi-1s*^{b} allotype and a normal dystrophin phenotype were injected into the TA of a dystrophin-negative *mdx* mouse of *Gpi-1s*^{a} allotype. To incapacitate host mpc, the right leg of this mouse had been X-irradiated 3–5 days previously. After several weeks, the injected muscle was removed for analysis. A portion of the injected muscle was analysed for its GPI and dystrophin content. The finding of dystrophin-positive muscle fibres and heterodimeric GPI means that the implanted mpc had made new muscle within the host muscle. The remainder of the muscle was tissue cultured; any myogenic cells were isolated and the clones expanded. Any myogenic clones of donor GPI allotype which fused into dystrophin-positive fibres must have been derived from the donor cells. This is strong evidence that the implanted mpc had given rise to stem cells within the host muscle.

stages of the experiment, i.e. 67 days after cell implantation, compared to those examined earlier, i.e. at 29 and 49 days after cell implantation (Morgan *et al.*, 1990). This result points to the fact that some of the cells introduced in the original cell implant must have remained within the host muscle as unfused proliferating mononuclear cells, still capable of fusing with the host muscle fibres during the latter stages of the experiment. Additionally, in very recent work (Morgan *et al.*, in preparation), where donor newborn mpc had been carefully injected only into the tibialis anterior muscle of the mouse, the neighbouring extensor digitorum longus and peroneus muscles were examined to detect the presence of muscle of donor origin. Donor and heterodimer GPI and dystrophin-positive fibres were only very rarely observed in neighbouring muscles at 7 weeks after implantation, but frequently so at 36 weeks. This finding indicates a further important feature required for successful myoblast transfer therapy, i.e. the ability of cells not only to remain as mononuclear stem cells as discussed above, but also to move into neighbouring muscles affected by the disease and disseminate their normal gene products throughout several affected muscles.

12.3.3 How many cells are required to treat a given mass of host muscle?

Ideally, when treating a single muscle, sufficient precursor cells to completely renew that muscle should be implanted. In murine experiments reported to date, between 10^5 and 10^6 cells have usually been injected into a single myopathic muscle (Morgan *et al.*, 1988a, 1990; Partridge *et al.*, 1989; Law *et al.*, 1988b; Karpati *et al.*, 1989a, b). Apart from estimating the amount of dystrophin present in injected muscles (Partridge *et al.*, 1989; Morgan *et al.*, 1990), the amount of new muscle which arises from a given number of injected cells has not been quantified. In irradiated *mdx* muscles, a single injection of 5×10^5 cells gave rise to up to 80% dystrophin-positive fibres. Many of these fibres were mosaic, showing that the donor cells indeed had become incorporated into the host fibres.

In the first experiments performed on boys suffering from DMD, suspensions containing 8×10^6 (Law *et al.*, 1990b) or 5×10^7 (Tremblay *et al.*, 1991) have been injected into a single muscle. The long-term results of such trials are awaited with interest.

12.3.4 Can multiple copies of the defective gene be implanted into the deficient muscle?

Once a highly proliferative stem cell population has been obtained, the efficiency of correction of the host muscle may be improved by introducing multiple copies of the gene which is missing or defective into each stem cell to be implanted. This could be achieved using retroviral vectors if the gene in question were not too large to be carried in such a vector. Subsequent to infection or transfection of stem cells with multiple gene copies, it would be desirable to expand the population *in vitro* prior to insertion into the target muscle. Blau and Hughes (1990), Hughes and Blau (1990) and Thomason and Booth (1990) have succeeded in using the replication defective mouse moloney leukaemia retroviral vector carrying the *lacZ* gene which encodes for the enzyme β-galactosidase to insert multiple copies of *lacZ* into hindlimb muscles of normal rats. In addition, muscle cell lines (Coleman *et al.*, 1991; Salminen *et al.*, 1991) and primary mouse mpc (Coleman *et al.*, 1991) have been infected *in vitro* and injected into skeletal muscle *in vivo*, giving rise to the protein product in the injected muscles. Human growth hormone has been produced in the serum of mice whose skeletal muscles had been injected with C2 cells into which the human growth hormone gene had been inserted by transfection (Barr and Leiden, 1991) or by infection (Dhawan *et al.*, 1991) with a retroviral vector.

There are of course both advantages and disadvantages to using such vehicles as retroviral vectors to insert multiple copies of genes into target muscles. First, this depends very much on the size of the gene, for genes in excess of 7 kb cannot be incorporated by retroviruses (Dickson and Dunckley, Chapter 11, this volume). Second, retroviral vectors may themselves be detrimental, for their introduction into target tissue may cause the activation of deleterious genes, such as oncogenes.

The injection of expression plasmids containing either full-length or Becker-like dystrophin cDNA into mouse skeletal or cardiac muscle led to dystrophin-positive fibres above background levels, but the efficiency of this gene transfer was very low (Acsadi *et al.*, 1991).

Lee *et al.* (1991) have transfected a full-length mouse dystrophin cDNA into non-muscle cells, choosing the COS cell, a kidney cell derived from the African green monkey, as their target tissue. Although dystrophin is not normally expressed within such a tissue these authors were able to show that when expressed within the target cell, the dystrophin product localized to a subsarcolemmal position, the normal location for this protein in the skeletal muscle fibre. Wolff and co-workers (1990) have successfully directly introduced pure *lacZ*, chloramphenicol acetyltransferase and luciferase DNA and RNA into the skeletal muscle fibres of host mice. The introduced gene products were detectable for up to two months after injection of the nucleic acids, although the number of fibres positive for the gene products was low; large amounts of DNA and RNA, i.e. in the order of 100 μg, were required to be injected to produce positivity in 1.5% of the muscle fibres.

In terms of myoblast transfer therapy, however, the advantage of injecting mpc into myopathic muscle, as opposed to genes alone, is that any injected cells which remain as unfused 'stem' cells will be under the normal biological control of the host muscle, and will be able to replicate and then to differentiate to form more new muscle, when and where appropriate.

12.3.5 Once incorporated into dystrophic muscle, can the precursor cells migrate far enough for the defective gene product to permeate throughout the muscle?

Certainly in the mouse there is good evidence that myogenic cells can migrate from an adjacent muscle into an area of regeneration. In cases where a whole extensor digitorum longus (EDL) muscle of a host mouse was autografted adjacent to a similar EDL allograft from a donor strain of mouse, donor and heterodimer GPI isoenzyme allotypes were detected, albeit rarely, within the autografted muscle (Watt *et al.*, 1987). In these instances, donor cells could only have been derived from the donor muscle, and, as mosaic fibres were formed, some of the host cells which had entered the regenerating muscle must have been myogenic. This indicates that myogenic cells are capable of migration between adjacent muscles, and in such cases it appeared likely that migratory muscle cells were crossing the connective tissue barriers between muscle bellies. The first evidence that migration of host mpc into an area of necrosis is a frequent occurrence was provided by experiments with freeze-killed autografted muscles; when no mpc were added to these grafts, new muscle was found in all instances and could only have been derived from the host. Where donor type mpc were added to the freeze-killed graft, donor and heterodimer GPI were found in 5/7 grafts; the host element of the heterodimer GPI can only have been derived from host mpc which had invaded the graft (Morgan *et al.*, 1987). Similarly, an irradiated *mdx* mouse muscle, injected myogenic cells rapidly permeate the target muscle (Morgan *et al.*, 1990); and myogenic cells of donor origin have been shown to move into the adjacent muscles of the *mdx* mouse (Morgan *et al.*, 1990, 1993; Watt *et al.*, 1993). The apparently greater movement of injected mpc in irradiated, rather than non-irradiated muscle may be due either to the absence of competing host mpc, or to some effect of the irradiation itself on muscle structure.

In the rat, evidence for migration is more controversial, some workers showing migration does occur, and others failing to find evidence of such a phenomenon. Lipton and Schultz (1979) were the first to produce conclusive evidence that myogenic cells could cross the endomysial sheath in the rat, when they reported the extensive migration and incorporation of autoradiographically labelled satellite cells throughout the normal tibialis anterior muscle of the rat subsequent to their insertion under the investing fascial sheath of this muscle. Using retroviral vectors containing the *lacZ* gene, Hughes and Blau (1990) also showed the willingness of myogenic cells to cross the basal lamina during development in the newborn rat. Following injection of *lacZ* retrovirus into a single region of the hindlimb muscles of the rat, several clusters of fibres positive for the marker were isolated throughout the muscle 2–3 weeks afterwards. These results however contrast with other work using the adult rat where myogenic cells failed to invade minced muscle grafts, devitalized by freezing and thawing (Ghins *et al.*, 1986) or EDL muscles killed *in situ* by freezing and thawing (Schultz *et al.*, 1986), although satellite cells will readily migrate from one muscle to another if the basal lamina is damaged (Ghins *et al.*, 1986; Schultz *et al.*, 1985) and do indeed migrate from an uninjured to injured area of the rat EDL muscle (Phillips *et al.* 1990).

All of these lines of evidence attest to the fact that muscle precursor cells are capable of migrating throughout and sometimes between, muscles. It is not known why the migration of mpc between muscles occurs in adult mice but not in adult rats. It is possibly due to the greater distances to be covered in the rat and the thicker epymysia; if this is the case, mpc may not readily migrate between muscles in man. Whether all myogenic cells are motile, or if such activity is confined to part of this population, is also unknown.

12.3.6 How much of a myofibre is affected by the incorporation of normal myonuclei?

Following implantation of normal cells into abnormal fibres, it is important to know just how beneficial such an injection will be, in terms of how widely normal gene products can be dispersed throughout the rescued fibre. There is evidence that some proteins become widely disseminated within the muscle fibre and some remain around the nucleus which encoded for them. In the case of the protein product of the *lacZ* gene, β-galactosidase, this protein is capable of dispersing along the length of the fibre in which one cell nucleus coding for the gene is incorporated (Hughes and Blau, 1990). Other proteins, including structural proteins such as sarcomeric myosin heavy chain proteins, remain localized in the vicinity of the nucelus which encoded it (Pavlath *et al.*, 1989). In the case of dystrophin, obviously the protein of interest in DMD, patchy distribution occurs in mosaic muscle fibres of *mdx* mice following the incorporation of normal donor precursor cells into such muscles (Partridge *et al.*, 1989; Karpati *et al.*, 1989b). This patchy distribution of dystrophin was also observed in young heterozygous *mdx* mice (Watkins *et al.*, 1989), and *xmd* dogs (Cooper *et al.*, 1990), again suggesting that the protein is confined to the vicinity of the nucleus that produced it. The distribution of dystrophin however differs in mosaic myotubes formed *in vitro* from normal rat and dystrophic mouse *mdx* nuclei, where the dystrophin is present over the entire subsarcolemmal face of the myotube (Huard *et al.*, 1991). Uniform dystrophin distribution is observed in older heterozygous animals, but it is not known whether this is due to the diffusion of the protein or to the fusion of 'normal' satellite cells to the myofibres. However, the debate as to whether a completely normal distribution of dystrophin, localized around the entire subsarcolemmal position of the skeletal muscle fibre, is necessary to restore the normal functioning of the dystrophic fibre after implantation of normal precursor cells, will have to wait until the actual role of dystrophin within the fibre has been finally elucidated.

12.3.7 How can rejection of the implanted cells be prevented?

The implantation of foreign cells into a non-histocompatible tissue would normally have the effect of causing a swift and strong immunological rejection within the target tissue. One would therefore expect that immunosuppressive measures are imperative if myoblast transfer therapy is to be contemplated. However, there have emerged several conflicting reports as to the immunogenicity of implanted cells and just how strong an immune reaction will be mounted by the host. From the point of view of the skeletal muscle itself, neither human nor murine fibres exhibit Class 1 major histocompatibility antigens (Ponder *et al.*, 1983), although they are very strongly manifest in certain human muscle diseases characterized by inflammatory and degenerative events (Appleyard *et al.*, 1985; Emslie-Smith *et al.*, 1989; Karpati *et al.*, 1988) and HLA Class I antigens are also expressed on regenerating fibres and on myoblasts and myotubes *in vitro* (McDouall *et al.*, 1989). HLA Class II antigens are not expressed on the muscle cell, although they may be induced in myoblasts in response to gamma interferon (Bao *et al.*, 1990; Hohlfeld and Engel, 1990). The lack of expression of these antigens on the muscle cell suggests that these cells may not raise such a violent immune response as might be expected when implanted into a target muscle fibre. It is possible that myoblasts, like keratinocytes (Bal *et al.*, 1990) may induce T cell tolerance.

In the experimental literature, there have been several cases cited where injected mpc have survived for prolonged periods of time in non-histocompatible animals, and certainly for times far in excess of that expected if the host were to mount an immune response. Karpati *et al.* (1989a,b) found muscle of donor origin in non-tolerant *mdx* mice which had been injected either with C2 cells or with human mpc. When grafts of whole or minced muscle are made into non-tolerant hosts, the grafts are rapidly rejected (Partridge and Sloper, 1977; Partridge *et al.*, 1978; Grounds *et al.*, 1980; Watt, 1982). By contrast, when suspensions of cells derived from disaggregated muscle, as opposed to minced muscle allografts, were injected into minced muscle autografts in non-tolerant hosts, donor and heterodimer GPI, indicating the presence of muscle of donor origin, were found frequently in strains compatible at the major histocompatibility locus but more rarely in strains which were incompatible (Watt *et al.*, 1991). Mosaic muscle fibres were found as late as 105 days after mpc implantation. Injected suspensions of cells prepared from regenerating adult muscle were however rapidly rejected, presumably due to the presence of inflammatory cells. This indicates that suspensions of muscle-derived cells are less immunogenic than is the whole muscle tissue, despite the fact that the cell suspension had not been enriched for myogenic cells (Watt *et al.*, 1991). There may be some immunological impediment to mpc grafting, even in histocompatible hosts which have been made tolerant. When donor mpc were added to dead muscle grafts, more new muscle was found in grafts made in *nu/nu* than in grafts made in neonatally tolerant mice (Morgan *et al.*, 1987). Similarly, fusion of donor mpc into host *mdx* muscle was more frequent when nude *mdx* mice, rather than neonatally tolerant *mdx* mice, were used as hosts (Partridge *et al.*, 1989).

Tolerance to grafts of minced (Watt *et al.*, 1981) and whole (Gulati and Zalewski, 1982) muscle and to implanted mpc (Watt *et al.*, 1984b; Law *et al.*, 1988) may be induced in rodents by the administration of cyclosporin A; survival of host muscle of donor origin has been noted up to 65 days after the cessation of cyclosporin A treatment (Watt *et al.*, 1984b). In human experiments, cyclosporin A has been used for immunosuppression (Law *et al.*, 1990, 1991; Gussoni *et al.*, 1992). Tremblay *et al.* (1991) found that their mpc implants survived and formed new muscle in non-immunosuppressed patients which were matched to the donor at both the MHC class 1 and 2 loci.

In the case of myoblast transfer therapy, there is the added problem of introducing synthesis of a new foreign protein in muscles of individuals which have not become developmentally tolerized to it. This is bound to be the case when either normal cells or dystrophin cDNA are introduced into dystrophin-deficient muscle. However, no evidence has yet been found to show that an immune response has been raised to dystrophin within the host muscle fibres.

12.3.8 How can the functional improvement of a treated muscle be measured?

To ascertain whether an implantation of normal mpc has been beneficial to the treated muscle, it is necessary not only to monitor the biochemical and histological changes which have occurred in that muscle, but also to measure changes in the functional state. In experiments performed in the *mdx* mouse to date, muscle physiology has not been measured, because untreated *mdx* muscles become bigger and stronger than controls

(Coulton *et al.*, 1988b). By contrast, in the *dy* mouse, Law *et al.* (1990a) have claimed improvement in both histological and physiological parameters.

In cell implantation experiments already performed in DMD patients, Law's group have reported that the extensor digitorum brevis muscles in three patients which had been injected with normal mpc contained dystrophin-positive fibres and displayed a more normal histology than did sham-injected muscles. The muscles injected with mpc had greater twitch tensions than they showed prior to the injection (Law *et al.*, 1991). Tremblay *et al.* (1991) found dystrophin-positive fibres in injected muscles in three patients, but have not examined the muscles physiologically. To ascertain that injection of normal mpc into muscles of DMD patients does lead to a marked and persistent functional improvement, many more controlled experiments need to be performed. Unfortunately, the better control experiment of injecting normal mpc into a muscle of one leg and dystrophic mpc into the muscle of the contralateral leg cannot be performed in humans for ethical reasons.

The problems of conducting trials of mpc injections in humans have been discussed in detail elsewhere (Karpati, 1991; Miller *et al.*, 1991; Partridge, 1990). These basically involve:

1. selection of the ideal patient/donor;
2. selection of the best muscle to be injected;
3. preparation of sufficient numbers of mpc for implantation;
4. eliminating infection risks and immunorejection;
5. control sham-injected muscles;
6. how to measure improvement in a treated muscle;
7. 'blind' assessment of the data.

Thus, myoblast transfer therapy is indeed one possible option for the treatment of myopathies such as DMD, but, prior to its therapeutic application, there are many parameters which need to be more fully investigated in animal models and in experimental trials in DMD patients, so as not to prematurely judge a technique which may have far-reaching and beneficial consequences, if developed by careful and logically-constructed investigation.

ACKNOWLEDGEMENTS

The authors' contributions to the work described in this chapter were supported by the Muscular Dystrophy Group of Great Britain, Action Research for the Crippled Child and the Leverhulme Trust. We wish to thank Mr Ron Barnett for photography.

REFERENCES

Acsadi, G., Dickson, G., Love, D.R. *et al.* (1991) Human dystrophin expression in mdx mice after intramuscular injection of DNA constructs. *Nature*, **352**, 815–18.

Alameddine, H.S., Dehaupas, M. and Fardeau, M. (1989) Regeneration of skeletal muscle fibres from autologous satellite cells multiplied *in vitro*. An experimental model for testing cultured cell myogenicity. *Muscle Nerve*, **12**, 544–55.

Anderson, J.W., Ovalle, W.K. and Bressler, B.H. (1987) Electron microscopic and autoradiographic characterization of hindlimb muscle regeneration in the mdx mouse. *Anat. Rec.*, **219**, 243–57.

Appleyard, S.T., Dunn, M.J., Dubowitz, V. and Rose, M.L. (1985) Increased expression of HLA ABC class 1 antigens by muscle fibres in Duchenne muscular dystrophy, inflammatory myopathy, and other neuromuscular disorders. *Lancet*, i, 361–3.

Bal, V., McIndoe, A., Denton, G. *et al.* (1990) Antigen presentation by keratinocytes induces tolerance in human T cells. *Eur. J. Immunol.*, **20**, 1893–7.

Bao, S., King, N.J.C., and dos Remedios, C.G. (1991) Elevated MHC class 1 and 2 antigens in cultured human embryonic myoblasts following stimulation with gamma interferon. *Immunol. Cell Biol.*, **68**, 235–42.

Barr, E. and Leiden, J.M. (1991) Systemic delivery of recombinant proteins by genetically modified myoblasts. *Science*, **254**, 1507–9.

Blau, H.M. and Webster, C. (1981) Isolation and characterization of human muscle cells. *Proc. Natl. Acad. Sci. USA*, **78**, 5623–7.

Blau, H.M. and Hughes, S.M. (1990) Retroviral lineage markers for assessing myoblast fate *in vivo*, in *Myoblast Transfer Therapy*, (eds R. Griggs and G. Karpati), Plenum Press, New York, pp. 201–3.

Bourke, D.L. and Ontell, M. (1986) Modification of the phenotypic expression of murine dystrophy: A morphological study. *Anat. Rec.*, **214**, 17–24.

Bradley, W.G. and Jenkison, M. (1973) Abnormalities of peripheral nerves in murine muscular dystrophy. *J. Neurol. Sci.*, **18**, 227–47.

Bulfield, G., Siller, W.G., Wight, P.A.L. and Moore, K.J. (1984) X chromosome-linked muscular dystrophy (mdx) in the mouse. *Proc. Natl. Acad. Sci. USA*, **81**, 1189–92.

Cardasis, C.A. and Cooper, G.W. (1975) An analysis of nuclear numbers in individual muscle fibres during differentiation and growth: a satellite cell-muscle fibre growth unit. *J. Exp. Zool.*, **191**, 333–46.

Carlson, B.M. (1973) The regeneration of skeletal muscle – a review. *Am. J. Anat.*, **137**, 119–50.

Carpenter, J.L., Hoffman, E.P., Romanul, F.C. *et al.* (1989) Feline muscular dystrophy with dystrophin deficiency. *Am. J. Pathol.*, **135**, 909–19.

Carnwath, J.W. and Shotton, D.M. (1987) Muscular dystrophy in the mdx mouse: histopathology of the soleus and extensor digitorum longus muscle. *J. Neurol. Sci.*, **80**, 39–54.

Chaudhari, N., Delay, R. and Beam, K.G. (1989) Restoration of normal function in genetically defective myotubes by spontaneous fusion with fibroblasts. *Nature*, **341**, 445–7.

Christian, C.N. and Nelson, P.G. (1977) Synapse formation between two clonal cell lines. *Science*, **196**, 995–8.

Coleman, M.G., Partridge, T.A. and Watt, D.J. (1991) Introduction of multiple gene copies into donor precursor cells as a therapeutic intervention in inherited myopathies, in *Muscular Dystrophy Research: From Molecular Diagnosis Toward Therapy*, (eds C. Angelini, G.A. Danielli and D. Fontanari), Excerpta Medica, Amsterdam, New York and Oxford, p. 199.

Cooper, B.J., Winand, N.J., Stedman, H. *et al.* (1988) The homologue of the Duchenne locus is defective in X-linked muscular dystrophy of dogs. *Nature*, **334**, 154–6.

Cooper, B.J., Gallagher, E.A., Smith, C.A. *et al.* (1990) Mosaic expression of dystrophin in carriers of canine X-linked muscular dystrophy. *Lab. Invest.*, **62**, 171–8.

Cooper, B.J. (1990) The role of the XMD dog in the assessment of myoblast transfer therapy, in *Myoblast Transfer Therapy*, Adv. Exp. Med. Biol. 280, (eds R.C. Griggs and G. Karpati), Plenum Press, New York, pp. 279–84.

Coulton, G.R., Morgan, J.E., Partridge, T.A. and Sloper, J.C. (1988a) The mdx mouse skeletal muscle myopathy: 1. A histological, morphometric and biochemical investigation. *Neuropath. Appl. Neurobiol.*, **14**, 53–70.

Coulton, G.R., Curtin, N.A., Morgan, J.E. and Partridge, T.A. (1988b) The mdx mouse skeletal muscle myopathy: II. Contractile properties. *Neuropathol. Appl. Neurobiol.*, **14**, 299–314.

Dean, M.F., Stevens, R.L., Muir, H. *et al.* (1979) Enzyme replacement therapy by fibroblast transplantation. *J. Clin. Invest.* **63**, 138–45.

Dean, M.F., Muir, H., Benson, P. and Button, L. (1980) Enzyme replacement therapy in the mucopolysaccharidoses by fibroblast transplantation, in *Enzyme Therapy in Genetic Diseases: 2. Birth defects: Original article series*, **Vol XVI**, no. 1, (eds R.J. Desnick, N.W. Paul and F. Dickman), Alan R. Liss, New York, pp. 445–56.

Dean, M.F., Muir, H., Benson, P.F. and Button, L.R. (1981) Enzyme replacement therapy by transplantation of HLA-compatible fibroblasts in Sanfillipo A syndrome. *Paediatr. Res.*, **15**, 959–63.

Dhawan, J., Pan, L.C., Pavlath, G.K. *et al.* (1991) Systemic delivery of human growth hormone by injection of genetically-engineered myoblasts. *Science*, **254**, 1509–14.

Emslie-Smith, A.M., Arahata, K. and Engel, A.G. (1989) Major histocompatibility class 1 antigen expression, immunolocalization of interferon subtypes and T cell mediated cytotoxicity in myopathies. *Hum. Pathol.*, **20**, 224–31.

Ervasti, J.M., Ohlendieck, K., Kahl, S.D. *et al.* (1990) Deficiency of a glycoprotein component of the dystrophin complex in dystrophic muscle. *Nature*, **345**, 315–19.

Frair, P.M. and Peterson, A.C. (1983) The nuclear-cytoplasmic relationship in 'mosaic' skeletal muscle fibres from mouse chimaeras. *Exp. Cell Res.*, **145**, 167–78.

Gearhart, J.D. and Mintz, B. (1972) Glucosephosphate isomerase subunit-reassociation tests for maternal–fetal and fetal–fetal cell fusion in the mouse placenta. *Dev. Biol.*, **29**, 55–64.

Ghins, E., Colson-van-Schoor, M. and Marechal, G. (1986) Implantation of autologous cells in minced and devitalised rat skeletal muscles. *J. Muscle Res. Cell Motil.*, 7, 151–9.

Gibbs, D.A., Spellacy, E., Roberts, A.E. and Watts, R.W.E. (1980) The treatment of lysosomal storage diseases by fibroblast transplantation: some preliminary observations, in *Enzyme Therapy in Genetic Diseases: 2. Birth Defects: Original Article Series*, **Vol XVI**, no. 1 (eds R.J. Desnick, N.W. Paul and F. Dickman), Alan R. Liss, New York, pp. 457–74.

Grounds, M., Partridge, T.A. and Sloper, J.C. (1980) The contribution of exogenous cells to regenerating skeletal muscle: an isoenzyme study of muscle allografts in mice. *J. Pathol.*, **132**, 325–41.

Gulati, A.K. and Zalewski, A.A. (1982) Muscle allograft survival after cyclosporin A immunosuppression. *Exp. Neurol.*, 77, 378–85.

Gussoni, E., Pavlath, G.K., Lanctot, A.M. *et al.* (1992) Normal dystrophin transcripts detected in Duchenne muscular dystrophy patients after myoblast transplantation. *Nature*, **356**, 435–8.

Hall-Craggs, E.C.B. (1974) Rapid degeneration and regeneration of a whole skeletal muscle following treatment with bupivicaine (marcaine). *Exp. Neurobiol.*, **43**, 349–58.

Hermanson, J.W., Moschella, M.C. and Ontell, M. (1988) Effect of neonatal denervation–reinnervation on the functional capacity of a 129ReJ *dy/dy* murine dystrophic muscle. *Exp. Neurol.* **102**, 210–16.

Hoffman, E.P., Brown, R.H. and Kunkel, L.M. (1987) Dystrophin: the protein product of the Duchenne muscular dystrophy locus. *Cell*, **51**, 919–28.

Hoffman, E.P., Morgan, J.E., Watkins, S.C. and Partridge, T.A. (1990) Somatic reversion/suppression of the mouse mdx phenotype *in vivo*. *J. Neurol. Sci.*, **99**, 9–25.

Hoffman, E.P. and Gorospe, J.R. (1991) The animal models of Duchenne muscular dystrophy: windows on the pathophysiological consequences of dystrophin deficiency, in *Topics in Biomembranes, vol 38*, (eds J. Morrow and M. Mooseker), Academic Press, New York, pp. 113–53.

Hohlfeld, R. and Engel, A.G. (1990) Induction of HLA-DR expression on human myoblasts with interferon-gamma. *Am J. Pathol.*, **136**, 503–8.

Huard, J., Labrecque, C., Dansereau, G. *et al.* (1991) Dystrophin expression in myotubes formed by the fusion of normal and dystrophic myoblasts. *Muscle Nerve*, **14**, 178–82.

Hughes, S.M. and Blau, H.M. (1990) Migration of myoblasts across basal lamina during skeletal muscle development. *Nature*, **345**, 350–3.

Jones, P.H. (1979) Implantation of cultured regenerate muscle cells into adult rat muscle. *Exp. Neurol.*, **66**, 602–10.

Jones, G.E., Murphy, S.J. and Watt, D.J. (1990) Segregation of the myogenic lineage in mouse muscle development. *J. Cell Sci.*, **97**, 659–67.

Karpati, G., Pouliot, Y. and Carpenter, S. (1988) Expression of immunoreactive major histocompatibility complex products in human skeletal muscles. *Ann. Neurol.*, **23**, 64–72.

Karpati, G., Pouliot, Y., Carpenter, S. and Holland, P. (1989a) Implantation of nondystrophic allogeneic myoblasts into dystrophic muscles of mdx mice produces 'mosaic' fibres of normal microscopic phenotype. *Cell and Molecular Biology of Muscle Development*, Alan R. Liss, New York, pp. 973–85.

Karpati, G., Pouiliot, Y., Zubrzycka-Gaarn, E. *et al.* (1989b) Dystrophin is expressed in mdx skeletal muscle fibres after normal myoblast transplantation. *Am. J. Pathol.*, **135**, 27–32.

Karpati, G. (1991) Myoblast transfer in Duchenne muscular dystrophy. A perspective, in *Muscular Dystrophy Research: From Molecular Diagnosis Towards Therapy*, (eds C. Angelini, G.A. Danielli and D. Fontanari), Excerpta Medica, Amsterdam, New York, Oxford, pp. 101–8.

Koenig, M., Hoffman, E.P., Bertelson, C.J. *et al.* (1987) Complete cloning of the Duchenne muscular dystrophy (DMD) cDNA and preliminary genomic organization of the DMD gene in normal and affected individuals. *Cell*, **50**, 509–17.

Kornegay, J.N., Tuler, S.M., Miller, D.M. and Levesque, D.C. (1988) Muscular dystrophy in a litter of golden retriever dogs. *Muscle Nerve*, **11**, 1056–64.

Law, P.K. and Yap, J.L. (1979) New muscle transplant method produces normal twitch tension in dystrophic muscle. *Muscle Nerve*, **2**, 356–63.

Law, P.K. (1982) Beneficial effects of transplanting normal limb-bud mesenchyme into dystrophic mouse muscles. *Muscle Nerve*, **5**, 619–27.

Law, P.K., Goodwin, T.G. and Li, H-J. (1988a) Histoincompatible myoblast injection improves muscle structure and function of dystrophic mice. *Transplantation Proc.*, **XX**, (3), Suppl. 3, pp. 1114–19.

Law, P.K., Goodwin, T.G. and Wang, M.G. (1988b) Normal myoblast injections provide genetic treatment for murine dystrophy. *Muscle Nerve*, **11**, 525–33.

Law, P.K., Goodwin, T.G., Li, H-J. *et al.* (1990a) Myoblast transfer improves muscle genetics/structure/function and normalizes the behavior and lifespan of dystrophic mice, in *Myoblast Transfer Therapy*, (eds R. Griggs and G. Karpati), Plenum Press, New York, pp. 75–87.

Law, P.K., Bertorini, T.E., Goodwin, T.G. *et al.* (1990b) Dystrophin production induced by myoblast transfer therapy in Duchenne muscular dystrophy. *Lancet*, **ii**, 114–15.

Law, P., Goodwin, T., Fang, Q. *et al.* (1991) Pioneering development of myoblast transfer therapy, in *Muscular Dystrophy Research: From Molecular Diagnosis Towards Therapy*, (eds C. Angelini, G.A. Danielli and D. Fontanari), Excerpta Medica, Amsterdam, New York, Oxford, pp. 109–16.

Lee, C.C., Pearlman, J.A., Chamberlain, J.S. and Caskey, C.T. (1991) Expression of recombinant dystrophin and its localization to the cell membrane. *Nature*, **349**, 334–6.

Lipton, B.H. and Schultz, E. (1979) Developmental fate of skeletal muscle satellite cells. *Science*, **205**, 1292–4.

Martin, H. and Ontell, M. (1988) Regeneration of dystrophic muscle following multiple injections of bupivicaine. *Muscle Nerve*, **11**, 588–96.

McDouall, R.M., Dunn, M.J. and Dubowitz, V. (1989) Expression of class I and class II MHC antigens in neuromuscular diseases. *J. Neurol. Sci.*, **89**, 213–26.

Meier, H. and Southard, J.L. (1970) Muscular dystrophy in the mouse caused by an allele at the dy locus. *Life Sci.*, **9**, 137–44.

Menke, A. and Jockusch, H. (1991) Decreased osmotic stability of dystrophin-less muscle cells from the mdx mouse. *Nature*, **349**, 69–71.

Miller, R.G., Steinman, L., Majumdar, S. *et al.* (1991) Methodologic considerations in clinical studies of myoblast implantation, in *Muscular Dystrophy Research: From Molecular Diagnosis Towards Therapy*, (eds C. Angelini, G.A. Danielli and D. Fontanari), Excerpta Medica, Amsterdam, New York, Oxford, pp. 117–122.

Morgan, J.E., Coulton, G.R. and Partridge, T.A. (1987) Muscle precursor cells invade and repopulate freeze-killed muscles. *J. Muscle Res. Cell Motil.*, **8**, 386–96.

Morgan, J.E. (1988a) Phosphorylase kinase activities in damaged mouse skeletal muscles. *J. Neurol. Sci.*, **86**, 149–58.

Morgan, J.E. (1988b) Myogenicity *in vitro* and *in vivo* of mouse muscle cells separated on discontinuous Percoll gradients. *J. Neurol. Sci.*, **85**, 197–207.

Morgan, J.E., Watt, D.J., Sloper, J.C. and Partridge, T.A. (1988a) Partial correction of an inherited defect of skeletal muscle by grafts of normal muscle precursor cells. *J. Neurol. Sci.*, **86**, 137–47.

Morgan, J.E. and Partridge, T.A. (1988b) Synergism between myogenic and non-myogenic cells during muscle formation *in vivo*. *J. Cell Biochem.*, Suppl. 12C, 331.

Morgan, J.E., Coulton, G.R. and Partridge, T.A. (1989) Mdx muscle grafts retain the mdx phenotype in normal hosts. *Muscle Nerve*, **12**, 401–9.

Morgan, J.E., Hoffman, E.P. and Partridge, T.A. (1990) Normal myogenic cells from newborn mice restore normal histology to degenerating muscles of the mdx mouse. *J. Cell Biol.*, **111**, 2437–49.

Morgan, J.E., Pagel, C.N. and Partridge, T.A. (1991) Implanted normal muscle precursor cells reconstitute and repair degenerating mdx muscles, in *Muscular Dystrophy Research: From Molecular Diagnosis Towards Therapy*, (eds C. Angelini, G.A. Danielli and D. Fontanari), Excerpta Medica, Amsterdam, New York, Oxford, pp. 235–6.

Morgan, J.E., Pagel, C.N., Sherratt, T. and Partridge, T.A. (1993) Long-term persistance and migration of myogenic cells injected into pre-irradiated muscles of *mdx* mice. *J. Neurol. Sci.*, in press.

Moschella, M.C. and Ontell, M. (1987) Transient and chronic neonatal denervation of murine muscle: A procedure to modify the phenotypic expression of muscular dystrophy. *J. Neuroscience*, **7**, 2145–52.

Moss, F.P. and Leblond, C.P. (1971) Satellite cells as the source of nuclei in muscles of growing rats. *Anat. Rec.*, **170**, 421–36.

Nicholson, L.V.B., Johnson, M.A., Gardner-Medwin, D. *et al.* (1990) Heterogeneity of dystrophin expression in patients with Duchenne and Becker muscular dystrophy. *Acta Neuropathol.*, **80**, 239–50.

Ontell, M. (1986) Muscular dystrophy and muscle regeneration. *Hum. Pathol.*, **17**, 673–82.

O'Reilly, R.J. (1983) Allogeneic bone marrow transplantation: current status and future directions. *Blood*, **62**, 941–64.

Partridge, T.A. and Sloper, J.C. (1977) A host contribution to the regeneration of muscle grafts. *J. Neurol. Sci.*, **33**, 425–35.

Partridge, T.A., Grounds, M. and Sloper, J.C. (1978) Evidence of fusion between host and donor myoblasts in skeletal muscle grafts. *Nature*, **273**, 306–8.

Partridge, T.A., Morgan, J.E., Moore, S.E. and Walsh, F.S. (1988) Myogenesis *in vivo* from the mouse C2 muscle cell-line. *J. Cell Biochem.*, Suppl. 12C, 331.

Partridge, T.A., Morgan, J.E., Coulton, G.R. *et al.* (1989) Conversion of mdx myofibres from dystrophin-negative to -positive by injection of normal myoblasts. *Nature*, 337, 176–9.

Partridge, T.A. (1990) Invited review: Myoblast transfer: A possible therapy for inherited myopathies? *Muscle Nerve*, **14**, 197–212.

Pavlath, G.K., Rich, K., Webster, S.G. and Blau, H.M. (1989) Localization of muscle gene products in nuclear domains. *Nature*, **337**, 570–3.

Peterson, A.C. (1974) Chimaera mouse study shows absence of disease in genetically dystrophic muscle. *Nature*, **248**, 561–4.

Peterson, A. and Pena, S. (1984) Relationship of genotype and in vitro contractility in mdg/mdg ↔ +/+ 'mosaic' myotubes. *Muscle Nerve*, 7, 194–203.

Phillips, G.D., Hoffman, J.R. and Knighton, D.R. (1990) Migration of myogenic cells in the rat extensor digitorum longus muscle studied with a split autograft model. *Cell Tissue Res.*, **262**, 81–8.

Ponder, B.A.J., Wilkinson, M.M., Wood, M. and Westwood, J.H. (1983) Immunohistochemical demonstration of H2 antigens in mouse tissue sections. *J. Histochem. Cytochem.*, **31**, 911–19.

Rahim, Z.H.A., Perret, D., Lutaya, G. and Griffiths, G.R. (1980) Metabolic adaptation in phosphorylase kinase deficiency. Changes in metabolite concentrations during tetanic stimulation of mouse leg muscles. *Biochem. J.*, **186**, 331–41.

Ralston, E. and Hall, Z.W. (1989) Transfer of a protein encoded by a single nucleus to nearby nuclei in multinucleated myotubes. *Science*, **244**, 1066–9.

Salminen, A., Elson, H.F., Mickley, L.A. *et al.* (1991) Implantation of recombinant rat myocytes into adult skeletal muscle: a potential gene therapy. *Human Gene Therapy*, **2**, 15–26.

Sanderson, R.D., Fitch, J.M., Linsenmayer, T.R. and Mayne, R. (1986) Fibroblasts promote the formation of a continuous basal lamina during myogenesis in vitro. *J. Cell Biol.*, **102**, 740–7.

Schultz, E., Jaryszak, D.L. and Valliere, C.R. (1985) Response of satellite cells to focal skeletal muscle injury. *Muscle Nerve*, **8**, 217–22.

Schultz, E., Jaryszak, D.L., Gibson, M.C. and Albright, D.J. (1986) Absence of exogenous satellite cell contribution to regeneration of frozen skeletal muscle. *J. Muscle Res. Cell Motil.*, 7, 361–7.

Sicinski, P., Geng, Y., Ryder-Cook, A.S. *et al.* (1989) The molecular basis of muscular dystrophy in the mdx mouse: A point mutation. *Science*, **244**, 1578–80.

Stockdale, F.E. and Holtzer, H. (1961) DNA synthesis and myogenesis. *Exp. Cell Res.*, **24**, 508–20.

Thomason, D.B. and Booth, F.W. (1990) Stable incorporation of a bacterial gene into adult rat skeletal muscle *in vivo*. *Am. J. Physiol.*, **258**, C578–C581.

Tremblay, J.P., Roy, R., Bouchard, J.P. *et al.* (1991) Human myoblast transplantation, in *Muscular Dystrophy Research: From Molecular Diagnosis Towards Therapy*, (eds C. Angelini, G.A. Danielli and D. Fontanari), Excerpta Medica, Amsterdam, New York, Oxford, pp. 123–30.

Valentine, B.A., Blue, J.T. and Cooper, B.J. (1989) The effect of exercise on canine dystrophic muscle. *Ann. Neurol.*, **26**, 588.

Wakeford, S., Watt, D.J. and Partridge, T.A. (1991). X-irradiation improves mdx mouse muscle as a model of myofibre loss in DMD. *Muscle Nerve*, **14**, 42–50.

Walsh, F.S. and Ritter, M.A. (1981) Surface antigen differentiation during human myogenesis in culture. *Nature*, **289**, 60–4.

Walsh, F.S. (1990) N-CAM is a target cell surface antigen for the purification of muscle cells for myoblast transfer therapy, in *Myoblast Transfer Therapy*, (eds R. Griggs and G. Karpati), Plenum Press, New York, pp. 41–6.

Walton, J.N. and Gardner-Medwin, D. (1974) Progressive muscular dystrophy and the myotonic disorders, in *Disorders of Voluntary Muscle*, (ed. J.N. Walton), Churchill Livingstone, Edinburgh and London, pp. 561–613.

Watkins, S.C., Hoffman, E.P., Slayter, H.S. and Kunkel, L.M. (1989) Dystrophin distribution in heterozygote mdx mice. *Muscle Nerve*, **12**, 861–8.

Watt, D.J., Partridge, T.A. and Sloper, J.C. (1981) Cyclosporin A as a means of preventing rejection of skeletal muscle allografts in mice. *Transplantation*, **31**, 266–71.

Watt, D.J. (1982) Factors which affect the fusion of allogeneic muscle precursors *in vivo*. *Neuropathol. Appl. Neurobiol.*, **8**, 135–47.

Watt, D.J., Lambert, K., Morgan, J.E. *et al.* (1982) Incorporation of donor muscle precursor cells into an area of regeneration in the host mouse. *J. Neurol. Sci.*, **57**, 319–31.

Watt, D.J., Morgan, J.E. and Partridge, T.A. (1984a) Use of mononuclear precursor cells to insert allogeneic genes into growing muscles. *Muscle Nerve*, **7**, 741–50.

Watt, D.J., Morgan, J.E. and Partridge, T.A. (1984b) Long term survival of allografted muscle precursor cells following a limited period of treatment with cyclosporin A. *Clin. Exp. Immunol.*, **55**, 419–26.

Watt, D.J., Morgan, J.E., Clifford, M.A., and Partridge, T.A. (1987) The movement of muscle precursor cells between adjacent regenerating muscles in the mouse. *Anat. Embryol.*, **175**, 527–36.

Watt, D.J. (1990) A comparison of long-term survival of muscle cell suspensions and minced muscle allografts in the non-tolerant mouse, in *Myoblast Transfer Therapy*, (eds R. Griggs and G. Karpati), Plenum Press, New York, pp. 35–9.

Watt, D.J., Morgan, J.E. and Partridge, T.A. (1991) Allografts of muscle precursor cells persist in the non-tolerised host. *Neuromuscular Disorders*, **5**, 345–55.

Watt, D.J., Karasinski, J. and England, M.A. (1993) Migration of lacZ positive cells from the tibialis anterior to the extensor digitorum longus muscle of the X-linked muscular dystrophic (*mdx*) mouse. *J. Muscle Res. Cell Motil.*, **14**, 121–32.

Webster, C., Pavlath, G.K., Parks, D.R. *et al.* (1988) Isolation of human myoblasts with the fluorescence-activated cell sorter. *Exp. Cell Res.*, **174**, 252–65.

Webster, C. and Blau, H.M. (1990) Accelerated age-related decline in replicative life-span of Duchenne muscular dystrophy myoblasts: Implications for cell and

gene therapy. *Somat. Cell Mol. Genet.*, **16**, 557–65.
Wolff, J.A., Malone, R.W., Williams, P. *et al.* (1990) Direct gene transfer into mouse muscle *in vivo*. *Science*, **247**, 1465–8.
Wysocki, L.J. and Sato, V.L. (1978) 'Panning' for lymphocytes: a method for cell selection. *Proc. Natl. Acad. Sci. USA*, **75**, 2844–8.
Yablonka-Reuveni, Z. and Nameroff, M. (1987) Skeletal muscle cell populations. Separation and partial characterization of fibroblast-like cells from embryonic tissue using density centrifugation. *Histochemistry*, **87**, 27–38.
Yablonka-Reuveni, Z., Quinnn, L.S. and Nameroff, M. (1987) Isolation and clonal analysis of satellite cells from chicken pectoralis muscles. *Dev. Biol.*, **119**, 252–9.

Index

Made in United States
Cleveland, OH
18 October 2026

46341975R00198